Luminos is the Open Access monograph publishing program
from UC Press. Luminos provides a framework for preserving and
reinvigorating monograph publishing for the future and increases
the reach and visibility of important scholarly work. Titles published
in the UC Press Luminos model are published with the same high
standards for selection, peer review, production, and marketing as
those in our traditional program. www.luminosoa.org

Making Our Beasts

CRITICAL ENVIRONMENTS: NATURE, SCIENCE, AND POLITICS

Edited by Julie Guthman and Rebecca Lave

The Critical Environments series publishes books that explore the political forms of life and the ecologies that emerge from histories of capitalism, militarism, racism, colonialism, and more.

1. *Flame and Fortune in the American West: Urban Development, Environmental Change, and the Great Oakland Hills Fire*, by Gregory L. Simon

2. *Germ Wars: The Politics of Microbes and America's Landscape of Fear*, by Melanie Armstrong

3. *Coral Whisperers: Scientists on the Brink*, by Irus Braverman

4. *Life without Lead: Contamination, Crisis, and Hope in Uruguay*, by Daniel Renfrew

5. *Unsettled Waters: Rights, Law, and Identity in the American West*, by Eric P. Perramond

6. *Wilted: Pathogens, Chemicals, and the Fragile Future of the Strawberry Industry*, by Julie Guthman

7. *Destination Anthropocene: Science and Tourism in The Bahamas*, by Amelia Moore

8. *Economic Poisoning: Industrial Waste and the Chemicalization of American Agriculture*, by Adam M. Romero

9. *Weighing the Future: Race, Science, and Pregnancy Trials in the Postgenomic Era*, by Natali Valdez

10. *Continent in Dust: Experiments in a Chinese Weather System*, by Jerry C. Zee

11. *Worlds of Green and Gray: Mineral Extraction as Ecological Practice*, by Sebastián Ureta and Patricio Flores

12. *The Fluvial Imagination: On Lesotho's Water-Export Economy*, by Colin Hoag

13. *The Low-Carbon Contradiction: Energy Transition, Geopolitics, and the Infrastructural State in Cuba*, by Gustav Cederlöf

14. *Unmaking the Bomb: Environmental Cleanup and the Politics of Impossibility*, by Shannon Cram

15. *The Taste of Water: Sensory Perception and the Making of an Industrialized Beverage*, by Christy Spackman

16. *The Pathogens of Finance: How Capitalism Breeds Vector-Borne Disease*, by Brent Z. Kaup and Kelly F. Austin

17. *The Nightcrawlers: A Story of Worms, Cows, and Cash in the Underground Bait Industry*, by Joshua Steckley

18. *The Plantation Ideal: Landscapes of Extraction in Mozambique*, by Wendy Wolford

19. *The Almond Paradox: Cracking Open the Politics of What Plants Need*, by Emily Reisman

20. *Making Our Beasts: Paleontology in the United States*, by Elana Shever

Making Our Beasts

Paleontology in the United States

Elana Shever

UNIVERSITY OF CALIFORNIA PRESS

University of California Press
Oakland, California

Suggested citation: Shever, E. *Making Our Beasts: Paleontology in the United States*. Oakland: University of California Press, 2026. DOI: https://doi.org/10.1525/luminos.256

Library of Congress Cataloging-in-Publication Data

Names: Shever, Elana, author.
Title: Making our beasts : paleontology in the United States / Elana Shever.
Other titles: Critical environments (Oakland, Calif.) ; 20.
Description: Oakland, California: University of California Press, [2026] | Series: Critical environments: nature, science, and politics ; 20 | Includes bibliographical references and index.
Identifiers: LCCN 2025028724 (print) | LCCN 2025028725 (ebook) | ISBN 9780520425668 (cloth) | ISBN 9780520416727 (paperback) | ISBN 9780520416734 (ebook)
Subjects: LCSH: Paleontology—United States. | Vertebrates, Fossil—United States. | Fossils—Collection and preservation—United States. | Paleontology—Social aspects—United States. | Science museums—Social aspects—United States. | Dinosaurs in popular culture—United States.
Classification: LCC QE746 .S44 2026 (print) | LCC QE746 (ebook) | DC 560.973—dc23/eng/20250625

LC record available at https://lccn.loc.gov/2025028724
LC ebook record available at https://lccn.loc.gov/2025028725

GPSR Authorized Representative: Easy Access System Europe, Mustamäe tee 50, 10621 Tallinn, Estonia, gpsr.requests@easproject.com

34 33 32 31 30 29 28 27 26 25
10 9 8 7 6 5 4 3 2 1

publication supported by a grant from
The Community Foundation for Greater New Haven
as part of the *Urban Haven Project*

CONTENTS

List of Illustrations — *ix*

Introduction: Meeting Prehistory Halfway — *1*

1. Becoming Stone — *34*

2. Making Charismatic Violence — *61*

3. Forging Connection Across Millions of Years — *90*

4. Paleontology, Colonialism, and Edutainment — *113*

5. The Paleontological Real — *142*

Conclusion: Remaking Our Beasts — *172*

Acknowledgments — *185*
Notes — *189*
Bibliography — *217*
Index — *243*

ILLUSTRATIONS

MAP

1. North America in the Mesozoic era and the present   *xi*

FIGURES

1. A sixty-seven-million-year-old *Triceratops horridus* skull   *36*
2. Excavating a paleontological quarry in the Dakota Badlands   *47*
3. Jacketing a fossil in the Dakota Badlands   *49*
4. Reconstruction of a decaying Triceratops skull   *52*
5. A Triceratops skull undergoing preparation   *53*
6. The cast of AMNH 5027 in the Denver Museum of Nature and Science   *62*
7. The cast of Stan's skull (BHI 3033) in the Morrison Natural History Museum   *78*
8. The Bone Bar exploration station in *Prehistoric Journey*   *84*
9. Docents and visitors gather around the Bone Bar   *85*
10. A docent shows students Pteranodon's five fingers   *95*
11. A puppeteer in a dinosaur skeleton greets a girl   *107*
12. Dino-Sue in Disney World's DinoLand U.S.A.   *114*
13. "Welcome to DINOLAND U.S.A. The Friendliest Fossils in America"   *122*
14. Riding Triceratops Spin in DinoLand U.S.A.   *123*
15. Climbing the *Tyrannosaurus rex* skeleton   *124*

16. Arthur Lakes's *Professor Mudge Contemplating Atlantosaurus Bones*   134
17. Arthur Lakes's *Digging Out Bones at Morrison, Quarry No. 10*   135
18. Sketch of the paleontological continuum   159
19. Matthew Mossbrucker teaches a child to use an airscribe   166
20. Matt in National Geographic's *T-Rex Autopsy*   167

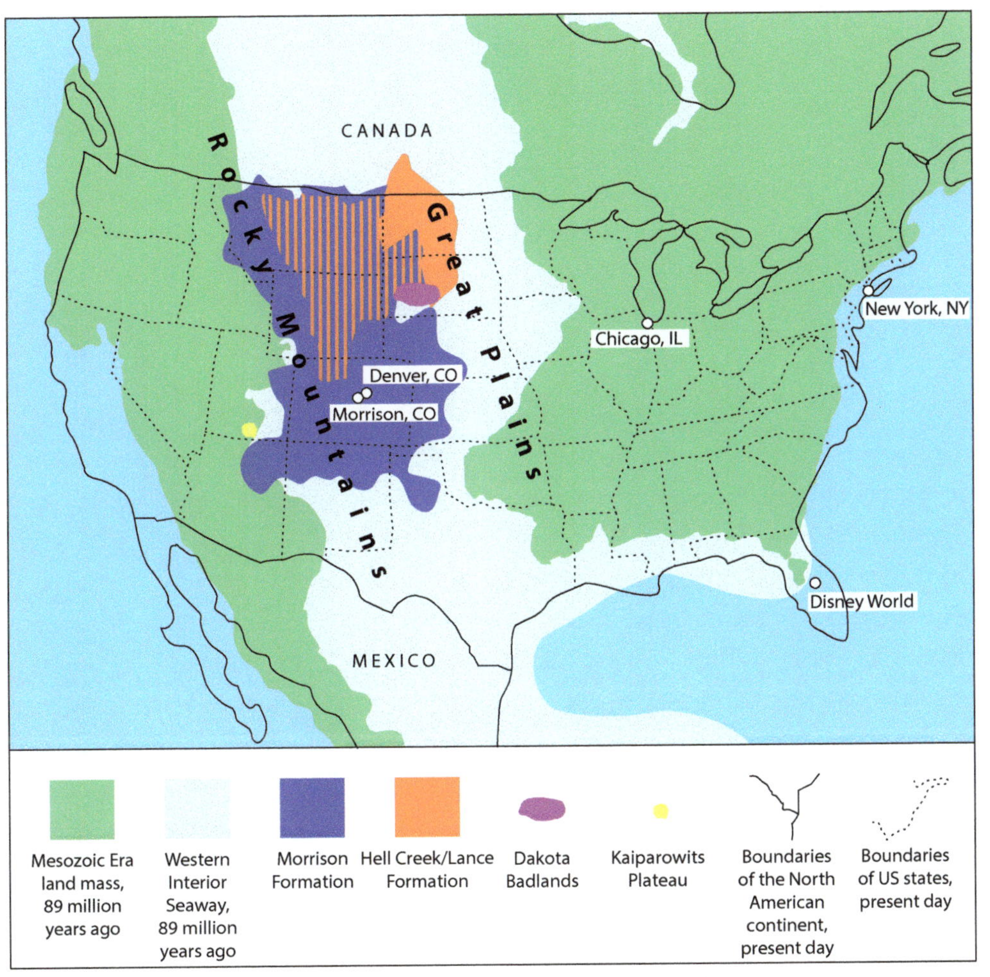

MAP 1. North America in the Mesozoic and the current eras. Map by Lara Scott.

Introduction

Meeting Prehistory Halfway

I sit with a group of adults and children on wooden bleachers facing a steep ridge in the Front Range of the Rocky Mountains. We look up at a rockface covered in hundreds of enormous footprints impressed in the light gray sandstone. A few straggly plants stick out from cracks in the rock, but other than that, the rockface is barren. It's a hot summer day, and I squint in the glare of the midday sun so I can see the slivers of shadow in the deepest of the footprints. The longer I stare at the rockface, the more footprints I can make out. In some places, I recognize prints with clearly defined edges and dark interiors that make their shape easy to discern, while in others I see only what might be the impression of a toe. I am not sure. Gradually, I discern a series of roughly parallel footprints that extend from the ground near the bench to the top of the ridge, more than a dozen feet above us. The toes, it seems, mostly point upward and slightly to my left.

Carl, the guide leading this tour, tells us that in the late Cretaceous period this spot was not a ridge, but flat "beachfront property" along the Western Interior Seaway, the body of water that split North America for millions of years (see map). Sixty-five million years later, the tracks were discovered when a road was built over the ridge in the 1970s. Carl is a regular volunteer guide at this park, called Dinosaur Ridge, where more than a hundred thousand tourists and locals come each year to see dinosaur fossils embedded in the rock.[1] His well-practiced tours weave stories and jokes into explanations of geology and paleontology. Like many of Dinosaur Ridge's volunteers, Carl is a retired industrial geologist who, after decades of working outdoors, could not imagine sitting at home. He is a tall, broad-shouldered, white man who is strong and fit despite his sixty-plus years. Today, as usual, he is wearing Dinosaur Ridge's unofficial uniform: a forest-green cargo vest, khaki cargo pants, work boots, and a wide-brimmed canvas hat with a leather strap. He holds a walking stick that he more often uses for pointing out features in the rock than to help him navigate the uneven terrain.

Carl calls our attention to two parallel sets of footprints on the rockface. Both have three wide toes, but one set is much larger than the other. He begins to recount a story of two herbivorous dinosaurs, a mother and child, strolling along the riverbank millions of years ago. Using his walking stick to indicate the trail of larger footprints, he tells us, "This is an adult walking on four legs." Pointing to a smaller line of footprints, he says, "This is her baby hopping around on two legs," adding, "babies are curious," to explain its circuitous path. He then points to where the adult's front footprints "disappear." Lifting his arms to imitate the dinosaur's posture, he asks, "She stood up, right?" Several in the group nod in agreement. "It looks like the baby's tracks separate from the mother's tracks. That's why she stood up. She's looking around for her baby." Turning to the children in the group, he says, with a wink, "It's just like if you guys go to the mall with your parents. You always stay right with them?" Children and adults nod their heads with knowing smiles. Carl shows us another trail of footprints in the rockface. It is the trackway of a carnivorous dinosaur, he explains, because these footprints have long, slender toes. "This predator is getting close to the baby," he says. He indicates where the juvenile's trackway parallels the adult's again near the top of the ridge and concludes that the baby "goes back to momma, and they walk off together." A happy ending to this prehistoric tale.

What is going on here? How does Carl know that the baby is hopping? That the adult is its mother? That she stood up to look for her baby who had wondered off? Is Carl objectively relating scientific facts about the prehistoric world in terms that his audience can understand?[2] Or is he imposing contemporary ideas and values onto the physical remains of prehistory? Is there another, better, way to interpret these tracks, or to understand why Carl told this story about them? *Making Our Beasts* scrutinizes moments like this one to reveal how the material remains of the prehistoric world become meaningful and valuable objects in our present world. It examines how fossils of dinosaurs and other prehistoric animals come into being, following a journey that began in the Mesozoic era and continues in the current era, the Anthropocene. In tracing this journey, I explore the numerous ways that prehistory is alive in the contemporary United States. In classical anthropological terms, I "make strange" the prehistoric creatures that are found in the United States today. Why do they show up in so many places? What are they doing there?

The moment at the rockface foreshadows several themes in the pages that follow. You might have noticed that Carl gendered the adult dinosaur as female, and presented her as playing a traditional feminine role of caring for a child. Carl also established a parallel between the relationship of these two dinosaurs and those of the adults and children on his tour that day. He suggested that there are universal characteristics of parents and children that transcend time and species: Children are curious, innocent, and irresponsible, so parents need to protect them from harm, especially from predators lurking in the shadows. As a retired geologist and volunteer paleontology guide, Carl, particularly his body, comportment, and dress, convey the gendering of science. Geology, and its subdiscipline of paleontology, are overwhelmingly occupied by men.[3] Moreover, these fields adhere to a specific masculine ideal. Carl wore the clothes typical of geologists, an outfit not only normatively male but, more specifically, one that reproduces the masculinity of those scientist-explorers who have traveled the North American continent since the nineteenth century. Carl also spoke with the authority and knowledge of a scientific expert, a position long gendered as male. I observed Dinosaur Ridge's female guides taking on the same dress and voice, sometimes deluging visitors with even more facts than their male counterparts, in order to assert their scientific authority.

The above vignette also reveals the confluence of science, education, and entertainment. Paleontologists conduct research at Dinosaur Ridge, but it is also a park. It is one of many "edutainment" sites that use paleontology to educate and entertain people at the same time. The idea behind edutainment is that people learn things painlessly when they are enjoying themselves. Entertainment is the "sugar" that makes the "medicine" of scientific knowledge go down. There is much research showing that young children learn best through hands-on play, but edutainment is aimed at adults, too. It is commonly used to teach science to people of all ages. You can find science edutainment in museums, classrooms, homes, and parks, whether national parks like Dinosaur National Monument or amusement parks like Disney World. On Carl's tour of Dinosaur Ridge, for example, he taught visitors that the geology and environment of the spot where they were standing had changed over millions of years, from flat riverine landscape to the dry mountainous landscape that now surrounded them. He also informed them that carnivorous animals have different foot shapes from herbivores. He did this by telling a clever story instead of lecturing, using narrative and humor rather than dense explanations and diagrams.

Carl's tour of Dinosaur Ridge exemplifies how paleontology is conceived as an adventure: the daring hunt for fossils, the thrill of touching the remains of an animal from millions of years ago, and the exhilaration of discovering a new species. The prehistoric world is portrayed as one full of excitement, most often in the form of fights between predators and prey. Carl invoked the thrill of paleontology and prehistory by recounting one such story of predation, heightening its drama by including a mother and child pair. He also helped his audience to experience the excitement of scientific discovery by leading them through the steps that paleontologists use to interpret trackways. He pointed to a set of tracks and asked, "What's that dinosaur doing?" Several kids answered, "walking." He corrected them, "I don't think he was walking. He was probably running. How do I come to that conclusion?" Carl's audience was stumped, so he explained by comparing this trackway with another one nearby. "Look at the size of the track.

This track is much smaller than that one, right? But the distance between the two [footprints] is longer." He showed them the difference in the two gaits by running and then walking a few steps himself. Here is the first of many instances in this book where people compare their bodies and behaviors with those of prehistoric animals that they cannot observe directly. Finally, after completing his story about the dinosaurs, he invited the people on his tour to get out of their seats and feel the rocks, even to climb part of the way up the rockface. Carl made the tour itself into a scientific adventure by first presenting the clues in the rock, then explaining them, and finally inviting visitors to physically imitate the paleontologists who discovered the footprints. This hints at another theme in the pages that follow: the importance of touch in paleontology. Beyond edutainment events like Carl's tour, the use of touch is widespread in vertebrate paleontological research and education. *Making Our Beasts* examines how paleontological fieldwork and labwork, as well as teaching and learning, are all highly tactile, sensual, and affective activities.

It is also important to pay attention to what is left out of Carl's tour. Most of Dinosaur Ridge's guides tell visitors heroic tales about the first fossil hunters who came to this part of Colorado, and the famous species of dinosaurs first discovered there. Missing from all the tours I took was any critical discussion of paleontology's involvement in white settler colonialism. Since the founding of the United States, the search for fossils in the Western territories has been framed as a scientific adventure, one that revealed treasures hidden within the country's vast and untapped land. As I discuss in chapter 4, Dinosaur Ridge is one among many places that perpetuate myths about the Western frontier and celebrate the scientific value of colonialism, while erasing its violence and paleontology's role in dispossessing native peoples. When Carl recounted how paleontologists benefited from road construction, his story depicted paleontology as working in harmony with the economic development of the Denver region, not harming anyone or anything.

Scientists tell stories as much as everyone else does. Anthropologist Thom Van Dooren writes that telling stories about extinct species is "not an attempt to obscure the truth of the situation, but to insist on a truth that is not reducible to" the data scientists collect.[4] I believe that Carl and many others at my field sites would agree. Storytelling is part of doing science, not a distortion of it, nor is it something done only after the work of science is finished. Stories are told in scholarly publications, in classrooms, in laboratories, and at fossil dig sites. Paleontology's storytelling illustrates the impossibility of isolating science from its social context. Carl's tour highlights how stories about the prehistoric world reveal prominent ideas, morals, desires, and assumptions of people in the contemporary world. For example, mothers worry about their children; children are venturesome. But paleontology does not merely reflect today's social meanings and values.[5] I argue that current ideas are not imposed onto the physical remains of prehistory; rather, meanings and materiality emerge together. More broadly, *Making Our Beasts* examines how the science of vertebrate paleontology and the materiality of fossils shape the social world and how they, in turn, are shaped by it. To accomplish this, I spent countless hours at paleontological research institutions, science museums, nature parks, and entertainment venues, where I observed, participated in, and interviewed people about their engagements with paleontological objects, research, and knowledge. I listened to many stories and shared some of my own.

Making Our Beasts is a science-in-action ethnography that examines vertebrate paleontology in the United States. I conducted most of my research in those parts of the US Great Plains and the Rocky Mountain foothills that comprised the Western Interior Seaway in the Cretaceous period (see map). This shallow inland sea, which stretched from what is today the Canadian Arctic to the Gulf of Mexico, is where the remains of many of today's iconic dinosaurs have been excavated. Much of my fieldwork took place in the booming Denver metropolitan area, where a thriving vertebrate paleontological community has sustained the region's importance to paleontology since the nineteenth century. There, I conducted ethnographic research at one large world-renowned science museum, the Denver Museum of Nature and Science, and at one small specialized museum, the Morrison Natural History Museum, as well as doing fieldwork at Dinosaur Ridge, the nature park in the Front Range of the Rocky Mountains (and the site of Carl's tour), and at an idiosyncratic chain hotel whose owners have turned it into a recreation of a fossil hunter's home. Like the early fossil hunters, I traveled from Denver to the Dakotas and Montana to excavate in the fossil-rich geological formations of the Badlands. I conducted additional fieldwork at other science museums and parks, research institutes and "edutainment" sites, both in the Western United States and on the East Coast, where tons of the fossils found in the Western United States have been collected and displayed. I interviewed paleontologists involved in television productions, including PBS Kids' long-running show *Dinosaur Train* and the National Geographic special *T-Rex Autopsy*. I also traveled to Disney's Animal Kingdom in Florida to further explore paleontology edutainment. At all my field sites, I talked with and worked beside scientists, fossil preparators, curators, educators, volunteers, tourists, teachers, and students.[6]

At the same time that I interacted with people, I engaged with prehistoric animals in the multiple forms they take in the world today. The contemporary world is full of the physical remains of ancient organisms, from the paving stones on my street to the gas in the tank of my car. But these are not *meaningful* embodiments of prehistory. The most meaningful and valuable forms of prehistoric matter are the fossils that represent singular beings, animals with whom you can imagine having a relationship like you have with a pet. But it is important to recognize that these fossils are extremely rare. Although it would be impossible to count them all, I would bet there are many times the number of plastic dinosaur toys than fossilized dinosaur bones in the world today. The extensive cultural elaboration of prehistory means that prehistoric animals take multiple additional forms: casts and reconstructions displayed in museums; the animatronic and other models at entertainment venues; the CGI dinosaurs in popular movies, television shows, and video games; books in schools and homes across the country. Through my field-work, I interacted with an even wider variety of matter essential to paleontology, including glues, putties, and plasters; simple hand tools and intricate power-driven ones; and images, graphs, and texts. I encountered these materials while engaging with the wide range of people who enjoy them, both on special occasions, such as a school trip or family vacation, and in the course of their daily lives, as scientists, curators, fossil preparators, educators, volunteers, students, parents, and caregivers. The resulting book analyzes the complex interface between vertebrate paleontology as a scientific field and as a popular interest. My research showed me that in some times and places the line between scientific and popular paleontology becomes so thin that it all but disappears.

SCIENCE IN ACTION

The prehistoric world intersects with our contemporary one in many ways, yet it remains strange to us. It is a world inhabited by unfamiliar plants and animals living on landmasses different from the current continents with distinct climatic conditions. Paleontology is the quest to understand the evolution of life on earth in prehistory. The *pre* in prehistory refers to the time before humans started making their mark on the world. Paleontology asks how life emerged and evolved before people arrived on the scene and began to manipulate it. Vertebrate paleontology, the subfield I explore most closely, studies the evolution of animals with spines, beginning with tiny wormlike swimmers with mineralized skeletons in the Paleozoic era, more than five hundred million years ago. Dinosaurs and mammals both emerged, more than two hundred million years after the first vertebrates, in the Mesozoic era. Although those first mammals are humanity's ancestors, dinosaurs have received more attention. They are a diverse group of species that evolved over a hundred million years—which, paleontologists like to point out, is far longer than homo sapiens has been on earth.

It is important to recognize that vertebrate paleontologists cannot study prehistoric animals in the same ways that other scientists study contemporary ones. Paleontologists cannot observe their behaviors, scan their brains, sample their tissues, or sequence their genes. Instead, they examine the traces of their lives that are left on earth today, and they try to reconstruct their bodies and activities, as well as the evolution of their species. It is especially challenging to study the iconic large dinosaurs that fascinate both scientists and everyday people because there are so few specimens, and none is complete. A popular paleontology textbook states that "every paleontologist has to act as a crime scene investigator/coroner/forensic pathologist/detective, determining what killed the victim and trying to reconstruct the events at the 'scene of the crime.'"[7] *Making Our Beasts* examines how the difficult work of vertebrate paleontology is done: in field stations and laboratories, in research institutes, universities and museums, and in entertainment venues and the media.

There is a common belief that science is an isolated or apolitical practice divorced from the power dynamics of the rest of the world, though this misconception has been repeatedly dispelled by critical scholars of science.[8] Although science is often conceived as a "culture of no culture," scientific inquiry is actually a social and political practice that is both formed by and formational of its context.[9] The social, political, and economic setting in which the sciences are practiced affects what questions scientists ask and how they get answered, and how scientific knowledge is transmitted and interpreted.[10] Paleontology is no exception. Following a science-in-action approach, I argue that science is best understood as a process of inquiry, rather than as a body of facts and theories. Science is knowledge-making, not knowledge itself. Bruno Latour and Steve Woolgar's classic study *Laboratory Life* asserts that scientific facts are produced in a laboratory, much as widgets are produced in a factory: materials—from organisms to test tubes to electricity—come in; scientific laborers turn them into facts about nature; then these are sent out in scientific journal articles. Other scholars have countered that scientific knowledge is not always produced in such circumscribed, self-contained, and orderly settings. Emily Martin has shown that even the most restricted laboratories are not "citadels" but entities with "porous and leaky" walls through which the outside world enters.[11] More recently, she has explored how scientific research is all around us—in shopping malls, workplaces, and on the internet.[12] Paleontology demonstrates that this is not new. Paleontology has been carried out in all sorts of settings, from farms and ranches to suburban backyards and urban construction sites. *Making Our Beasts* challenges the persistent belief that the best science is what people with PhDs do in laboratories and that everything else is something other than, or less than, science.

The scholarship about science in action has thoroughly documented how scientific inquiry produces much more than facts and theories. Science produces scientists and their habitus, desires, and selfhood.[13] It produces technicians and volunteers.[14] Moreover, science produces things: samples, specimens, instruments, data, models, documents, images, and waste. This means that scientific research does not only study nature to create knowledge about it; it makes nature too. In chapter 1, I describe how paleontological fieldwork and labwork generate both valuable specimens and useless dust. More broadly, *Making Our Beasts* contributes to the science-in-action scholarship by demonstrating that vertebrate paleontology creates fossils and knowledge together through a process that involves physical and affective interactions among people and matter. It expands scientific knowledge creation to include not only the production of fossil specimens but also the casts of those specimens. The science-in-action scholarship understands objects as agentive, yet still tends to give short shrift to other forms of matter. *Making Our Beasts* shows how a wide variety of materials are involved in every stage and facet of fossil making. They include organisms and the traces they left. They include deeply buried sedimentary rock layers and their outcrops, mountains, buttes, and plateaus, and other geological formations. They include rock hammers, pickaxes, dental picks, microscopes, and miniature jackhammers called airscribes. They include glues, putties, plasters, and other adhesives. The people who do paleontology do not simply *use* things to investigate prehistoric nature. To say that would be to claim that people maintain control over the scientific process. It is more accurate to say that paleontologists work *with* matter, be it an outcrop of Mesozoic rock, a dental pick, or a flowing glob of latex. Yet the roles of humans and nonhumans are not the same. *Making Our Beasts* reveals the unequal power dynamics involved in paleontology's creative processes. It unpacks how paleontology's ideas and objects are made and transformed through the movements of people and matter in and out of geological formations, fossil quarries, laboratories, collections, exhibits, media, and other sites.

A science-in-action approach that is focused on the practices and processes of scientific research encourages questioning who is a scientist and who is not. My fieldwork quickly showed me that most people who carry out tasks in vertebrate paleontology do not hold the title of scientist. Vertebrate paleontology is unusual among the sciences because several of its most accomplished figures do not have doctoral degrees. This was common a hundred years ago, but is uncommon in most sciences today. In the early days of paleontology in the United States, field research depended on the knowledge and skills of Native Americans and white colonists living in the Great Plains, of the US military stationed in the Western territories, and of college students coming from the East Coast.[15] Chapter 1 shows that the field continues to depend on a multitude of amateurs, with white urbanites engaged in volunteer-tourism now replacing the rural people who worked for earlier generations of paleontologists.[16] Likewise, paleontological laboratories need the expert labor of legions of volunteers and technicians. As I learned to do paleontology, I quickly realized how much this science depends on the specialized skills of car mechanics, construction workers, electricians, dental hygienists, sculptors, illustrators, and writers, as well as scientists with degrees in various subfields of geology, biology, chemistry, and physics. Paleontology is a "community of practice" engaged in processes that involve—really, require—people with a wide range of skills and experiences.[17] Its practitioners are engaged in energetic debate about who should be named as an author at the top of a research article and who should be thanked in the acknowledgements at the end. I do not attempt to adjudicate the weighty question of who is and who is not a scientist, but simply describe people as they describe themselves.

DELIMITING PALEONTOLOGY

Paleontology's practitioners are also engaged in the "boundary-work" of defining, contesting, and negotiating what counts, and does not count, as science.[18] Here I avoid the a priori separation of scientific research from science education and communication. Instead, I explore how the boundaries of paleontological science have been delimited and redefined, especially in relation to commercial fossil collecting and the media and entertainment industries. On the one hand, paleontology is more open to recognizing the expertise of those without formal educational credentials in science (although I discuss limits to this in chapter 2 and the conclusion.) On the other hand, the work of commercial fossil collectors is hotly contested. While some argue that commercial operations contribute invaluably to the fossil record, others claim that they make it more difficult for scientists to do research. The media industry is less controversial than commercial fossil collecting. While scientists in many fields risk losing prestige when they appear in the media, vertebrate paleontologists do not. In addition to working behind the scenes as consultants to movies and television shows, renowned dinosaur paleontologists such as Robert T. Bakker and Scott Sampson have appeared onscreen themselves. *Making Our Beasts* examines the continual boundary-work that occurs in paleontology to define what is and is not science.

Scientists often dismiss paleontology as lesser than other sciences because it shows up in popular media, because it appeals to nonscientists, and because some of its practitioners have become household names. Some of the sites where paleontology is conducted are also widely seen as completely *un*scientific spaces, which reinforces this reputation. For instance, before being studied and mounted at the Field Museum of Natural History in Chicago, the famous *Tyrannosaurus rex* "Sue" was partially prepared in a laboratory housed at Disney World's Animal Kingdom in Florida. Disney's laboratory was not so different from the ones at the Denver Museum of Nature and Science, or the Paleontological Research Institute (PRI), an affiliate of Cornell University in Ithaca, New York. As I learned early in my fieldwork, besides holding one of the largest and most acclaimed fossil collections in North America, PRI also hosts dinosaur egg hunts for children at Easter. Neither the places where we find paleontology nor the doing of paleontology itself can be neatly divided between science and entertainment. I agree with the scholars who counter that "science as it occurs within popular culture is not simply a more diluted, less sophisticated, error-plagued imitation of 'real' science: it, too, is 'real' science."[19] In the chapters that follow, I present the teaching of paleontology as part of the process of doing science, not as something else carried out afterwards. The "dissemination" of science is best conceived as "the linked play associating observation, experiment, testing, confirmation, replication, description, measurement, demonstrations, representation, interpretation and presentation."[20] In paleontology, in particular, dissemination includes much more than writing peer-reviewed journal articles. It involves talking with museum-goers while cleaning a fossil (chapter 1), constructing exhibits (chapters 2 and 3), giving tours of a fossil quarry (chapter 1), leading a classroom simulation (chapter 4), and appearing on television (chapter 5). Eminent paleontologists perform all these knowledge-making practices, but not by themselves. They are always working with other people and matter.

Paleontology is not merely tied to the entertainment industries, but inexorably enmeshed in global capitalism. Even before paleontology's development into an academic discipline in the nineteenth century, fossils were taken, bought, and sold alongside other natural resources. As paleontology developed from a gentlemen's hobby into a science and a profession, its practitioners participated in both the global circuits of knowledge production and the transnational markets for luxury items. Paleontology has depended on imperial explorers, colonists, military officers, government officials, and business magnates who have turned natural materials into incredible wealth.[21] Furthermore, it has played a special role in colonialism and native dispossession in the United States. In this, paleontology is not so different from mining and petroleum geology, two commercially focused subfields of the discipline. It has joined in the dynamic common among extractive enterprises in which educated experts—usually white men—travel to rural places in search of "buried treasures." Once they find them, they remove them and send them off for the use and enjoyment of people elsewhere. This dynamic has shaped how fossils have been displayed and interpreted.[22] In paleontology today, these dynamics have been challenged but have not disappeared. Some fossils are now kept and displayed closer to where they were found. Some landowners receive compensation when fossils are found on their land. Paleontology has had resource rushes, booms, and busts, not unlike those in the gold, coal, and petroleum industries. Dinosaur fossils, like fossil fuels, are often believed to "magically" generate wealth from nature.[23] This has led to contentious, often litigious, struggles over the ownership of fossil quarries and access to them, as the saga of the *Tyrannosaurus rex* called "Sue" dramatically illustrates.[24] Although the prices for spectacular dinosaur fossils have dramatically increased of late, it is important to remember that fossils have never been priceless.[25]

The political-economic and social dynamics of prospecting, extraction, and exchange in paleontology often resemble those in other subfields of geology. They share common technical expertise, discourse, and habitus. Paleontology, like mining and petroleum geology, hinges on the unpredictable hunt for the earth's hidden treasures, with its cycles of boom and bust. There are important differences from these more lucrative subfields of geology, however. Paleontology does not operate through an extractivist logic, according to which fossils' value is assessed solely in economic terms.[26] In paleontology, fossils are also valued for their ability to increase knowledge, to educate, and to be aesthetically pleasing. Moreover, the most prized fossils are the ones that are different from those already known, not those that match the standard and are thus easier to commercialize. Because paleontology is seen as less exploitative than other forms of geology, it is sometimes used (or, one might say, abused) to lend a magnanimous aura to commercial extraction. Oil and mining companies often sponsor paleontological research, exhibits, and education programs.[27] Many of the volunteers I encountered at paleontology labs, field stations, and museums were retired from careers in commercial geology and were dedicated to using their expertise magnanimously in research and education. Instead of feeding the desire for plentiful cheap energy in the United States, as much geology does, paleontology feeds the desire for the newest and the largest.[28] Barely a week goes by without a news story about the discovery of a new species, or a trove of fossils, or a novel theory about dinosaurs.

A FULL-BODIED SCIENCE

The critical study of science in society is being increasingly eclipsed by science and technology studies (STS) because, as STS scholars argue, science and technology are now inseparable.[29] A technology can be any object or procedure that extends human capacities to intervene in the world or that becomes people's "full partners in . . . the dance of world-making."[30] While technology includes ancient stone axes and methods for making fire, STS scholars have tended to focus on the novel biotechnologies and digital technologies in "cutting-edge" laboratories. Studies of science in action in these spaces show how high-tech machines are themselves actors (or "actants" in actor-network-theory terms) intimately involved in generating the "natural" objects they study, as well as generating the knowledge about them. Instead of understanding specimens "as things to be discovered," these studies suggest that "the process of 'discovery' is increasingly one of active production."[31]

Making Our Beasts shows that "active production" did not begin with the advent of high-tech laboratory science. As chapter 4 highlights, fossil making has been going on in paleontology since its emergence in the nineteenth century. Today, the making of scientific specimens and knowledge together occurs not only in the newest high-tech laboratories, but also in ostensibly old-fashioned labs and field sites far from prestigious research institutes. Vertebrate paleontologists increasingly use technologies such as CAT scans, CGI modeling, and GIS mapping— yet fossil excavation, preparation, and observation are largely conducted with the same technologies that were used in the nineteenth century. Paleontological mapping today employs both remote sensing technologies and pencil, paper, and measuring tape. In the shift toward ever more complex technoscience, the ongoing scientific research conducted with human muscles and simple hand tools has been disregarded.[32] *Making Our Beasts* illustrates its continued relevance.

I quickly realized the importance of "low-tech" methods in paleontology while conducting field research in the Hell Creek Formation of the Dakota Badlands. Although this region has changed tremendously since the first fossil hunters started digging there, it remains a difficult place to conduct scientific research. After driving many miles from the closest town, first on a desolate highway, then on unpaved roads and cow paths, we still had to hike several miles across sandy hills to get to our excavation site. After being carried in trucks and on human backs for miles, our equipment had to be able to withstand high heat and blowing sand at the quarries. There was no way we could connect that equipment to electric lines in the field, nor could we create a controlled environment. Severe storms arose with little notice. Even in the North Dakota town where we stayed, these storms would flood streets and cut off the electricity.[33] For reasons like these, paleontology continues to rely on durable nonelectrical technologies that have withstood the test of time: rock hammers, scratch awls, dental picks, and whisk brooms. Across the Western United States and beyond, doing paleontological research requires working under conditions people cannot regulate. While I was undertaking field-work in Montana, a storm uplifted my tent and blew it against a barbed-wire fence. The Kaiparowits Plateau in southern Utah, where several paleontologists I know conduct research, is an even more challenging location (see map and conclusion). Even the paleontology laboratories in climate-controlled facilities where I have worked or visited do not resemble the high-tech theater portrayed in *Jurassic Park*. They are best described as "a hardware store, a tinkerer's workshop, and an artist's studio" rolled into one.[34] This book counterbalances the tendency of STS scholars to examine the development of new scientific technologies and to ignore the ways that so much of science continues to rely on seemingly archaic technologies and grueling human labor.

Participating in paleontological fieldwork taught me the importance of touch in this science. It led me to question the claim that the sciences privilege vision above other senses, showing instead how paleontology relies on touch and sight together.[35] Despite the increased use of remote-sensing technologies, the basis of paleontological research still involves physical contact between human bodies and earthly matter.[36] Chapter 2 illustrates how the handling of fossilized *Tyrannosaurus rex* teeth has reinforced the species' exotic and vicious nature. Chapter 3 demonstrates how people's intimate physical contact with other fossils forges personal connections across millions of years, countering the pervasive conception of prehistoric animals as vicious beasts wholly unlike modern animals.[37]

Paleontology challenges not only the notion that doing science today necessitates high-tech equipment but also the idea that science requires its practitioners to maintain a dispassionate disposition and a detached relationship to their objects of study. Paleontological research requires both curiosity about prehistoric animals and tender care of their remains. Paleontology, like environmental conservation, is "passionate practice, energized by the enthusiasm of scientists, volunteers, and other publics."[38] In chapter 2, I examine the role of charisma in paleontology, focusing on the interplay between the charisma of paleontologists and that of the prehistoric animals they research. Then, in chapter 3, I describe a fossil excavator who developed such a personal and passionate relationship with the remains of the prehistoric animal she was excavating that she cried over its death. Whether professionals or volunteers, many of those who work in paleontology form affective relationships with fossils that defy the scientific archetype of the unfeeling researcher who objectively probes specimens in search of new knowledge. By *affect*,

I mean the sensory experiences and somatic feelings that complement thinking, but that are too often misrecognized as the opposite of thought and as outside science. I build on Natasha Myers's assertion that life-science labwork is a "full-bodied practice" in which "seeing, feeling, and knowing are entangled."[39] I extend this line of argument both to research spaces outside laboratories and to people who do not have (and are not pursuing) doctoral degrees. Paleontology is sometimes looked down on by scientists in other fields for being too old-fashioned, too popular, too impassioned. Paleontology is all these things, but that does not make it inferior to other sciences.[40] In fact, I argue, science requires affective engagement with the material world. Paleontology offers a paradigm for how it can be done across the sciences.

MEANING, VALUE, AND MATERIALITY EMERGE TOGETHER

Dinosaurs are "evocative objects" that impart strong "emotions and ideas of startling intensity."[41] While extinct dodos may "linger in our imaginations," dinosaurs inhabit many people's everyday lives.[42] You can find them in so many places across the United States: in classrooms, playrooms, and living rooms; in science museums, amusement parks, and gas stations; in books, television shows, and social media sites; in corporate logos and song lyrics; stamped on clothing and formed into foods, from breakfast cereals to chicken nuggets. They are used, especially, in educating and socializing children. Many children in the United States know more about some dinosaur species than they do about the myriad mammalian ones alive today.[43] I have bathed my children, quite literally, in water filled with (toy) animals understood to be the most ferocious ever to live on earth. I have seen *Tyrannosaurus rex* on my nieces' and nephews' burp cloths, towels, sheets, pajamas, and room decorations. The number of dinosaur-themed toys is astounding. They include everything from plush "stuffies" to sophisticated video games. Dinosaurs have become "childhood idols," equivalent to teens' and adults' movie and sports stars.[44] Although dinosaurs are often associated with childhood, my fieldwork showed me that they are not only a children's phenomenon. Plenty of adults participate, too. Dinosaurs are one of the few interests that bridge political divides in the United States.[45] If you take a minute to think about the proliferation of dinosaurs in the United States today, it is quite strange.

I realized the immense cultural elaboration of paleontology in the United States after I noticed its absence in Argentina. While conducting fieldwork in fossil-rich Patagonia, I saw that the "discovery" of some of the world's most remarkable fossils incited far less excitement among Argentines than do less impressive finds in the United States.[46] Although I found dinosaurs in natural history museums in Argentina, they were not displayed with the same prominence as in the United States. Dinosaur toys and media were popular inasmuch as they were *Estadounidense* (of the United States), not because they were dinosaurs. Although dinosaur fossils have been found throughout the world, dinosaurs are symbols of Americanness in Argentina. Appreciating that got me thinking about their prominence within the United States, something I had taken for granted having grown up there.

Dinosaur books, films, television shows, toys, and internet sites are often aimed at "hooking" children and adults on the excitement of science. Rather than privileging scientific accuracy, this media often uses dinosaurs' popularity to instill the value of science. But not the value of *all* science. It is most often the heroic adventure of scientific discovery, which celebrates fieldwork and ignores labwork. In the chapters that follow, I uncover the multiple ways that, in the United States, dinosaur fossils, casts, and other objects come into being infused with messages about scientists and scientific research, about gender and kinship, about hierarchy, violence, progress, nature, and about what is human and what is not.

Making Our Beasts demonstrates that dinosaurs' material characteristics, physical forms, scientific importance, social uses, cultural and moral significances, and economic value have formed and transformed together. Materiality, meanings, values—these cannot be separated or temporally sequenced. By *materiality*, I mean the physical and sensual qualities of matter that cannot be reduced to abstractions, symbols, or language. Materiality names the combined characteristics that make a difference in how matter interacts in the world.[47] Scholars often contrast the materiality and the sociality of objects, but here I show that materiality and sociality develop together and transform together. Like other natural resources, fossils are not just sitting "'out there' ready to be seized upon and utilized." They come into being within the particular networks of "substances, technologies, discourses, and the practices deployed by different kinds of actors."[48] Chapter 2, for instance, explains how *Tyrannosaurus rex*'s enormous gaping mouth with gigantic jagged teeth came into being together with the notion that the species was a tyrannical king that ruled life in the Cretaceous period. Of course, there were very large animals living millions of years ago, some of whose bones turned to rock. In addition, there were centuries-old ideas about giant beasts and tyrannical kings in the Euro-American tradition. But it was not until the twentieth century that a species of prehistoric animals with common anatomical features, behaviors, and charismatic appeal came into being.[49] In other words, *Tyrannosaurus rex* could not be "big, fierce and . . . scary" until remains of these prehistoric animals were excavated, prepared, studied, named, and reconstructed in a specific social, political, and economic context.[50] Key elements of this context include exploration, conquest, and settler colonialism in the rural Western United States (chapter 4), scientific philanthropy and volunteer-tourism among urbanites on the East Coast (chapter 1), and Euro-American ideas about beasts, the real, and nature versus culture (chapter 5).

When the anthropology of science was still new, Emily Martin challenged her colleagues and students to "wake up the sleeping metaphors in science"—that is, to uncover the powerful assumptions, values, ideas, and ideals that underlie scientific fields.[51] *Making Our Beasts* does this for vertebrate paleontology in the United

States. Yet, at the same time, it counters the portrayal of dinosaurs as a tabula rasa on which ideas about the social/natural world are inscribed. Fossils and stones are *not* "blank containers of values or meanings that can be swiftly replaced."[52] Fossils are made, but they cannot be made into anything or everything. *Making Our Beasts* traces the mutual creation and transformation of the guiding ideas and material manifestations of paleontology together.

My ethnographic methodology for studying dinosaur paleontology provides a counterpoint to the approach of cultural studies scholars who interpret dinosaurs as icons or "texts" that reflect the dominant ideals, assumptions, and values of the United States. This scholarship has insightfully revealed how contemporary understandings of dinosaurs, and their prehistoric world, reflect dominant American ideas about nature, masculinity, colonialism, and savagery, authenticity, progress, capitalism, modernity, and postmodernity.[53] This scholarship on dinosaurs makes too sweeping claims about the meanings of dinosaurs, and unlike most recent work in cultural studies, tends to neglect the ways that race, gender, and other differences shape peoples' experiences with science and popular culture. Even more problematically, the cultural studies' approach presents the physical remains of dinosaurs as infinitely malleable social objects and portrays the science of paleontology as nothing more than a reflection of society. For example, Susan Willis argues that "extinction has turned dinosaurs into vessels for all sorts of encoded meanings." Unlike living species whose "characteristics restrict the imagination," dinosaurs can be "molded and shaped to embody a huge variety of cultural meanings and serve a range of personal needs."[54] This neglects the power of materiality and the rigor of scientific practice. W. J. T. Mitchell's tongue-in-cheek *The Last Dinosaur Book* exemplifies the cultural studies approach to dinosaurs. Mitchell argues that dinosaurs are American cultural icons that mirror *the* American national identity, smoothly morphing to reflect each historical era in the United States. In contrast, *Making Our Beasts* argues that dinosaurs come into being through unpredictable more-than-human processes that are thoroughly physical *and* conceptual, creative *and* factual. I show that neither scientists nor nonscientists can treat the material world as a blank screen onto which they can write anything they want; they work with its materiality in forging objects and their meanings. Despite my differing approach to meaning and materiality, I share with cultural studies scholars an interest in how the contemporary manifestations of long-dead animals shape the contemporary world.

FOSSILS ARE MADE, NOT DISCOVERED

One of the central arguments of *Making Our Beasts* is that fossils are made, not discovered. It can be difficult to comprehend this because fossil making is so strongly concealed by the myth of scientific discovery. Paleontologists, fossil excavators, preparators, museum docents, teachers, and nearly everyone else I met during my research talked about "discovering" fossils. To discover is, in essence, to make public what is already there but not yet known. For something to count as a discovery, it must be announced by a person with the right status in the right time and place, and taken by influential others to be both true and unprecedented.[55] The "fairy tale" of discovery depends on a genius who recognizes what others have not seen.[56] In paleontology's fairy tale of discovery, fossils are the hidden ladies passively waiting to be rescued and taken to the scientific palace by the heroic explorer prince. Since the founding of the United States, the search for fossils has been framed as a scientific adventure to discover the treasures of the country's vast, rich, and untapped lands. These treasures were not only impressive skeletons to be displayed in museums and minerals to fuel industrial production. They also generated scientific knowledge about the earth. Thomas Jefferson, among other "founding fathers," used both this knowledge and these materials in efforts to prove the country's greatness and denounce claims of European superiority.

This discourse of paleontological discovery is exemplified today by DinoLand U.S.A. in Animal Kingdom, part of Disney's Florida entertainment complex (described in chapter 4). Its "Boneyard Playground" is an elaborate oversized version of the sandbox dig pits I found at nearly every public paleontology site I visited. Disney's version creates a paleontological excavation site next to an Old West town, where children and adults can effortlessly "discover" dinosaur skeletons—with all bones intact and articulated in anatomical position—hidden beneath the sand. As they do so, they are encouraged to imagine themselves as the saviors of dinosaurs, or at least of their fossil remains. This simulation is far from the reality of paleontological fieldwork that I describe in early chapters of the book. For one thing, this fieldwork requires days and days of prospecting to find a promising excavation site. Even at the best sites, paleontologists never find a complete and articulated skeleton. More importantly, as I describe in chapter 1, fossils come into being through an unpredictable and complex process of making.

My research has shown me that scientific discovery is a powerful fable that illustrates cultural values, not the actual process of scientific inquiry. The notion of *discovery* is used in and beyond science to suggest that vast stretches of pristine wilderness in western North America remained untouched by people until white explorers arrived.[57] The ideal of scientific discovery has been deployed to dispossess Indigenous people of their homelands, material resources, and knowledge. Disney's elaborate dig pit simulates a scene that has never, and could never, occur. Yet it powerfully communicates that the colonists who extracted dinosaur fossils (much like gold, coal, and oil) from the ostensible no-man's-land of the Western United States were the first to find and know these valuable objects. It further implies that anyone who locates and removes fossils is doing good by making them available for generating scientific knowledge and, therefore, contributing to societal progress. The ubiquitous talk, images, and simulations of scientific discovery that are used within and about paleontology obscure the complex and uncertain process by which fossils come into being. The somatic practices of the people I worked with in paleontological quarries and laboratories tell a different, more complicated, story about how scientific specimens and knowledge come into being together.

Fossils are not stable and fully formed objects hidden in the ground, just waiting to be found. They emerge through a complex and unpredictable process that involves much more than human acts of locating and reconstructing prehistoric nature, or ascribing meaning and value to inert matter.[58] The role of the scientist is not to be a "totalitarian ruler over a world of parts whose integration is already given by Nature," but instead to be a "go-between, in amongst a world of non-parts that must be coaxed into getting along together."[59] For example, chapter 1 describes how, in the Denver Museum's paleontology laboratory, fossil preparators "coax" pieces of rock "into getting along together," with the indispensable assistance of dental picks, brushes, compressed air, and plastics, in order to form whole fossils. However, long before these pieces reach the lab, the emergence of a fossil requires a fortuitous combination of numerous human and nonhuman forces. These include the growth and decay of a prehistoric animal's bones, as well as the sedimentation, mineralization, and erosion of the rock in the area it lived and died. The expertise of paleontologists, fossil excavators, and preparators, the labor of volunteers, and the capital of wealthy patrons of science are also needed to form a fossil. The prehistoric animals that inhabit the world today would not exist without the unintended collaboration of these and other heterogeneous actors and forces, spread over millions of years.[60]

Paleontological practitioners work arduously and creatively to bring fossils into being, but this does not mean that people make them in any way they please. Although I illuminate the human involvement in fossils' coming into being throughout the book, I refuse the cultural constructionist (and creationist) claim that people make dinosaurs into "mirrors" of their social world. As any craftsperson will tell you, they do not impose their will onto inert matter, but cooperate with it to bring something new into being.[61] Fossil making requires that fossil excavators and preparators cooperate with rock, not the most pliable of materials. This approach to fossil making as a more-than-human craft reveals that the conventional distinction between specimens (natural objects created by nonhumans) and artifacts or artworks (cultural objects created by people) is untenable. I elaborate this argument in chapter 5 as I scrutinize the conventional dichotomy between "real" fossils and casts, and I suggest that paleontological objects form a continuum rather than a binary opposition.

DINOMANIA!

The famed paleontologist Stephen Jay Gould coined the term "dinomania" for the immense popularity of dinosaurs in the United States among children and adults alike. Paleontologist José Luis Sanz elaborates this further, claiming that dinomania is the practice of "surround[ing] oneself with all forms of dinosaur iconography" and amassing encyclopedic "information about their appearance, size, and way of life," which is found most extensively, but not exclusively, in the United States.[62] I encountered many people—children and adults, scientists and nonscientists—who do this. What Gould and Sanz do not specify is that the majority of these enthusiasts are white and male. This points to the fact that dinomania is not explained by universal human psychology but by a particular form of socialization. It may be true that learning about dinosaurs and playing with models of them help children to learn to manage their fears, as some have suggested, but both the nature of these fears and this way of dealing with them are culturally specific phenomena. Moreover, dinomania is not sui generis. It did not arise either spontaneously or accidentally. We need to look for explanations of dinomania in the confluence of social and political-economic forces in the United States, not in brain chemistry. This means that dinomania has a history.

Historians, paleontologists, cultural studies scholars, and others have debated where and when dinomania emerged, with claims ranging from England in the 1850s to the United States in the 1970s.[63] By the time the English gentleman-scientist Richard Owen coined the term "dinosaur" in 1842, geology was already the most popular science in the English-speaking world.[64] The "discovery" of large numbers of strange and enormous animals embedded in the geological formations of North America fed dinomania's development on both sides of the North Atlantic (chapter 2). As I elaborate in chapter 4, the remains of enormous prehistoric creatures took on far greater significance as they became intertwined with the founding of the United States and the effort to establish its global reputation.[65] Before the United States declared independence from England, Thomas Jefferson already was promoting the American *incognitum* or "mammoth" as proof of North America's natural greatness against European claims of America's natural inferiority.[66] After he became president, he sent expeditions to explore the territory and to assess the promise for white settlement, farming, mining, and industry. Jefferson hoped that the explorers would also find more evidence of animals that would match, or even best, the size and strength of those found in Europe.[67] They did not find living mammoth, but they did find fossils of enormous extinct animals. These charismatic prehistoric megafauna have been deployed as symbols of American greatness, for audiences both inside and outside the United States, ever since.

North American scientists joined European ones in developing the gentleman's hobby of natural history collection, transforming it into the scientific field of paleontology. The American geologist and doctor Ferdinand V. Hayden is credited with discovering the first North American dinosaur fossils in Montana in 1855, just one of several influential remains that established the United States as a dinosaur fossil "treasure trove." The reputation of the United States was solidified by the famous "bone wars" that took place from the 1870s to the 1890s between paleontologists Othniel C. Marsh and Edward D. Cope, each of whom viciously competed to amass more fossils and name more species than the other. In the United States, the new science blossomed in museums as much as, if not more than, at the country's fledgling universities (chapter 2). Marsh's and Cope's capacious collecting rapidly filled collection rooms with tons and tons of dinosaur and other fossils, some of which have yet to be cataloged.

These early American paleontologists focused on identifying, describing, and naming new species more than on understanding the evolution of life on earth. As I discuss further in chapter 2, charismatic paleontologists helped to make the new science a public affair, with nonscientists attending talks and following newspaper accounts of the latest paleontological "discoveries." Following Charles Willson Peale's popular display of a mastodon skeleton at the American Philosophical Society in Philadelphia in 1802, Cope put a Hadrosaur skeleton on display at the city's Academy of Natural Sciences in 1878. Visitor attendance at the museum nearly doubled. Anthropologist Franz Boas, who worked at the American Museum of Natural History in New York in the beginning of the twentieth century, quickly recognized that "mounted dinosaurs had an uncanny power to capture people's imaginations and thus ensured a steady stream of visitors to the museum."[68] Some scientists "complained that huge crowds were forced to 'move in nearly continued streams'" quickly through the exhibition, "affording little opportunity for the examination of the specimens."[69]

Paleontologists have had a complicated relationship with dinomania. The first generation of paleontologists in the United States was more interested in prehistoric mammals and marine animals than in dinosaurs. They bemoaned the fact that the public interest in mounted dinosaur skeletons took attention away from their mission to study the fossils. One could argue that it was not these scientists but others—fossil prospectors and collectors, World's Fair visitors and museum-goers, newspaper owners, journalists, and their readers—who pushed paleontology to focus on the giant land animals of the Mesozoic era, prioritizing them over other species of prehistoric fauna, and over all prehistoric flora. The wealthy capitalists who funded paleontological expeditions and exhibitions, and who even bankrolled whole natural history museums, pushed paleontology further toward dinosaurs.[70] The scientists' grumbling notwithstanding, a plethora of museum exhibitions and traveling expositions were erected across the United States, drawing enormous crowds. The 1933–34 Chicago World's Fair included two exhibitions featuring animatronic Tyrannosaurus, Triceratops, and Stegosaurus. Chapter 4 explores the metamorphosis of this history into DinoLand U.S.A. at Disney World in Florida. While plenty of people around the world are interested in dinosaurs, dinomania is highly elaborated in the United States.

Stephen Jay Gould's influential essay naming "dinomania" credits the commodification of dinosaurs by the American entertainment industries for its emergence and growth.[71] For him, Steven Spielberg's film version of Michael Crichton's book *Jurassic Park* (1993) was a turning point. In his review of the movie, Gould admitted that "the boy dinosaur enthusiast still dwells within me" and that he had seen all the dinosaur films. *Jurassic Park*'s combination of compelling story and technological innovation hit "a spectacular new level of achievement."[72] Yet *Jurassic Park* grew out of popular American traditions in writing and cinema that date back to at least the beginning of the twentieth century.[73] It took less than twenty-five years from the time the very first film was shown in a public theater in France until dinosaur-themed films began to charm large audiences in the United States. While dinosaur movies fit within established American print and cinematic genres, like science fiction and cartoon, they also can be seen as a genre of their own.[74] From *The Lost World* (1925) and *King Kong* (1933) to the *Jurassic Park* series (1993–), movies featuring dinosaurs have ranked among the most popular of all time.

American dinosaur films have some distinctive themes that have recurred over more than a century. None is more prominent than ferocious conflict between people and dinosaurs.[75] The violence of these movies has continued the convention of portraying prehistoric animals as primarily struggling for survival, which reaches back to the Crystal Palace in England, where, beginning in the 1850s, sculptural reconstructions of prehistoric creatures were exhibited in violent poses and competitive interactions.[76] Most American dinosaur films fit a pattern that José Luis Sanz identifies as "dinosaurs versus civilization."[77] In these, dinosaurs threaten human life, disrupt the social order, and wreak havoc on its physical environment. Conversely, one of the first animated films ever made was *Gertie the Dinosaur* (1914), in which the dinosaur is a sweet and shy, if gigantic, pet.[78] While many American dinosaur filmmakers aim to impress their audience with the latest moving-image technologies, these movies also try to impart a message to viewers. For a long time, their lesson was that human intelligence and creativity, represented by scientists, will triumph over the brute force of nature, represented by prehistoric beasts. Sometimes, though, paleontologists are "mad scientists" who defend dinosaurs' lives despite the risks to humans. Regardless, dinosaurs terrorize everyone in these films, including the scientists. More recently, the message of dinosaur movies has shifted away from celebrating human control over nature and toward warning against the hubris of the belief that humans can manage natural forces and bend them to their desires.[79] However, it is important to remember that dinomania does not guarantee a dinosaur-themed movie's success. There have been many flops! It is even more important to recognize that media-makers cannot control what it is that viewers take away from their creations. The anthropology of media has demonstrated the surprising ideas that viewers find in the media they consume. For instance, chapter 5 explores how a National Geographic special about *Tyrannosaurus rex* communicates unexpected lessons about evolution alongside its predictable celebration of violence.

Scholars and popular critics have debated whether science or the arts has been the driving force in the development of dinomania. Some argue that paleontologists generate knowledge, which is then adopted by artists and commercialized by entertainment industries. For example, *Tyrannosaurus rex* in media shifted their posture, and then gained feathers, after paleontologists revised their understanding of the species' anatomy and physiology.[80] Conversely, historian of science Hugh Torrens insists that "the growth of dinomania has occurred almost independently of the growth of proper scientific knowledge."[81] This either/or debate misses the many ways in which science and art are intertwined in paleontology. For instance, in the eighteenth and nineteenth centuries, it was common for earth scientists to incorporate poetry into their scientific publications and to put on dramatic fossil shows.[82] Moreover, artistic works have themselves inspired scientific explanations.[83] Today, science and art are sharply divided fields of study, with science hailed as the acme of objectivity and art as the height of subjectivity. Yet, paleontology demonstrates the untenability of this separation.[84] Paleontological restoration is a good illustration of this. People with expertise in anatomy, physiology, the fossil record, and artistic representation work together to reconstruct the bodies and behaviors of prehistoric animals and to create two- and three-dimensional models. While the collaborations involved in paleontological restoration may be unusual today, all science involves creativity and interpretation.

The entertainment industries have been undeniably important in sustaining dinomania, but so too have the US government and its institutions, though these have received far less attention. The United States would not have risen rapidly to become a global leader in paleontology were it not for its government.[85] Since the nation's founding, paleontologists have been involved with government exploration expeditions and geological surveys. Among them was Othniel C. Marsh of the "bone wars," who served as chief paleontologist at the United States Geological Survey (USGS) for a decade, while also a professor at Yale University.[86] Marsh's career was enabled by his wealthy uncle George Peabody, yet his voluminous fossil collecting, and that of his less affluent rival Cope, would not have been possible without government support. Funding through agencies, such as the Bureau of Indian Affairs, accompanied other essential government assistance. The US military enabled copious fossil collecting in the midst of its bloody colonization of the Western territories. The army provided Marsh's fossil-hunting expeditions with bases of operations and supplies as well as military escorts. White soldiers and Pawnee scouts provided protection to the Yale fossil hunters and took part in searching for fossils themselves.[87] Although the fossils collected during government-funded expeditions were supposed to enter national museum collections, many ended up in private and university collections instead.[88] The Department of the Interior still holds millions of fossils in nonmuseum facilities.[89] When the Petrified Forest in Arizona was declared a national monument in 1916 (later becoming a national park), that act laid the groundwork for the paleontology program at the National Park Service, which today manages more than twenty national parks and monuments that protect and display fossils for research, education, and popular enjoyment.[90] These national parks receive millions of visitors yearly.[91]

Even after surveying the development of dinomania, a question remains to be answered: Why do dinosaurs captivate so many people? When I asked the people

I met during fieldwork, I usually received an answer like one that a psychologist offered to Stephen Jay Gould: Dinosaurs are popular because they are "'big, fierce, and extinct'—in other words, alluringly scary, but basically safe."[92] There is much truth in this answer. Yet it rests on an assumption of universal human emotions, making it both ahistorical and decontextual. Gould remarks that "dinosaurs were just as big, as fierce, and as extinct" when he was a child in the mid-twentieth century, "but only a few nerdy kids [like himself], and even fewer professional paleontologists, gave a damn about them."[93] As any anthropologist will tell you, interests and desires are culturally constructed, learned, and changing, no matter how ingrained they feel. There is nothing in human nature that automatically makes people fascinated, or scared, by long-dead animals. Even in the most intense moments of dinomania, these charismatic prehistoric megafauna, like charismatic religious leaders, strongly draw in some people but do not appeal to others (chapter 2). There is another problem with the big-fierce-extinct trifecta: it fails to recognize that the materiality of dinosaur fossils (and casts and models, too) changes as its sociality does.[94] The materiality of an object cannot be reduced to a list of the physical properties irrespective of time and place any more than a person's affect can be reduced to a list of psychological characteristics or sociality reduced to static cultural features. There is no short and easy answer to why dinosaurs are so popular in the United States. It is explained as much by the country's cultural, political-economic, and historical context as by the biological characteristics of the prehistoric animals themselves. In the chapters that follow, I reveal a plethora of social and material reasons for dinosaurs' popularity, from ongoing settler colonialism in the country, to the ways that jaw and teeth fossils are prepared in paleontology laboratories, to the history of popular representations of beasts and the scientific naming of new species.

LITHIC AGENCY

While some scholars view dinosaurs as malleable cultural constructions, most people who visit parks like Dinosaur Ridge assume that the fossils they see entombed in rock are traces of prehistory that have not changed in millions of years. Most visitors to the Denver and Morrison museums seeing fossilized bones believe they represent the animals' bones as they were in life, that the mounted skeletons depict their bodies caught in the midst of a struggle for survival. I thought this too before I began this research. How did I not know better? Like most readers, I grew up with powerful ideas about rock and its uses: monuments, buildings, sayings, music, emblems, and logos that use rock to communicate strength, stability, endurance, and immutability. Here, rock is perceived as "ahistorical, nonpolitical and inorganic."[95] In the Euro-American tradition, the earth's lithic crust is conceived as the passive ground on which human history stands. Earthquakes and volcanoes sometimes challenge this understanding, yet humans and rocks are largely perceived as opposites: humans are sentient and agentive, lithic bodies are insentient and inert. My fieldwork showed me how wrong these ideas are. I learned—from geologists, paleontologists, fossil excavators and preparators, from sandstones and shales, from the Dakota Badlands and the Rocky Mountain Front Range—that rock is transient and active. While rock forms on a time scale beyond my perception, I saw how lithic matter disintegrates if it is not protected from wind and water. As I recount in chapter 1, I saw rock turn to dust before my eyes.[96] I felt it break my skin and crumble in my hands.

My analysis of paleontology in this book brings together political-ecological and posthumanist STS approaches to the material world. Political ecology has developed sharp tools for probing the making of natural resources under the highly unequal conditions of capitalism.[97] While most of this work focuses on industrial resources, several recent ethnographies of gemstones are especially helpful for highlighting how lithic matter is much more than raw material for capitalism.[98] Uniting a political ecology approach to resource extraction with a cultural biography of things, these studies reconceptualize gemstones as complex assemblages of matter, labor, knowledge, meaning, and value.[99] They demonstrate the intertwined construction of stone's symbolic meaning and economic value within specific social and political-economic contexts that are, furthermore, embedded in uneven global commodity circuits.[100] Chapter 1 discusses some of the insights from the ethnography of gemstones that pertain to fossils. Both gems and fossils have long been exchanged through transnational markets that capitalize on a combination of specialized knowledge and expert discernment, on the one hand, and popular appeal, on the other. Large and unusual gems and fossils have been subject to elaborate narrations of their discovery and conflicts over their ownership, which have increased their financial value.[101] For North American consumers in particular, both gems and fossils inspire awe and wonder at "nature," conceived as prior to and independent of human history and social relations (chapter 5). Yet fossils are objects of scientific study—specimens—as much as they are elite commodities. I therefore fuse this political ecological approach to precious stones with STS's insights into the mutual creation of scientific specimens and knowledge, and its conception of more-than-human agency.

Work in posthumanist STS challenges scholars to reconsider not only how natural resources become commodities, gifts, and symbols but, more radically, how these materials *act* in the world. STS scholarship moves beyond appreciating how the physical properties of materials can assist or stymie people's efforts to mold them; it also asserts that nonhuman entities partake in making the social world. In other words, rock has agency. This conception of agency breaks away from its standard dictionary definition as the human ability to exert power over other people or things. By *agency*, I mean the capacity of human and nonhuman bodies to effect change collectively in the world. No one, and no thing, can act alone. Jane Bennett calls this conception of agency "distributive" because it "depends on the collaboration, cooperation, or interactive interference of many bodies and forces."[102] Scientists not only rely on a wide range of people in order to create knowledge but also work together with numerous nonhumans. Their collective achievements are not the result of intentional and concerted efforts to alter the course of events. Most scientific "discoveries" are the result of a host of purposeful and accidental actions fortuitously coming together in the right time and place. However, to argue that entities other than humans exert agency in science is not to claim that people and things have equal ability to control the course of history. There are significant inequalities within distributive agency. Just think of mass extinction's different causes in the current Anthropocene and in the Mesozoic era, when humans did not yet exist.

The agency of things I describe here is closely identified with how agency is theorized in Actor Network Theory (ANT), but my approach differs in several important respects. Latour began to elaborate nonhuman agency through his close readings of mundane objects, such as door-closers and keys.[103] However, his and other classic ANT studies neglect the unstable coming into being and ongoing transformations of these things in a world of highly unequal power dynamics.[104] Latour presents the door-closer, for instance, as a stable object without a history. Many anthropologists in the material cultural tradition make a similar mistake. Daniel Miller contests "our common-sense opposition between persons and things, the animate and the inanimate, the subject and the object," through his study of commonplace things, such as clothing and houses.[105] While he analyzes how these objects are interpolated in the formation of human personhood, he does not examine the socio-material and political-economic formations and transformations of the objects themselves. The things seem static. Therefore, while Miller and other scholars who piloted the materialist *return* in the twenty-first century have brought renewed attention to the agency of objects, they have not fully integrated either the global political-economic context of capitalism or the instability of objects into their analyses. Nonetheless, this scholarship has helped me to challenge the conventional distinction between specimens, artifacts, and artworks, a theme I return to in chapter 5.

To draw attention to the vitality of matter, some posthumanist STS scholars have developed a materialism focused on material flows rather than on the agency of fixed objects. Inspired by Gilles Deleuze and Felix Guattari's *A Thousand Plateaus*, these scholars examine the continual material regeneration of bodies in a way that questions the dichotomy between organic and inorganic, between growing and making.[106] They assert that inorganic matter is, in some sense, alive. In this approach, lively (or vital) matter refers not to an inherent agency within a bounded object but to its ongoing becoming—that is, its continual formation, transformation, and disintegration. More-than-human ethnographies have begun to explore a range of such vital "existents," and their complex relationships to people, communities, and environments.[107] Although stones and fossils certainly fit Elizabeth Povinelli's notion of existents, they have not been analyzed in these terms.[108] *Making Our Beasts* does this: it analyzes fossils as objects that are continually coming into being through more-than-human processes.

Recognizing the continual emergence, vibrancy, and agency of rock is especially challenging because the lithic serves as an emblem of immutability in the dominant Euro-American tradition.[109] Bennett, who has been central to the development of a posthumanist approach to matter, as well as to distributive agency, writes that stones "greet us as stable bodies" but are actually "mobile, internally heterogeneous materials." Rock is misrecognized as immutable because its "pace of change [is] *slow* compared to the duration and velocity of the human bodies participating in and perceiving them." All objects, but especially lithic ones, seem static "because their becoming proceeds at a speed or a level below the threshold of human discernment."[110] Bennett employs the concept of "becoming" to highlight the distributive agency involved in making things, including rocks. In her reinterpretations of scientific texts against the grain, there is no individual actor, or even set of discrete actants, who creates things.

The geographer and sculptor David Paton provides an illustration of rock's vibrancy and agency through his work at an English granite quarry.[111] Despite their apparent hardness, granite objects are "radically ephemeral" and represent only a "brief stop on a much longer passage through time."[112] Paton shows that granite is not simply a raw material that people transform into useful or beautiful objects, such as a countertop, by applying human labor and creativity to the properties of this stone. Neither is a granite countertop an actant that does the work of a human butler by, say, holding a coffee pot. Paton, writing with fellow geographer Caitlin DeSilvey, contends that granite is best understood as "worked matter, generated through recurring encounters between persons and stone."[113] Put differently, stone objects are "impermanent by-products" of "the world's ongoing generation and regeneration."[114] In short, rock is not made once but is continually coming into being. The same can be said of humans. Donna Haraway has developed the allied notion, "become *with*," to highlight the continual mutual emergence of humans and "companion species," such as dogs.[115] Yet it is harder to perceive how people and fossils help make each other than to see how humans and living animals do.

The life story of Tyler Lyson, one of the paleontologists whom I got to know at the Denver Museum, provides a powerful illustration of how one paleontologist and some fossils have raised each other. Tyler grew up in Marmarth, North Dakota, a ranching town of fewer than 150 people that sits atop the Hell Creek Formation in the Badlands. While other kids in Marmarth honed their rodeo skills, Tyler spent his free time wandering his family's ranchlands in search of fossils. Even before he was in high school, he made some impressive finds—Triceratops horns containing the teeth of its predator, a group of prehistoric turtle shells, Hadrosaur remains. When he was sixteen years old, he excavated a Hadrosaur that ended up on display at the North Dakota Heritage Center. By the time Tyler began formal education in geology at an East Coast college, he was already an expert fossil surveyor, excavator, and preparator. He regularly spotted fossils that professionals missed. Kirk Johnson, a former Denver Museum paleontologist who went on to direct the Smithsonian's National Museum of Natural History, noticed this when Tyler was twelve years old. Johnson told a *Smithsonian Magazine* reporter, Tyler is "remarkable in the field because he's trained his eye since he was small. He can see everything."[116] More accurately, the Hell Creek Formation trained Tyler. It taught him to see the landscape, feel the rock, and sense its fossils far better than a textbook or professor could. Hundreds, maybe even thousands, of fossils would not exist today were it not for Tyler, and Tyler would not be the accomplished paleontologist he is without the Badlands rock. He did not discover these fossils alone, but worked with lithic matter and other materials to bring these fossils into being.

This approach to the world makes clear that rock is not merely the platform undergirding human action. Fossils, gemstones, and pieces of granite all pulse with activity. They interact, exchange with, and affect the matter around them. They each have their own history and development. They do things that affect others, but not in the same ways that people do. They are different, and act differently, from living organisms in many ways. While I draw inspiration from Bennett and allied scholars' dynamic conception of lithic matter, I join the growing critique of posthumanist scholarship's tendency to equate living and nonhuman bodies.[117] Rocks are vibrant but they are not alive in the way that humans are.[118] As Povinelli writes, they "extrude into their environment, changing wind patterns and leaving soil deposits, and they ingest the living that changes their geochemical imprint."[119] Because of this, I am careful to describe fossils as vibrant, but not lively, in the chapters that follow.

Paleontologists have taught me to understand that rock is an energetic substance. In posthumanist STS terms, it is agentive. It changes, and it changes things around it, too, albeit on a timescale that is incomprehensible to most nongeologists. As part of lithic formations, fossils are dynamic objects. Their matter undergoes incredible transformations as an animal grows and dies, as ground water permeates bone cavities, as organic structures are replaced by inorganic ones, as minerals cement and compact into sedimentary rock. Before becoming part of a fossil, its lithic matter moves from the surface of the earth to deep below it, and then rises to the surface once more. Where a floodplain once lay, a mountain now stands. Most of us only think of rock as agentive in the drama of an earthquake or volcano, but geologists understand that it is agentive all the time.

A fossil's "dance of agency" is extremely slow at some points, and elapses in less than a second in others.[120] Humans do not always lead this dance, and nonhuman matter does not always cooperate with the scientific choreographers' plans for it.[121] Yet the material transformation of fossiliferous rock accelerates when humans enter the scene. They help to disintegrate some parts of it while stabilizing other parts. A piece of rock only coheres into a fossil—an object that can move from a field quarry to a laboratory and then to a collection or display—in collaboration with glues, putties, and plasters, rock hammers, airscribes, and microscopes, human hands, eyes, muscles, and sweat. Like archeological and ethnographic artifacts, paleontological specimens only come into being when people "define, segment, detach, and carry them away" from the context of their creation.[122] Only when fossiliferous rock comes into contact with human bodies does a coherent individual fossil emerge. With people's help, plastic consolidants seep into the rock, while plaster strengthens and supports pieces of it. Adhesives fuse fragments together, and putties fill in holes. The composite objects that result from fossil-making processes not only comprise bone turned rock, putties, glues, and other materials. They also incorporate human labor, skill, knowledge, meaning, and value. Like the fossil excavators mentioned earlier, preparators told me about instances when, after spending countless hours carefully removing the sediments around a valuable fossil, the fossil itself crumbled to dust in their hands. Although people influence the form that rock takes, they cannot arrest its vibrancy. Scholars too often forget this when analyzing pieces of rock that have temporarily taken the form of religious icons, commodities, gifts, artifacts, or even specimens. *Making Our Beasts* attempts to amend this.

Understanding how fossils come into being requires integrating a political-economic approach that pays attention to power dynamics and inequality with a posthumanist STS approach that attends to the agency, vibrancy, and instability of matter. In other words, it requires fusing the two meanings of "materialism"—Marxist and posthumanist—in the social sciences. For Marxist cultural scholars such as Raymond Williams, materialism explains how capitalism shapes the ways that everyday things and works of art are manufactured, distributed, consumed, and, moreover, valued—aesthetically, morally, and monetarily.[123] For posthumanist STS scholars such as Karen Barad and Jane Bennett, materialism means that nonhuman matter is an active force in making the social and material world.[124] I unite these two conceptions. *Making Our Beasts* shows how scientists and specimens, amateur fossil hunters, dinosaur enthusiasts, and charismatic prehistoric animals, come into being together, made by the same forces, yet not in the same ways. It emphasizes the unequal relations among social and material forces and human and nonhuman bodies. Moreover, tracing specific instances of fossil making historically and ethnographically helps to move us beyond critiques of the Euro-American dichotomies of animate/inanimate, artifacts/specimens, geos/bios, growing/making, and human-creators/nonhuman-objects. I replace these with a contextual understanding of the formation and transformation of some iconic, and also iconoclastic, bodies that inhabit the contemporary social/material world today.

THE VIBRANCY OF ROCK IN ANTHROPOLOGY

One of the most renowned scholars of religion in the twentieth century, Mircea Eliade argued that people make pilgrimages to sacred stones the world over because they are "that which [the human] is not."[125] Eliade was writing about ostensibly nonmodern people, but you probably know people who have traveled far to visit extraordinary rocks: Mount Rushmore, Noble Rock, and the Western

Wall, the Rosetta Stone, Banalinga, Stonehenge. Scientists, too, make pilgrimages to see, touch, and even taste special stones. Geologists travel to an outcrop in northern Italy where you can clearly discern the boundary between the Mesozoic and Cenozoic Eras (long known as the K-T boundary, now called the K-Pg boundary) that indicates the mass extinction that wiped out most species of dinosaurs. Paleontologists revisit specific spots in the Western United States, where iconic dinosaurs were unearthed, in the hope of discovering a new prehistoric species or otherwise figuring out another piece of the evolutionary puzzle. But we usually distinguish scientists' expeditions from nonscientists' pilgrimages to rocks. The paleontological record contains dinosaur fossils that have names, genders, biographies, and the power to reveal truths. The ethnographic record is thick with sacred stones that share these same attributes. Anthropologists have done a great deal to challenge the classic distinction between *their* magic and *our* science, yet we usually think differently about lithic specimens and sacred stones.[126] We still tend to separate sacred stones that speak *to* people from scientific specimens that are studied *by* people. *Making Our Beasts* refuses this division. It brings together anthropological insights on the meanings and uses of rock with the critical study of science to understand fossils and people's relationships with them.

For far too long, anthropologists portrayed people who have relationships with stones and mountains as premodern atavists with false beliefs, while they saw themselves as relating to the material world with scientific reason.[127] For example, a few decades before I undertook fieldwork in the Great Plains, Lee Irwin conducted a very different kind of fieldwork in the region. In his 1994 ethnography, *The Dream Seekers: Native American Visionary Traditions of the Great Plains*, Irwin discusses the importance of sacred stones for the religious practice of the region's Indigenous peoples. He presents these stones as inanimate objects that native people imagine to be animate beings. Recently, anthropologists have begun to reconceive Indigenous and other peoples' relations to stones and mountains as beings in terms of subaltern ontologies—that is, oppressed conceptions of existence and ways of being in the world. Roger Sansi Roca explains a striking ontology found in Candomblé, a religion of African origin practiced in Brazil, which approaches lithic and human bodies in a way that is probably unfamiliar to most readers. In this context, sacred stones "almost become an exterior organ of [the practicant's] body, a part of her 'distributed person.'"[128] The agency of Candomblé stones does not emanate from them independently but from their physical relationship with Candomblé's human practitioners. Because the body of a sacred stone is "radically different from the human body," their agency is different, too.[129] While Candomblé's human saints eat and dance, sacred stones sit and host feasts. Candomblé presents an alternative to the hegemonic Euro-American ontology, yet it does not exist in a bubble untouched by the world around it. Candomblé practitioners must contend with a dominant ontology that rejects lithic agency. Therefore, it is not enough to recognize multiple overlapping understandings of personhood, being, and matter;

one must also acknowledge the unequal power of different ontologies in the current configuration of the world.[130] While Sansi Roca and other anthropologists explore alternatives to the Euro-American understanding of the earth's materiality, I examine a physical science that embodies the Euro-American ontology.[131]

At the same time that anthropologists have taken renewed interest in the diversity of ways that people relate to the earth's materiality, the Anthropocene has prompted social scientists to scrutinize the hegemonic Euro-American understanding of the material world.[132] Anthropologist Elizabeth Povinelli and geographer Kathryn Yusoff argue that a distinction between *bios* (life) and *geos* (nonlife) is central to European-based ontology.[133] They assert that geology has played a central role in separating lifeless matter, such as rocks, mountains, and deserts, from living beings, and making the former available for exploitation. Drawing on Michel Foucault's conception of power, Povinelli names "geontopower" as the special form of power that governs nonlife, accompanying biopower's governing of human life. She rightly critiques posthumanist scholars for paying insufficient attention to the differences in the agency of living and nonliving bodies, or in how they are governed. Their work thus elides the distinct ways that geontopower and biopower work. However, I believe that Povinelli and Yusoff overstate the division between *bios* and *geos* in science. Paleontology, for one, does not abide by this dichotomy.[134] The language of science separates the physical from the biological, the organic from the inorganic, and living bodies from inert matter. But the everyday practice of paleontology tells a different story. Paleontologists study animal and plant life *in* rock, not life as separate from rock. We might even call their object of research "life/rock." Today, many paleontologists prefer to call themselves paleobiologists, paleozoologists, or paleobotanists in order to capture the interdisciplinarity of a field that crosses the life and physical sciences.[135] In fact, paleontology has challenged and defied the boundary between life and nonlife since it first developed out of natural history in the nineteenth century. In short, paleontology has always refused to stay within "the inorganic slot."[136]

As an anthropologist, I felt at home in paleontology because anthropology also sits at the crossroads of ostensibly different ways of knowing that are usually separated in academia. Yet anthropology, too, has maintained the divorce between *bios* and *geos*. Anthropologists have given far more attention to investigating the biological sciences than the physical sciences, and almost none to fields like paleontology that bridges them.[137] While there are many ethnographies examining the changing and contextual meanings of life within science, there is scant examination of the intersections among the life and physical sciences, and the ways that the *bios/geos* dichotomy has been breached as well as reinforced.[138] I counter this trend by exploring vertebrate paleontology as a scientific field at the intersection of the physical and life sciences that links questions about life and nonlife though the study of fossils. *Making Our Beasts* also identifies and analyzes key claims, assumptions, and stories in the Euro-American way of thinking about, and relating to, matter. This requires taking ideas that are understood to be universal features of the world and showing that they are culturally specific ways of relating to it.

WHY BEASTS?

Why, you may ask, is the book titled *Making Our Beasts?* Why *beasts?* Because, I argue, dinosaurs and many other prehistoric animals are experienced in the United States through a particularly European-based conception of the beast. Animals are "good to think with," declared the anthropologist Claude Lévi-Strauss long ago.[139] The beast is a particularly interesting kind of animal to think with because it blurs the usual divisions between the imaginary and the real, the artistic and the scientific, and the human and the nonhuman, as I discuss further in chapter 5. While monsters are conceived as mythical threats, beasts are understood to be natural ones. Beasts are a class of animals that share certain features: enormity, strength, and ferocity. In contrast to domesticated animals, beasts elude people's management and control.

In today's video games, "beast mode" favorably describes the situation when a player temporarily takes on a stronger and more aggressive form that allows them to overpower their opponents.[140] Since long before video games, people have been called beasts when they are perceived as "brutal, coarse, contemptible, cruel" and believed to be missing the "intelligence, morality, reason, and self-control" expected of people.[141] Beastly humans are assumed to lack the self-discipline to thwart their instinct for violence and the desire for sex. Beasts and beastly people may be strong and smart in a cunning way, but they are perceived as lacking the intelligence, rationality, and morality expected of civilized people. This characterization of beasts has been racialized, gendered, and then deployed to claim that some people are not fully human and are undeserving of political representation or social inclusion. The equation of Indigenous, Black, and other colonized groups with beastliness has enabled their genocide, dispossession, and enslavement, and continues to justify colonialism, incarceration, social exclusion, and massacre. As Sylvia Wynter has persuasively argued, the seemingly universal terms used to describe humanity and animality in science and beyond are drawn from a white masculine European tradition that portrays full humans as the mirror of the authors' own image, and Black and Indigenous peoples as less than fully human, and therefore, more animalistic.[142] Within this context, prehistory is usually imagined as the original state of nature, the time when beastliness was everywhere. The Mesozoic Era is thus the antecedent to the Holocene, the era during which society emerged to tame humans' animal instincts and guided some to become "civilized," while others remained atavistically "savage."[143] As chapter 4 discusses, colonialism has dehumanized Native Americans as "prey" for white settler cowboys as well as "beasts of burden." Positioned as uncontrollably violent, beastly people have themselves been subjected to extreme violence by colonizers who act in the name of civilization. To suggest that someone is beastlike makes them unworthy of rights, care, or full social participation. It even justifies their slaughter. To kill a person is murder, but to kill a beast is heroism. I hope the "our" in the book's title leads you to notice, and to question, who among "us" gets likened to beasts, and who gets positioned as the very opposite of beastliness.

In Europe and the places Europeans colonized, beasts have long been treated as objects of curiosity, exploration, preservation, and examination, as well as extermination or taming. There is some evidence that the fossilized remains of large animals that European people found in their environments inspired the beasts in their folktales.[144] In medieval Europe, the interest in beasts was expressed in "bestiaries," collections of stories about observed and imagined animals, as well as plants and stones, which identified each one's special qualities and moral message. As chapter 2 discusses, the eighteenth-century Swedish botanist Carl Linnaeus tried to tame the European tales of beasts by placing them within an orderly system encompassing all life on earth: taxonomy. He divided beasts into different groups of fauna, his Latin invention for animals. Linnaean taxonomy did not entirely replace existing ideas about beasts. Rather, the notion of the beast became embedded within dinosaurs' scientific names. The term Richard Owen invented for the taxa—*dinosauria*—means "fearsome lizard." This association of dinosaurs with beasts has remained strong, despite the decline in scientific use of taxonomy. The concept of the beast has shaped more than just the words we use to label and describe dinosaurs. It has been incorporated into paleontological practice and the physical coming into being of dinosaur fossils and mounted skeletons. Chapter 2 explains how the expert labor of scientists, fossil excavators, and preparers, together with an assortment of nonhuman actors and forces, came together to make the beast we call *Tyrannosaurus rex*, the "tyrant lizard king" of the prehistoric world. In my fieldwork, I witnessed moments in which people reinforced the common notion that dinosaurs are inherently bestial others, wholly unlike themselves, but also moments in which people found a commonality with prehistoric animals that defied this conception. When people connect with prehistoric animals that are usually considered beasts, they often question the dichotomy between beasts and humans, a dichotomy that has been harmful to so many people and other animals.

The ongoing legacies of the beast bring us back to the footprints at Dinosaur Ridge, where this chapter opened. The park's visitors, volunteers, and staff do not encounter these tracks in isolation but as steeped in Euro-American conceptions of the world and traditions of storytelling, whether or not they identify the Euro-American folklore tradition as their own, whether or not their people have been called beastly. Scholars know little about what the people living on the land thought about the giant fossils there, but they may have inspired legends.[145] The white colonists who moved west with the expanding United States in the nineteenth century brought with them a tradition of folk stories that included numerous accounts of giant footprints. I see a continuation of the beast-tale tradition in the people who travel long distances from their homes to visit the dinosaur footprints at Dinosaur National Monument on the Utah-Colorado border, and those who make daytrips to Dinosaur Ridge on the outskirts of Denver. Scientists are not immune from the fascination with giant footprints. I met one paleontologist, Martin Lockley, who has made a career out of studying prehistoric trackways at Dinosaur Ridge. By this time, he had retired from his position at the University of Colorado, he had identified thousands of previously unstudied prehistoric footprints across the Western United States (and elsewhere) and had written more than eight hundred publications.[146] In his retirement, Lockley continued documenting trackways, training students, and working to preserve the footprints in Dinosaur Ridge that have awed scientists and nonscientists alike. The interest in beasts is carried on in multiple different forms that continue to be racialized and gendered.

CHAPTER OUTLINE

In this introduction, I lay out some of the themes and arguments that carry throughout *Making Our Beasts*. In the chapters that follow I elaborate them through ethnographic storytelling and analysis. Chapter 1 narrates the life of a fossil, from prehistoric animal to either scientific specimen or dust. Stories about how particular Triceratops fossils came into being counter the narrative about "discovery" that dominates the discourse of paleontology, both within and beyond scientific communities. The chapter identifies the numerous human and nonhuman forces needed to create these Triceratops fossils as seemingly stable, coherent, and individual objects. I also demonstrate the inequalities among the human and nonhuman forces involved in fossil-making.

Chapter 2 elaborates how fossils' materiality and meaning emerged together to form *Tyrannosaurus rex* into the iconic, enormous, and vicious dinosaur species so well-known today. Here, I bring together the scholarship on charismatic people and charismatic megafauna to analyze the intertwined charisma of "rock star" paleontologists and prehistoric animals.

Chapter 3 counters the focus on the charisma of violence and examines how intimate and affective connections are forged between people and fossils through touch. Exploring moments in which scientists, technicians, students, and volunteers recognize shared experiences and emotions with prehistoric animals, I scrutinize the role of anthropomorphism in vertebrate paleontology, and in science more broadly. I argue that anthropomorphism needs to be recognized as a long-standing and common, even inescapable, practice in scientific research. The condemnation of scientific anthropomorphism rests on the assumption of human exceptionalism, and on the unquestioned belief that there is a fundamental difference between humans and all other animals, a belief that posthumanist scholars have smartly critiqued. Chapter 3 shows how anthropomorphism has aided vertebrate paleontology, not detracted from it.

Chapter 4 examines key moments in the history of paleontology to reveal how this science relied on, and contributed to, settler colonialism in the Western United States. It explores how this legacy is sustained at science edutainment sites, such as DinoLand U.S.A. in Disney World's Animal Kingdom, which use paleontology to perpetuate the myth of the American West. I argue that dinosaur edutainment promotes a national origin story that replaces the horrors of colonialism with a heroic narrative of giant beasts, and also replaces "Indian" hunters with fossil hunters who bring civilizing science to the "wild West."

Most people think that fossils are immutable stones and that casts and reconstructions are not real fossils but human-made things. Yet, when people have extended tactile encounters with these objects, they do not experience them in this way. In chapter 5, I scrutinize what "real" means in paleontology and consider its relation to several dichotomies that are central to the Euro-American tradition: nature versus culture, growing versus making, specimen versus artwork, and science versus both entertainment and belief. Counter to the assumption that casts are "fake" fossils, this chapter shows how they can be more useful for scientific study than "real" specimens. To capture people's actual engagements with fossils, casts, reconstructions, and models, this chapter lays out a multidimensional continuum of paleontological objects that replaces dichotomies such as real versus fake. Chapter 5 connects this seemingly philosophical question about the real to the ongoing debate over evolution and creationism in the contemporary United States. I demonstrate how a National Geographic special called *T-Rex Autopsy* cuts through this Gordian knot by adopting the sort of continuum approach that this chapter proposes.

The pervasive physical and symbolic presence of dinosaurs in the United States today says more about the contemporary world than it does about the prehistoric one. I end *Making Our Beasts* with a call for greater thoughtfulness and deliberateness in what we—scholars, educators, students, consumers, and citizens of an endangered world—do with the traces of these long-dead animals, both their physical remains and their life stories. The conclusion looks to the future and envisions a more inclusive scientific field and a conception of prehistory that dismantles gender, racial, and other hierarchies of life on earth.

1

———

Becoming Stone

One Friday in May 2019, a construction crew building an expansion of the Wind Crest retirement complex in a Denver suburb stopped their digging. A construction worker saw something in the dirt that he thought might be dinosaur fossils and convinced his boss to investigate. By Monday, the news media announced that he was right; he had discovered the remains of a large dinosaur, probably a Triceratops. Denver's television and print news organizations covered the story extensively as paleontologists, fossil preparators, and volunteers from the Denver Museum excavated the site for eight weeks. As residents of the retirement community and other people in the area watched, the team uncovered ribs, limbs, and other bones that amounted to more than 30 percent of a large adult Triceratops's skeleton. One of several syndicated articles in the *Denver Post* stated:

> Bones discovered recently at a Highlands Ranch construction site were those of a large adult triceratops dinosaur, Denver Museum paleontologists have determined.
>
> The partial skeleton included a limb bone and several ribs of the three horned dinosaur, Maura O'Neal, spokeswoman for the Denver Museum of Nature & Science said Friday . . . The fossils were found in a rock layer which dates back 65 million to 68 million years, a museum news release said.
>
> Of all the dinosaur bones discovered in Colorado, the remains of triceratops have been among the most common.[1]

The many news accounts, like this one, repeat the word "discovery" over and over. They announce that a construction worker *discovered* the first fossils. They describe paleontologists from the Denver Museum *discovering* more of the skeleton. They explain that numerous Triceratops have been previously *discovered* in the Denver area. They document myriad people's excitement about the *discovery* of a dinosaur in their backyard. "This is a remarkable discovery that our team takes great pride in unearthing," stated the project director for the construction company in a press release. "On behalf of the residents and employees of Wind Crest, we are thrilled to be part of such an incredible scientific discovery," the executive director of the retirement community echoed.

"Discovery" is widely celebrated, but it is often a misleading name for the process of creating scientific specimens and knowledge. As I note in the introduction, to discover is to find something that is already there but unknown. Unknown to whom? Just think of Columbus's supposed discovery of America. Centuries later it is still widely held that North America was a treasure trove of flora, fauna, minerals, and fossils only discovered when whites arrived on the scene.[2] "Discovery" implies an already existing natural specimen—whether human or nonhuman—passively waiting to be found. It concomitantly acclaims the unique accomplishment of an individual subject who alone finds the specimen, thereby disregarding all the other people, nonhumans, institutions, and forces that formed the object and made its discovery possible. In nineteenth-century North America, writes the historian Roy MacLeod, "discovery became the ambition of the scientific traveler," who journeyed across ostensibly unknown lands with "the unremitting desire to be 'first.'" These travelers, and the consumers of their enthralling accounts, saw the Indigenous peoples living there as "an artifact; perhaps an opportunity, at most a distraction."[3] "Discovery" has continued to involve the removal of materials from their surroundings and their control by someone—often with the title of scientist—who claims intellectual mastery as well as physical ownership of them.[4] "By the act of discovery, we lay claim to possession," MacLeod writes.[5] Fossil specimens thereby become private property, which is routinely taken and kept far away from the people and places where it was found. Fossils are often claimed to be held "in trust" for science that will benefit all people, but its benefits are not equitably distributed. At the same time, nonscientists' claims to these objects are discredited, and the more-than-human process by which specimens come into being is obscured. "Discovery" thus seems an extraordinary act by an exceptional individual. This chapter challenges this widespread understanding of scientific "discovery" by closely examining how fossil specimens come into being through a more-than-human process that is shot through with asymmetrical power dynamics.

Despite the continual portrayal of the "discovery" of a dinosaur skeleton as an exceptional event, the news media noted that unearthing the remains of a Triceratops in May 2019 was far from unusual. Many have been found in the Denver area in the last few decades. The population explosion that began in Denver in the 2000s led to a construction boom across the metro region. Construction workers in Denver's expanding suburbs have routinely dug into fossil-rich geological layers dating to the Mesozoic era. At the end of this era, a large inland sea, known as the Western Interior Seaway, divided the continent. While most of the geological formations from this time were subsequently buried under other layers, the rise of the Rocky Mountains brought some of them to the surface, particularly in the

FIGURE 1. This sixty-seven-million-year-old *Triceratops horridus* skull is on display in the Denver Museum's *Prehistoric Journey*. The fossil was found by construction workers building a housing development in a Denver suburb in 2003. Photo by author.

eastern foothills of the Rockies, where Dinosaur Ridge park and the Morrison Museum are located (see map). This is also one of the most active areas of new construction in the Denver region.

Another Triceratops skull, this one unearthed by construction workers in 2003, sits on display in the Denver Museum. It is not one of the museum's most spectacular specimens as it shows signs of weathering and breakage (see fig. 1). To experts, the fissures indicate its complex history. For most museumgoers, the skull appears to be a static stone, an ahistorical trace of a prehistoric animal. A senior museum staff member named Robin explained to me that people think the fossils they encounter in museums are as they came out of the ground: "There's this misconception in the public that when you find a dinosaur, you dig it up, and you put it together, and you're done." During fieldwork at the Denver and Morrison Museums, I too noticed that few museumgoers realized the long and arduous process that went into making a fossil on exhibit. Fossils' materiality is perceived as fixed.[6] This is not surprising because the Euro-American tradition, in which most museumgoers are raised, employs stones as icons of solidity and durability, and emblems of timeless permanence. Dinosaurs, moreover, are used as symbols of things that do not change with the times. As I discuss further in chapter 4, dinosaur dig pits—a mainstay of science edutainment venues and museums—reinforce this misconception. There, children and adults simulate the "discovery" of dinosaur fossils beneath a shallow layer of sand or old tire chips. These plastic fossils are in the form of whole bones, or even complete and articulated skeletons.

This chapter explores what the notion of scientific discovery hides—that is, how fossils come into being through a complex and unpredictable process that involves much more than heroic acts of uncovering prehistoric nature. I came to understand this process, firsthand, while volunteering as a fossil excavator at the Badlands Paleontological Station and shadowing fossil preparators in a laboratory at the Denver Museum. Yet the process of making a Triceratops skull fossil begins, more than sixty million years before people arrive on the scene, with the life and death of an animal.

A FOSSIL IS BORN

Imagine a large quadruped animal, a Triceratops if you can bring one to mind. Imagine her grazing on the gently sloping banks of one of the rivers that feed the Western Interior Seaway. The Triceratops gets closer to the fast-moving river water to take a drink. She slips in the soft sand and falls over. As she struggles to get back on her feet, she hits her head on a rock sticking out just above the river water. The current pulls her under, and she drowns. Her body floats down the river as it rushes to meet the sea. Once in this inland seaway, her corpse hits another rock and is pushed onto a sandy bank at a bend in the river. Her heavy body starts to sink into the clay, silt, and sand. The tide is coming in now. The animal's body is smacked by waves of water carrying sand and gravel. As the tide rises, the body becomes fully submerged in the dark muddy water. Sticks, leaves, ferns, and gravel collect around it as its soft tissue decomposes. Soon, all that is left of the animal are its bones swaddled in a blanket of thick mud. As more river sediments settle on top of it, the bones and its protective cover, very slowly, turn into stone.

The fossilization—that is, the conversion of bone into stone—of an animal is a rare occurrence. The vast majority of organisms that once lived do not undergo it. The animal needs to die in a sandy or muddy environment, and then its bones need to quickly become submerged in the muck. A skull is among an animal's heaviest bones, so it is more likely than others to get buried without much harm. Still, the bone must rapidly fill with sediment and be permeated by groundwater so that the minerals in the water precipitate into the bone's pores and holes. To preserve a bone, these soft sediments must convert into rock. This occurs when certain minerals cement under pressure from layers of additional sediment above them. In many instances, a second mineralization process occurs in which less stable minerals are replaced by more stable ones. This permits the bone's structure to be preserved in a chemically altered state for millions of years. Yet lithic matter does not stay still. It gets buried further below ground as additional rock layers form above it, or it gets lifted up as a mountain emerges. The animal's bone structures are preserved through their conversion to stone, but paleontologists often find that many of them, including Triceratops skulls, become flattened by the pressure of rock layers above them.[7] Moreover, fossil-rich sedimentary rock, made from layers of cemented sand, silt, mud, and the occasional animal bone, often disintegrates into small particles once again. Fossilization is thus an uncertain process. Few animals turn into fossils, and even fewer survive as fossils into the present.

This is one version of a common story told today in science museums, documentaries, and textbooks. Its model of fossilization indicates that, from paleontologists' perspective, only nonhuman forces are responsible for fossilization. People play no role in it. Paleontologists, like most others, describe fossils as "discovered" by people. As the term "discovery" implies, only one person receives the credit for finding a new dinosaur fossil. In addition, the people involved do not find a dinosaur, but the partial remains of one or more than one animal. In nearly all cases, the animal's soft tissues are not preserved, and there are hard parts missing too. Vertebrate paleontologists further emphasize that fossilized bones are not usually found articulated—that is, lying in the anatomical positions they were in when the animal was alive. Either right before or right after death, the spines of many dinosaurs contort into an arc shape, with their necks bent so far backwards as to nearly touch their hips. The lower jaw and then the rest of the head usually separate from the body first. Some bones are carried off or broken to bits by scavengers. Others are swept away, broken or destroyed by rainstorms and sandstorms. While paleontologists emphasize that it takes enormous amounts of expert labor to locate, excavate, clean, repair, and assemble dinosaur skeletons, they also argue that they discover fossils that already exist; they do not create them. My experience at the Badlands Paleontological Station complicates this claim.

THE BADLANDS PALEONTOLOGICAL STATION

The Badlands Paleontological Station is one of myriad places where novices can try their hand at excavating prehistoric fossils. It is run by museum- and university-based paleontologists, and largely supported by private philanthropy.[8] Other field stations are operated by educational institutions, government agencies or private businesses, and offer everything from half-day excursions for children to multiweek programs for dedicated amateurs. Since the nineteenth century, fossil hunting has attracted college students looking for adventurous summer vacations.[9] This remains true today. Contemporary paleontological field stations run the gamut from conducting rigorous scientific research to providing "edutainment" (as discussed in chapter 4).

The Badlands Paleontological Station is a scientific research operation that has resulted in numerous scholarly publications, in addition to being featured in the news media and documentaries. Its research projects focus on better understanding the fauna, flora, and ecology of the late Cretaceous Period by studying a sedimentary rock layer known as the Hell Creek Formation. Extending across Montana, North Dakota, and South Dakota, this formation is where the enormous dinosaurs from the end of the Cretaceous, including Triceratops and Tyrannosaurus, are found.[10] As its name suggests, the Hell Creek Formation, in the Dakota Badlands, is not an easy place to conduct research. In the eyes of Ferdinand V. Hayden, the geologist who surveyed the area for the US government in the nineteenth century, the Badlands is "one of the wildest and most desolate regions on this continent." He commented that it earned its name, supposedly from the native Dakotans' term for the place, "not only from the ruggedness of the surface, but also from the absence of any good water, and the small supply of wood and game." Hayden further remarked:

> It is only to the geologist that this place can have any permanent attractions. He wends his way through its wonderful cañons among some of the grandest ruins in the world . . . and not unfrequently the rising or the setting sun lights up these grand old ruins with a wild, strange beauty . . . It is at the foot of these ruins that the fossil treasures are found.[11]

While much has changed since then, the ruggedness of the landscape endures. Today, *badlands* refer to any high and arid landscape featuring rocks that have been carved into tall and steep conical peaks by erosion. With little vegetation, wind whips across the landscape and rainstorms temporarily turn the dry terrain into a treacherous web of mud, streams, and gullies. These events erode the rockface further, exposing the fossils within it. If they are not spotted and excavated, they will crumble into sand within a few years.

The Badlands Paleontological Station is based in a ranching town whose heyday was in the 1920s. In 2017, it had fewer than 150 residents. They call the people who descend on it each summer "the bone diggers." The paleontological research station owns a house in town, where the scientists and volunteers eat their meals and socialize in the evenings. The scientists, fossil preparators, and other staff sleep in the house, while the volunteers stay in tents or rented rooms around town. The house has a barn behind it that has been turned into a laboratory for preliminary preparation of fossils before they are transported to a better equipped space, like the lab at the Denver Museum that I describe later in this chapter. Surrounding the town are miles and miles of grasslands punctuated by outcrops of the Hell Creek Formation. This part of the United States is a patchwork of privately owned land and Bureau of Land Management (BLM) property, which adds legal complications to the other difficulties of excavating fossils there. Research in this area requires a BLM permit as well as the consent of ranchers who are concerned with whether paleontologists' vehicles will destroy the precious grass for their livestock. Since 2009, collecting fossils on public land has been governed by the Paleontological Resources Preservation Act (PRPA), which requires permits for the excavation of rare or unusual fossils, including the remains of any vertebrates. The law has strengthened scientists' privileged access to public lands along with their claims to expertise, while restricting commercial collectors' access and mistrusting their expertise.[12] Paleontologists are deeply divided over whether PRPA helps to protect science or contributes to the loss of valuable specimens.[13] The debate has crystalized opposing ideas about the compatibility, or incompatibility, of science and capitalism. While the history of paleontology demonstrates that they have never been separable, PRPA has nonetheless shifted their dynamics.

The summer-long influx of professional and volunteer paleontologists is tolerated by the town residents to different degrees. Some tension stems from the fact that the visitors provide welcome business to the town's lone restaurant-bar and sundries shop, but the field station does not otherwise benefit the town's economy. The Badland Station's few employees and legions of volunteers come from afar, so no locals are hired. The station's thrifty cook purchases food at a Walmart more than a hundred miles away to keep costs as low as possible. So, while volunteers' labor and capital make the paleontological research possible, they do not substantially contribute to sustaining the town that hosts the Badlands Station.

Research at the Badlands Station is led by a vertebrate paleontologist named Darrell. Like most of the other paleontologists I know, he grew up in a rural setting, where he spent his free time roaming the countryside in search of fossils. While few of the people he grew up with have ventured far from their childhood homes, Darrell turned his youthful passion into a career that has taken him to prestigious universities and museums in big cities. But he has never felt at home there. He yearns to return to open spaces far from dense urban buildings, so the field season at the Badlands Station is the highlight of his year. Darrell supervised two professional fossil preparators, two graduate students, one undergraduate intern, and a rotating corps of volunteers, who helped carry out his research there. Other scientists and fossil preparators visit throughout the summer. All the staff and volunteers are white. Only one staff member, a preparator named Lisa, is female. Unlike in the field's early years, when nearly all paleontologists came from upper-class families, most of the scientists I know today come from middle-class backgrounds, including some who are first-generation college graduates. Still, paleontology produces knowledge "out of Western nature" through a division between "those who publish or otherwise culturally display" science and those who labor "as field assistants or local collaborators—without gaining such rewards."[14] Notwithstanding the volunteers' eager participation, most of the fruits of their labor—like that of field and laboratory preparators and other technicians—accrue to the institutions and scientists that run the research station. In paleontology, like in the environmental sciences that Jamie Lorimer has examined, the passionate and skilled labor of volunteers has not been fully acknowledged because it "threaten[s] to undermine the credibility of the objective natural knowledge they have helped to generate."[15] This is changing, but only slowly.

Over the course of a typical summer field season, approximately fifty people volunteer, for one or more weeks, to assist Darrell and the professional staff in prospecting and excavating fossils from the Hell Creek Formation. The volunteer corps I worked with consisted of nine men and four women, who ranged in age from teenagers to octogenarians. Some had been working at fossil digs every summer for more than twenty years, while for others, like me, this was their first time. Together we covered the breadth of white middle-class North America, from an autoworker to a corporate lawyer. The group included several retirees, a number of whom volunteered at natural history museums during the rest of the year, and several teachers, a few of whom taught earth science. The group also included the home-schooled girl and her mother I discuss in the beginning of chapter 3. While some of the volunteers had saved all year to afford to participate, for others the struggle was not with finding the money but with taking time away from work. Two participants stood out for having little expendable income, but they possessed technical skills, such as construction and vehicle repair, that are crucial for paleontological fieldwork. As volunteers, they were not paid for their labor, but unlike the others, they were not required to pay the $1,000 weekly participation fee. Most of the volunteers had college degrees.

The volunteers and I all had the luxury of being able to spend a week or more conducting the unremunerated labor on which paleontology depends. At the Badlands Station, "hard labor is a commodity consumed [mostly] by vacationing postindustrial workers."[16] The Badlands Station volunteers' unpaid labor can be understood as part of "a turn to 'inconspicuous consumption'" in which the upper-middle class prefers to consume "experiences, services, and personal enrichment rather than material goods."[17] Unlike Darrell and the Badlands Station's staff who came from the rural Western United States, most of the volunteers come from cities and suburbs on the East Coast, as I did. Visiting the Badlands was an unfamiliar adventure for them. Like tourists visiting Indigenous people in West Papua, the volunteers at the Badlands Station were motivated by the desire to have a meaningful experience that was more immersive than a "sight-seeing" trip.[18] They sought somatic engagement in another world—a prehistoric one—as well as the adventure of scientific "discovery." Many were motivated by a fascination with prehistoric charismatic megafauna, just as their nineteenth-century predecessors had been. In many ways, the volunteers reenacted the American frontier myth I discuss in chapter 4 of going West to a place "thin with population and thick with peril," in order to tame nature and make it productive—in this case, for science.[19]

A GEM IN THE ROUGH

I expected my week of volunteer-tourism at the Badlands Station to be research "lite." I was badly mistaken. Early each morning, we loaded our stuff into weathered vans and traveled over unpaved roads and cow paths, across private cattle ranches, and into the Badlands. From the place we left the vans, we had to hike several miles, through brush and over sandy hills, to reach the excavation site. There, we hid our bag lunches and jugs of water in the shade under the lone cottonwood tree. Facing sandstorms and temperatures above a hundred degrees, we spent long days excavating fossils in the unforgiving environment. Then, we reversed our journey and headed back to town.

This excavation site was, in Darrell's words, "a gem." He had identified it on a prospecting walk he took with his graduate student, Drew, at the end of the previous field season. The loose sedimentary rock of the Hell Creek Formation was strewn with so many fossil fragments it was hard not to step on them in some places. Well-preserved skeletons are still rare, although Darrell had spotted the characteristic horns of a Triceratops sticking out of the rock on the side of a steep butte. It turned out to be the tip of a remarkably unweathered and complete skull. Despite being nicknamed "the Trike," there was nothing juvenile about it. The skull alone was more than six feet long. However, before the Badlands Station's staff and volunteers could begin removing it, Darrell needed to obtain government permits and the agreement of the ranchers whose land would have to be crossed to get there. The following summer, with the permissions now in hand, Darrell led a team of paleontologists and volunteers in excavating the Trike, along with other Triceratops on the same butte and additional fossils nearby.

Although the Trike was a spectacular prize that may end up in a museum display, the Badlands Station's research was guided by the current debate among vertebrate paleontologists over what happened at "the end of the reign of the dinosaurs," as Darrell put it. Did the giant dinosaur species flourish until the major extinction that marks the end of the Cretaceous Period, or were they already in decline before that? To answer this question, a transnational team has been mapping the diversity of dinosaur species in the geological layer closest to the Cretaceous/Paleogene (K-Pg) boundary—often still called the K-T boundary—at sites around the world.[20] As part of this collaboration, Darrell directed fieldwork in one section of the Hell Creek Formation. One of Darrell's colleagues on this research project explained their approach to me this way:

> We have a very holistic approach compared to some. When we have a dinosaur expert or a crocodile expert, we don't just go out after dinosaurs or crocodiles. When we go out and collect whatever is representative of that ecosystem, from its vertebrates, to its plants, to its megafauna . . . If we don't have expertise [in a taxa] then we work with other colleagues to study those organisms, as well as the ones we have expertise on. So we're not very selectionist. We look for well-preserved specimens, whatever taxa, whatever group.

Another paleontologist involved in the research similarly told me that they "look beyond the dinosaurs" to understand prehistory. They both contrasted the ecosystem approach used at the Badlands Station to the "selectionist" one used elsewhere, in which fossil hunters go out in search of spectacular specimens—almost always enormous dinosaurs—leaving behind the other fossils they come across in the field. These other research teams, they implied, have acted more like commercial fossil hunters aiming to "discover" a spectacular specimen than like scientists aspiring to understand the evolution of life on earth.

On my first day at the quarries, Darrell likewise made clear to the volunteers that we were engaged in science, not treasure hunting. Not only articulated skeletons, but also broken, weathered, and disarticulated bones were useful for research, he told us. He gave us an overview of our work for the week. Darrell had five quarries open on two adjoining buttes. Since they were slightly different distances from the K-Pg boundary, they probably were from different moments at the end of the Cretaceous Period. He would not know for sure until after the fossils were excavated, prepared, and studied. That would be a long time off. For this week, we would be divided into five teams, each led by a paleontologist, fossil preparator, or experienced graduate student. One group would prospect the area for other potential quarries, as wind and rain continually exposed new fossils, while the rest of us would work on excavating in the existing ones. A few volunteers and I were assigned to the quarry at the top of the butte next to the one with the Trike, where the frill from the back of the skull of a second Triceratops was exposed. We were all eager to know if there were more parts of the animal entombed there. Another group was sent to work on a quarry below me, which had some badly weathered leg bones to excavate. The rest of the volunteers were divided between two quarries at the base of the buttes. Only Drew, Darrell's other graduate student, and the undergraduate interns were assigned to work on excavating the Trike.

Despite the valuation of science over spectacle at the Badlands Station, before we began our work, Darrell led the volunteers up the butte to "meet the Trike" in small groups. We carefully climbed up the steep and crumbly rockface to where part of a skull was emerging. As I stood on the narrow cliff with one of the groups and we looked at the oddly shaped stone, an experienced volunteer said, "This is the coolest Triceratops skull I have ever come face to face with." Others called it "beautiful." A gregarious volunteer named Bonnie asked Darrell to tell us the full story of the Trike's discovery. Darrell obliged her, explaining that he and his graduate student Drew were "walking along here, mapping the K-Pg boundary on this hill." Darrell extended his muscular, tanned arm to indicate where, as he continued, "We were walking on the back side, then we walked over around there. We were coming near the gully here and I looked up and I saw this." Darrell indicated the stone we were standing around. He was about to recount his first reaction, when another volunteer cut him off to ask, "What was poking up?" Darrell answered, "The frill, this horn, and that horn, the beak . . ." "Oh my goodness!" interjected Bonnie. Darrell went on, "I'm thinking, well, that's way too much. Right away, I'm thinking it could be, well, it sort of looks like a skull. But I wasn't *that* excited because I saw the petrified wood down there, so I think it *has* to be petrified wood.

It must just be some sort of weird white root, like that big root between you guys." We looked down at the petrified wood at our feet. Darrell continued, "So I climb up here. Drew was just a little way behind me. I get up here. Holy—!" he exclaims, but does not finish the sentence.

Bonnie called over Drew, who was nearby, to join Darrell in recounting the moment when they realized what they saw on the butte was really a dinosaur. Darrell reports that Drew said, "Oh, man, it's hot!" We laugh. We then learned that they did a lot of "chest bumping and high fiving" as they repeatedly said to each other, "look at him!" then, "no, *you* look at him!" As the volunteers admired the fossil and the story of its discovery, Darrell maintained his usual calm composure, as Drew beamed with pride.

Darrell told us how, just before they closed the Badlands Station for the season, he and Drew came back to the spot and dug out the "beak" (center lowest horn) and more of the Trike's skull, but then covered it back up with dirt. An experienced volunteer asked if they protected the site with a tarp for the winter. "No, I just left it," Darrell answered. He explained, "I didn't want to because people hunt here. Anything like a tarp might attract them. We use gray tarps because they blend in, but, still, I didn't want to risk attracting people." He later told us that he feared hunters "might look through a [rifle] scope, think 'what is that?'" and decide, "Well, let's shoot it and see if it moves." We all laughed, but this was no joke. At the same time, Darrell was concerned about what the winter weather would do to the exposed fossil. To try to prevent them from breaking off before he could return, he explained, "We poured a bunch of glue out because this [horn] was the only part that I was worried about." Darrell showed us the cracks in the rock with his finger as he elaborated, "We poured a bunch of glue on this, and on this, and on anything that was showing." Later that week I learned that this clear plastic glue, which paleontologists call "consolidant," plays a crucial role in both excavating and preparing fossils.

The story of the Trike's discovery complete, and the volunteers duly impressed, Darrell recommended we take turns posing for photos "spooning" the skull before we left for our assigned quarries. However, most of us—myself included—were hesitant to bring our bodies that close to the Trike, for fear of accidentally damaging it. So we gingerly "petted" it instead. After a few more minutes admiring the quarry and asking Darrell questions, the volunteers and I descended the butte. None of the volunteers, except for me, went back to that quarry to work on excavating the Trike later in the week. Yet the volunteers' excitement at working near it was palpable. At the end of the week, several told me they expected to remember touching the Trike for the rest of their lives. Darrell is well aware of this strong appeal. His specialty is a contemporaneous but uncharismatic organism that is far more abundant in the prehistoric record than dinosaurs. Yet he knows the importance of excavating these enchanting animals for the viability of his research program, particularly for its dependence on volunteer labor. Spectacular dinosaurs like the Trike recruit volunteers to paleontological research stations and motivates their hard work once there. Volunteers' encounters with Triceratops skulls at the Badlands Station give meaning to their arduous efforts and the mundane tasks assigned to them.[21]

A STONE EMERGES

I conducted the majority of my paleontological/anthropological fieldwork at the uppermost quarry where the edge of the frill of a second Triceratops was exposed. Erosion had begun the process of exhuming the fossilized matter from the rock formation of which it had become part millions of years prior. Our job was to complete this separation before wind and rain disintegrated the rock into sand. We worked to "remove a mountain with a paintbrush," as one volunteer put it. Like the others without prior experience, I was taught how to do this by sitting alongside experienced excavators and asking lots of questions. I learned a great deal from Ray, a tall and wiry retired science teacher, who was among the most experienced volunteers in the group, having participated in paleontological digs every summer for more than three decades.[22] I also learned from the rock itself.

The process of learning fossil excavation is a lot like learning a craft, such as weaving or blacksmithing, through apprenticeship. These crafters learn their trade through repeated practice under the guidance of experts. Mastering a craft requires embodied knowledge that cannot be written down in a manual. In paleontological fieldwork, students and new volunteers spend a lot of time observing and assisting experts, beginning to do simple tasks and then gradually taking on more complex ones.[23] The hardest lesson for new fossil excavators like me to learn was how to distinguish fossilized bone from the "matrix," the rock sediments surrounding it. We were taught to do this by looking for clues in the rock's color, shape, and texture. Fieldworkers often first recognize fossilized bone because its distinctive mineral composition makes it a different color from the surrounding sandstone. At this site, most of the fossilized material is darker and redder than the grayish sedimentary rock. Furthermore, the color change between fossil and matrix has a definitive line, unlike most mineral variations within rock that change gradually. Another clue to distinguish fossil matter is its organic shape. Teeth are the easiest to spot, but pieces of leg, arm, and rib bones are also relatively straightforward to recognize. While visiting this site with a school group the week prior, a student found a dinosaur vertebra this way. She was able to identify it because it had the same shape as vertebrae of animals alive today, although it was much larger. Additionally, fossilized bone usually retains the texture of the living matter it once was. Pieces of limb bone, for instance, have smooth surfaces and spongy-looking ends full of tiny holes where the bone once joined cartilage.

The other novice excavators and I also learned to recognize fossilized bone because it is usually harder than the sedimentary rock around it. In this section of the Hell Creek Formation, the matrix rock is known as "gumbo" because of its loose gravel-like and almost slimy texture that crumbles when pressed between fingers. Yet the gumbo is full of concretions—hard rock masses within the softer sedimentary rock. Although several of the large fossils found at the site were encased in concretions, most concretions do not contain fossilized bone. I mistook pieces of gumbo and concretions for fossils more times than I could count. When all else fails, we could lick the rock, since fossil matter has a distinctive metallic taste owing to its high concentration of iron. Thus, fossils' specific materiality enabled even novice volunteers to distinguish them from other rock.

These initial lessons in paleontological fieldwork illustrate how the emergence of fossils requires people to isolate one kind of lithic matter from another—fossils from matrix—visually and tactilely. Darrell was a master at seeing the landscape as a trove of fossils hidden amid the expanse of sedimentary rock, as his spotting of the Trike revealed. Looking up at a butte, he saw a horn where others saw only gumbo. The paleontologists with whom I have worked further perceive the prehuman history of the rock: how the high and dry butte was once the muddy side of a prehistoric waterway, and how organic matter turned into inorganic matter that still retains signatures of the creatures buried within it. They also feel it between their fingers more skillfully than most volunteers do. This expertise is not easily acquired. Darrell developed it through years of fossil hunting in the Badlands, first as a child, then as a student, and finally as a professional scientist. Most of the volunteers lacked this "skilled vision."[24] Ways of seeing are not only learned but also embedded in particular historical, spatial, and political-economic contexts, including hierarchies of knowledge.[25] In order for researchers to conduct paleontological fieldwork, their senses are "enlisted, honed, cultivated, and trained to engage with the material world in a corporeal manner.[26] Challenging the claim that Euro-American science privileges vision above other senses, paleontology demonstrates how sight and touch are intertwined. The volunteers at the Badlands Station learned to use eyes, hands, and sometimes even tongues, in concert, to differentiate fossil rock from matrix.

This begins to show that fossils only come into being with the arrival of trained experts—who have developed a particular way of seeing, ideas about geological processes, excavation techniques, and bags of tools—and volunteers with the able bodies and vacation time needed to separate fossil from other rock. Yet our ability to extricate fossils depended on the rock itself as much as on our knowledge, skill, and patience. First, biological and geological forces lent the fossil matter different mineral content, shape, and texture than the matrix matter. Next, wind and rain eroded part of the matrix to expose some fossil matter at the surface. Only then could eyes, tools, and hands, together, deploy the subtle differences between fossil and matrix to disengage the two.

FIGURE 2. As pieces of rock broke away from the butte at a paleontological quarry in the Badlands of North Dakota, the excavators paused to assess if they were fossil or matrix. Photo by author.

CONSOLIDATING ROCK

Once fossil rock has been separated out from the other lithic materials that make up a butte, the next step in the fossil-making process is fusing the fossil matter into stable and transportable individual stones. My fellow excavators and I first worked with handheld whisk brooms to brush off gumbo, then with scratch awls to remove concretion. When we reached something that seemed like it might be fossil matter, we switched to paintbrushes to expose its shape and texture without chipping off pieces (see fig. 2). This was the stage at which there was the greatest risk that the fossil matter would return to sand. While the fossil rock was harder than the matrix, it was still quite fragile. We tried to keep it together as much as possible, but it was abundantly clear we were not in control.[27] Our eyes, tools, and hands again did the work, but the rock did not always cooperate in our efforts to create discrete fossils. It unhelpfully broke in places that did not fall along fossil/matrix lines.

At the butte Ray sat closer to the edge of the ledge than my fellow inexperienced volunteers and I dared to, and he worked back from the exposed part of the frill to figure out the Triceratops skull's shape and extent. To prevent the frill from crumbling as he exposed more of its surface, Ray infused it with "consolidant," the strong, fast-acting, and clear plastic adhesive similar to the superglue that Darrell had used to protect the Trike's horns for the winter. I watched the plastic seep into the cracks in the fossils and helped to prevent them from disintegrating as Ray and the other excavators worked. The consolidant left the surface with the slight shine of a wet stone, which distinguished it from the surrounding rock. While there is controversy over the use of consolidants because they cannot be removed entirely, we used it liberally in the quarries. None of the paleontologists seemed concerned. Consolidant is a crucial element in the mix of matter that becomes a fossil. This plastic arrests its disintegration into undifferentiated and worthless sand, and thus creates a stable and coherent stone that is no longer part of the "current of materials" within the rock formation.[28] Yet the glue did not act alone. Ray put it in some spots and not others to hold some rock together, while other rock crumbled and fell away. Excavators thus intervene in rock in ways that are not completely different from what forces like wind and rain do: cleaving it in some places and not in others.[29] Yet Ray and the other excavators at the quarry interceded in order to direct the lithic flow toward a specific end: making a Triceratops skull, a stone with a specific scientific, aesthetic, and financial value.

By the end of my week as a volunteer excavator, the uppermost quarry looked like a maze with embedded but stabilized fossils sticking out of the butte. We had bagged many pounds of fossil fragments, but the skull itself was far from ready to be moved from field to laboratory. On my last day in the field I went to the quarry midway up the butte where some weathered leg bones of undetermined species were about to be sheathed in field jackets to prepare them for transportation to the lab. Lisa, one of the preparators and the lone woman on the scientific staff, had led some volunteers in removing most of the concretion around the fossils. They had left them standing on matrix pedestals. As Lisa put it, she aimed for a delicate balance between exposing enough of the fossil to be sure they had isolated the whole thing, while saving as much of the preparation as possible for the laboratory where a more careful job could be done. Additionally, she told me, they needed to leave sufficient concretion around the fossil to protect it in transit, while removing enough rock to make it light enough to carry. Once that was done to her satisfaction, Lisa guided me and two other volunteers in jacketing the fossils.

Jacketing further separates the animal-object from its lithic context and enables it to become a mobile object. This work felt like play in comparison to the grueling labor of isolating fossils from matrix that I had done the rest of the week. We first covered each fossil in damp paper towels, and then encased it and the remaining concretion in strips of burlap soaked in plaster to create a cast. The atmosphere turned jovial as we got covered in the soft white goo. Although we placed the burlap strips haphazardly, Lisa made sure the cast extended as far under the fossil as possible to ensure that no valuable material would break off or fall out when we flipped it over. She then smoothed it over with more plaster to prevent it from cutting the hands that would carry it down the butte. After the plaster dried, and she labeled it, we gently turned the package off its rock pedestal and created a bottom for the field jacket (see fig. 3). The boundaries of the fossil were now defined,

FIGURE 3. A professional fossil preparator and two volunteers flip a fossil wrapped in its field jacket off its rock pedestal, at a paleontological quarry in the Badlands of North Dakota. Photo by author.

and its physical wholeness was produced, by a successful collaboration between human and nonhuman forces. What was previously a mix of fossilized bone, concretion, gumbo, and consolidant had become a discrete object. Some rock had been removed to isolate distinct stones, while other parts of the matrix had been left to keep fossil fragments together as one. Some rock matter had been glued together, while other pieces had been separated and tossed away. As Adrian Van Allen writes, "each specimen's life history can be seen as a collection of moments, important events as it is collected, prepared, accessioned, and sampled."[30] By the end of the week, several fossils sat apart, each in its own hard white case with its own identification number, ready to be moved to the Badlands Station's laboratory, then to a more sophisticated one—like the laboratory at the Denver Museum described below—and then finally into a research repository.

The scholarship on museum archives and collections unpacks the human labor involved in bringing museum objects into being. While most of this work has scrutinized anthropological artifacts and not natural history specimens, many of their insights hold for fossils too.[31] Barbara Kirshenblatt-Gimblett, for example, points out that the artifacts in anthropology exhibitions do "not begin their lives as ethnographic objects," but were made into them when people "define, segment, detach, and carry them away." She describes the "surgery" fieldworkers perform whereby they cut objects out of one context and move them to another.[32] However, my experience at the Badlands Station shows that humans do not create collections alone. They do it together with biological, geological, and chemical forces, employing a motley assortment of tools and glues. A Triceratops skull, like any other fossil, is a physical materialization that mixes skilled vision and manual labor, wind and rain, tools and glues, capital and expertise, among other forces. Fossils, like other natural resources, are made through the "separation, parting, simplification, and reduction (and, occasionally, addition) on both material and conceptual levels."[33] Excavators do not control how rock coheres or falls apart, but they do regulate what ends up as a bounded object on its way to a laboratory and what does not. The process by which a fossil leaves a rock formation and comes into being as a discrete stone thus illustrates the distributed agency involved in temporarily converting some of the incessant, albeit incredibly slow, flows of rock into a discrete animal-object that is both a specimen and an artifact.

As I am writing this, the Trike and the other fossils we excavated at the Badlands Station that summer sit in a storage room, waiting their turn to be cataloged, prepared, and studied. I pick up the story of how fossils come into being at the Denver Museum, where another Triceratops skull was being prepared in the publicly visible fossil preparatory laboratory while I was conducting fieldwork there.

PREHISTORIC JOURNEY

Many Coloradans know the Denver Museum of Nature and Science as "the dinosaur museum." Here, as at natural history museums elsewhere, dinosaurs have had an "uncanny power to capture people's imaginations and thus ensured a steady stream of visitors."[34] The Denver Museum was established at the turn of the twentieth century by a group of Denver businessmen, most of whom had accumulated their wealth in mining, when Colorado was a new state and paleontology a new science. These tycoons joined those on the East Coast in financing museum collections in natural history as signs of their social distinction, largesse, civic-mindedness, and economic acumen.[35] The Denver Museum soon began excavating, preparing, studying, and mounting the fossils of dinosaurs and other prehistoric animals for an eager public. Its collection, focused on the Western United States, gained a national, then an international, reputation.[36] While dinosaurs have remained among the most popular stops for museumgoers, how they are displayed has shifted several times since the nineteenth century.[37] Yet creating new exhibits is costly, so it occurs infrequently and with significant deliberation and planning.

In the early 1990s, new staff in the paleontology division led the Denver Museum in dismantling its historic dinosaur hall and incorporating its prized specimens within the new exhibition *Prehistoric Journey*. This exhibition was especially designed to show off the museum's spectacular eighty-foot-long Diplodocus, whose tail had been shortened to fit into the old hall. They also wanted to place the dinosaurs within the context of evolution. Robin, a staff member mentioned earlier in this chapter, recounted, "It was the first exhibit to walk you through the history of life on earth chronologically. Previous to that [the approach] had been to put all the dinosaurs in one room, put all the mammals in one room, and put all the fish in one room." *Prehistoric Journey* thus fits within the larger trend of organizing natural history exhibitions around ecosystems and evolution, while not always making this explicit in an attempt to avoid creationist protests.[38]

One of the innovative elements introduced when the dinosaurs moved into *Prehistoric Journey* in 1995 was "enviroramas." This is the term the museum staff coined for their modern take on the nineteenth-century dioramas that adorn most natural history museums. Traditional displays clearly mark a physical boundary between stable objects and evolving interpretations. As Ella Butler notes, inside the display case "is the realm of the 'fact,'" while the signage—and I would add docents' speech—outside of it "is the interpretative counterpart."[39] Enviroramas dissolve this boundary. Instead of looking at specimens behind protective glass or a railing, visitors walk through three-dimensional, multisensory displays that use sound and movement to recreate a particular location at a particular moment in time.[40]

Two of *Prehistoric Journey*'s enviroramas feature dinosaurs. First, as visitors descend the narrow staircase that takes them from the Cenozoic into the Mesozoic era, they find themselves in the "Cretaceous Creekbed" envirorama. This three-dimensional immersive display recreates what Marmarth, North Dakota was like sixty-six million years ago. Visitors hear birds chirping and water running through a creek. The walls are painted with murals of vegetation, and the low lighting adds to the sense that you are in a forest. The creek is filled with moving water, and the trees made of branches gathered not far from the museum. The animals are all sculpted and painted reconstructions created to show what prehistoric animals looked like when they were alive. Two Stygimoloch are locked in a fight, apparently over the attention of a female. Near them, a decaying Triceratops skull lies on its side. Unlike the Triceratops skull displayed in the main exhibit hall (see fig. 1), the one in the envirorama was sculpted out of synthetic materials (see fig. 4). Despite the absence of rock, this skull illustrates the process that turns a bone into a stone more clearly than the fossil does. It has separated from the rest of the animal's body, which is not visible in the envirorama. The rough brown skin that once covered its head is peeling off to reveal the white bone beneath, and its eye sockets are empty, its soft tissue presumably removed by scavengers.

The final envirorama in *Prehistoric Journey* is the museum's main fossil preparation laboratory. Its three large bay windows allow visitors to watch as volunteer preparators work inside. Like the other fossil prep laboratories I have visited, this one resembles "a hardware store, a tinkerer's workshop, and an artist's studio" rolled into one.[41] It is, therefore, decidedly unlike the laboratory Brian Noble describes as a technologically sophisticated theater mimicking the one portrayed in *Jurassic Park*.[42] The museum claims this was the first publicly visible fossil prep

FIGURE 4. As visitors to the Denver Museum's *Prehistoric Journey* descend the staircase from the Cenozoic into the Mesozoic era, they find themselves in the Cretaceous Creekbed envirorama. This 3D version of a diorama, complete with sound and running water, contains a reconstruction of a decaying Triceratops skull lying on its side in a stream. Photo by author.

lab ever built, and it certainly has been imitated by museums around the world. Its "fishbowl" design eschewed the separation of front and back stage usually found at sites from hotels to cell phone repair shops to gemstone prep labs.[43] Between walls lined with shelves full of fossils at various stages of becoming, two professional preparators circulate among a dozen or so volunteers answering questions, troubleshooting problems, and making suggestions about the best technique to use for a particular task. Everyone in the laboratory is carefully trying to stabilize pieces of fossil rock that are at risk of disintegrating into dust. While most of the volunteers sit at long tables running the length of the laboratory, intently laboring with their heads down, a few volunteers like to work under the bay windows at the front so they can interact with the people who stop to watch. While most volunteers prepare the decidedly uncharismatic organisms, the ones working under the windows prepare the fossils of large dinosaurs and other charismatic prehistoric megafauna (see chapter 2). This living envirorama reveals to visitors the process that creates the seemingly static stones on display in the rest of the exhibition, even more than the Cretaceous "Creekbed envirorama" described earlier.

FIGURE 5. The public laboratory at the Denver Museum presents the messiness of fossil preparation work to visitors. In the foreground is a field jacket containing what the museum's paleontologists preliminarily assessed as a sixty-six-million-year-old Triceratops skull. Photo by author.

REPAIRING FOSSILS

On most days, there is a steady stream of people standing at the windows watching the preparators working inside the laboratory. Visitors, especially children, often ask the volunteers at the window what they are "making." The volunteers gently correct them, telling them that they are "cleaning" or "repairing" fossils, not making them. I want to suggest that the visitors' questions were not wrong, as long as "making" is understood as a collaborative human/nonhuman process and not a human invention from whole cloth.[44]

When I shadowed a long-time volunteer named Henry, he frequently faced questions about making because he liked to open the lab windows as wide as possible to encourage people to talk to him.[45] "Hey," Henry greeted a girl who had been silently watching him work for a while. She responded by asking, "What are you making?" Henry patiently told her, "I'm actually not making anything. What I'm doing is cleaning and repairing a fossil. This is a fossil that we took out of the ground in Wyoming." Henry pointed to the field jacket in front of him. It contained a mess of broken fossil fragments embedded in loose matrix, which one of the museum's paleontologists had preliminarily assessed as a sixty-six-million-year-old Triceratops skull (see fig. 5). Like the *Triceratops horridus* skull on display in the main exhibit (see fig. 1), this skull had been found by a contractor during a government-required survey prior to the construction of an oil platform on BLM land. It was no accident that, despite its poor condition, it was placed in the lab window. Just as Triceratops attracted philanthropists and volunteers to the Badlands Station, it attracted museumgoers to *Prehistoric Journey*'s final envirorama.

The visible laboratory was built to communicate to visitors the enormous amount of skilled labor that goes into fossil preparation. It seemed to do this effectively. After watching Henry work for a few minutes, one man commented, "When you see all the digs [on television], you don't realize just how bad shape they're in." He marveled at the labor that needed to be done. Henry reinforced this point when a woman inquired, "Can I ask what you're doing?" Henry answered:

> When we take [the fossils] out of the ground and bring them back to the museum, they come to the lab first. This is where we clean them. You've got to remove all of the sedimentary rock that is still on them to completely expose the fossil so the scientists have a complete view of it. [We also need to] repair the parts that are broken. That's what I'm in the process of doing.

Henry's explanation not only underscores the tedious labor of fossil preparation; it also indicates how labwork extends the fossil-making process that began in the field, with similar power dynamics among the human and nonhuman participants. Yet, paleontologists were infrequent visitors to the fossil prep lab, where technicians and volunteers did crucial scientific work largely on their own. Fossil preparators, like excavators, arrest the flow of matter in rock in order to consolidate certain parts into stable and coherent stones, while helping other parts to disintegrate into the undifferentiated rock they throw away. Henry's narration of the fossil-making process emphasizes preparators' agency, while omitting other forces, including the tools in his hands. In his terms, people "clean," "repair," and "reassemble" fossils. Other preparators analogized that fossil preparation is like putting together a jigsaw puzzle without the picture on the puzzle box.[46] Their explanations portray preparators as in control of the repair process and able to manipulate tools and materials to return fossils to their "natural" whole form.

My experiences at fossil quarries and prep labs have shown me that preparators' bodies tell a different story than their discourse does. Instead of "totalitarian ruler[s] over a world of parts whose integration is already given by Nature," preparators act as "go-between[s], in amongst a world of non-parts that must be coaxed into getting along together."[47] Like the volunteers at the Badlands Station, the ones in the Denver Museum's laboratory worked in partnership with paintbrushes and scratch awls to separate fossil from matrix. They also worked with other smaller and finer tools, like dental picks and toothbrushes, and electric ones, like microscopes and tiny pneumatic drills, which only functioned in a lab setting. Much like the eroding force of wind in the Badlands, airscribes—tiny jackhammers that generate miniature windstorms—are utilized in fossil prep labs. In both places, blowing air does much of the labor of removing the matrix around fossilized bone, thus making them into discrete and clean objects called specimens. In the fossil prep laboratory, people and tools, together, isolate fossils at a minute level that brings out subtle features, such as striations, where muscle once attached to bone. Yet even under the controlled conditions of the laboratory, the fossil material does not cooperate with the preparators' aims and tends to break apart in inopportune places.

Consolidants play an even larger role in creating fossils in the laboratory than they do in the field. Here, glues, putties, and clays not only prevent the disintegration of fossils but assemble fossil fragments together into solid wholes. Looking at the jumble of materials inside the field jacket in front of Henry, another visitor commented, "There is a million pieces here. How are you going to stick them together?" Henry replied with a chuckle, "Ahhh, glue!" He elaborated:

> [We use] acetone with little acrylic beads dissolved in it. . . . You can vary the viscosity of it to suit your needs. In the places where it's cracked but hasn't really broken, I can use a very thin version and drip it down in the cracks. Where it's actually broken, I can use a thicker version of it as a glue.

In the lab, as in the field, consolidants are a crucial part of the mix of matter that becomes a fossil. As the conversation across the lab window continued, the visitor observed, "I see a lot of pieces that are separated, in some cases by an inch. Are you going to stick those back together?" Henry pointed to a particularly broken-up section and answered:

> Yeah, all of these [fragments] here are sitting on a bed of sandstone that is holding them in place from the back. But all that sandstone is going to have to be removed. So, yeah, we will. We'll reassemble these as best we can. If we can't get them to fit entirely together, we may put some clay in there, or putty.

The making of fossils thus depends on the fortuitous cooperation of numerous materials.

The physical properties of each kind of matter are crucial to the fossil's coming into being as a stable object. It matters that the Triceratops in the Denver Museum's laboratory came from the Badlands, rather than the extremely hard sandstone in the Rocky Mountain foothills near Denver, where I also conducted fieldwork. The Badlands rock made the process of fossil making both easier and more difficult. Both in the field and in the laboratory, matrix separated from fossil without tremendous effort, but lots of labor was needed both to prevent breakage in the wrong places and to assemble the fragments into a solid object. That is where the properties of the different consolidants and fillers came in. As Henry explained to the visitors, some are good at seeping into tiny fissures in the rock, while others fill in large gaps between fossil fragments. The importance of nonhumans in fossil making goes far beyond these physical properties. Without biological and geological forces there would be no fossilized animal remains for people to find, isolate, and consolidate into objects. At the same time, appreciating the role of nonhuman forces in creating dinosaur fossils should not come at the expense of ignoring their asymmetrical entanglement with human ones.

While much of the archiving of life undertaken at natural history museums involves disarticulating individual creatures into smaller units, vertebrate paleontology is dedicated to putting disarticulated pieces together to generate whole animals that, in turn, stand in for entire species, even genera.[48] I explore this further in chapter 2. Here, I want to highlight how tools, consolidants, and their human handlers separate and then reassemble rock to form the contours of a fossil. In the laboratory, as in the field, some rock is excluded while other rock is fit together with glues and fillers. The context of the Denver Museum—particularly its substantial financial resources, infrastructure, skilled staff, and dedicated volunteer corps—enables the fossil-making process. But that cannot guarantee its success. No matter how much glue, labor, and money are poured into it, stone does not always coalesce into stable fossils. The last time I was at the Denver Museum, the materials inside Henry's field jacket still had not become a coherent animal-object and might never become one.

Many vertebrate paleontologists and fossil preparators spend significant amounts of energy trying to prevent fossil pieces from disintegrating. They regard this work as returning fossils to their natural state. But to which "natural state" do fossil excavators and preparators restore fossils? Fossils are not static objects, but organic-turned-inorganic matter undergoing a continual—though usually incredibly slow—process of change. Paleontologists envision a moment after a bone has turned to stone but before it has become broken or weathered, and they try to recreate that moment. Fossil preparation is similar to the finishing of gemstones, which anthropologist Elizabeth Ferry has studied. She explains that gem preparation is considered legitimate when it "restores the [stone] to . . . what it was 'supposed to' look like," and leaves no "indication of the human labor that went into its creation as such."[49] Fossil preparation too aims to hide the enormous amount of work involved in excavating, cleaning, and repairing fossils. I, like Ferry, argue that we must pay close attention to that human labor.

I conceive of human labor as integral to fossil making, not as simply discovering, restoring, or preserving the material traces of prehistory. Yet, as I discuss elsewhere in the book, I contest the cultural constructionist (and creationist) claim that scientists make dinosaur fossils into reflections of their own social position and ideology.[50] Instead, I join scholars who have reconceived craftwork as a process in which people work together with nonhuman matter.[51] In other words, people do not make objects, such as clay pots, by molding the clay to actualize their desires for it. Instead, people collaborate with clay to produce a pot. Neither one has total control over the process or the outcome. The potter's fine-motor skills and the clay's elasticity both play a part. The same is even true of making granite building stones.[52] I found that scientists, fossil excavators, and preparators cooperate with rock rather than imposing their vision onto it. There is, however, yet another set of important forces that have enabled the coming into being of Triceratops in the Denver area, forces that have significantly shaped the dynamics of fossil making since the beginning of the nineteenth century.

THE CREATIVE FORCE OF PHILANTHROPIC CAPITAL AND UNPAID LABOR

With its characteristic three horns on the front of its head, Triceratops is among the most widely recognized and beloved dinosaur species in the United States and beyond. Triceratops stars frequently in entertainment, advertising, and education. It has special importance in the Denver area, where its fossils were first unearthed by white colonists. Triceratops fossils continue to be found in this region. After a seven-foot-long Triceratops skull was uncovered during construction of a new baseball stadium for the Colorado Rockies in the early 1990s, the team's mascot became "Dinger" the purple Triceratops. Media coverage of this and similar "discoveries" celebrate gentrification as advancing science but neglect how they revive internal colonialism.

I heard the story of the discovery of Triceratops in Colorado many times during my fieldwork: It begins in 1887, when a high school science teacher named George Cannon found an enormous horn and skull fragments while on a walk in the Rocky Mountain foothills not far from what is today downtown Denver. Cannon dug them up and sent them to the paleontologist Othniel C. Marsh, who was amassing an impressive collection of fossils at Yale University. Marsh initially classified the fossils as belonging to an ancient bison, but after he received more complete skulls of this "horned beast" from other fossil hunters, he reclassified Cannon's fossils as part of a ceratopsian dinosaur.[53] Marsh coined the genus Triceratops in 1889. This episode was part of the "bone wars," in which Marsh and his rival Edward Drinker Cope competed to collect more fossils and name more species.

Cannon's luck in finding fossilized Triceratops bones in the Rocky Mountain foothills was not as innocent as it seems. His hobby of hunting for fossils was made possible by the massive white settlement and Indigenous dispossession of the Western United States. As I explain in chapter 4, the history of paleontology is part of the larger story of settler colonialism in the Western United States. Like the mining of gold, uranium, and oil, fossil hunting drew on tropes of discovery, progress, and the frontier that have characterized—and motivated—resource extraction in these territories.[54] Since the 1870s, the Western United States has been viewed as among the world's richest dinosaur graveyard and fossil hunters have been seen as noble soldiers fighting to advance the nation's science. With help from the US government, paleontological expeditions ignored treaties and tribal sovereignty in their pursuit of fossils and, in the process, provided geographic information that aided the military and the colonists. "Like the settlers seeking Indian lands for farming or ranching, and the miners seeking precious metals on Indian lands, [paleontologists] staked out their own claims to the natural resources in which they were most interested, and 'opened up the West' in their own way."[55] Fossil "trophies" were deployed to justify conquest as advancing scientific discovery, as well as to justify military, political, and economic missions.[56]

William H. Goetzmann's Pulitzer Prize-winning history of exploration in the American West, *Exploration and Empire*, provides a vivid illustration of how early paleontology has been portrayed as an intrepid quest to civilize the Western frontier. Goetzmann claims that the "collection, classification and organization" carried out by state-sponsored explorers, including Hayden and Marsh, "was the essence of civilization come to the wilderness."[57] While it is important to recognize that "western places were not simply created, colonized, and generally 'acted upon' by outside forces," it is also true that the extraction of natural materials from the US "internal periphery" benefitted those outside the region far more than those within it.[58] In the case of fossils, those who benefited the most were East Coast collectors, museums, and universities.[59]

Paleontology in the Western United States illustrates the persistence of a close and mutually beneficial relationship among science, colonialism, and capitalism in other ways as well. Most of the vertebrate paleontologists I know repeatedly receive offers from private dealers who want to buy their fossils. Despite their refusals, they are part of a discipline that has been shaped since its inception by the interests, values, and aesthetics of the wealthy. While fossil hunting in the United States has been supported by multiple federal agencies, the most successful paleontologists have been able to draw on private riches as well as state coffers.[60] Marsh, for instance, was bankrolled by his maternal uncle, the financier George Peabody. He paid for Marsh's education, from boarding school to master's degree, then endowed the Peabody Museum of Natural Science at Yale, where Marsh became faculty chair in paleontology.[61] Once installed at Yale, Marsh ceased leading fossil-hunting missions to the Western United States. With the support of his uncle, wealthy friends, and government agencies, Marsh paid others to collect fossils for him and bought ones already excavated with little regard for how they were acquired. In a letter to the director of the US Geological Survey shortly before his death, Marsh asserted that he spent $200,000 (equivalent to $4 million today) "at [his] own expense" on fossil collecting while head of the Division of Vertebrate Paleontology for the Survey.[62] Marsh and his contemporaries thereby made it worthwhile for colonists to spend time fossil hunting.[63] In the last decades of the nineteenth century, tons (literally!) of stone made its way from fossil beds in the West to museum and private collections in the East.

Paleontology, like anthropology, transitioned from a gentlemen's hobby into an academic discipline and occupation on both sides of the North Atlantic in the second half of the nineteenth century.[64] Yet US paleontology continued to be directed by affluent collectors' desire for fossils as marks of social distinction.[65] Peabody was joined by other business magnates, including J. P. Morgan and Andrew

Carnegie, who spent part of their fortunes building natural history museums and filling them with fossils.[66] Underwriting museums and their collections extended the legacies of these men by associating them with the prestige of science and the romance of discovery. High-end collectors and donors prized the skeletons of prehistoric charismatic megafauna over other fossil remains. Carnivorous *Tyrannosaurus rex* and its valiant vegetarian adversary Triceratops have been especially highly sought. These fossils offer an investment and an object of conspicuous consumption unlike almost any other: they are unrivaled in their rarity and physical enormity. In the hunt for the largest and most ferocious dinosaurs, other evidence of prehistoric life has been neglected, even destroyed.[67] Upper-class capital, aesthetics, and values have thus shaped paleontology from its beginnings. With the decline in state funding in recent decades, corporate and individual funders have become even more important.[68]

While I was visiting the Badlands Station with a school group prior to my stint as a volunteer there, four museum philanthropists coincidentally came, by private jet, to visit one day. I found myself standing with them at the bottom of a butte with the Trike. These potential funders were two heterosexual white couples, probably in their sixties, who knew each other from their involvement with both the petroleum industry and the board of a major natural history museum.[69] One of the paleontologists later told me that these oil magnates "could buy a dinosaur skeleton with the stroke of a pen." They were invited to visit the Badlands Station's quarries in the hope that "meeting the Trike" would motivate them, following in the footsteps of business tycoons who had bankrolled the paleontologists who excavated these lands a century earlier, to support the Badlands Station. As we looked up at the Trike, Darrell talked to the philanthropists, the students, teachers, and me about the scientific importance of the site in terms of its proximity to the extinction of the nonavian dinosaurs. The philanthropists nodded their heads with interest, but really they had come to get an up-close look at the Trike. Two of them even climbed up the butte in their immaculate outfits to touch its horns. It was clear that without this dinosaur, they would not have taken an interest in Darrell's research or made the day trip to the quarry. These contemporary paleontology supporters shared with their predecessors a taste for prehistoric charismatic megafauna. The visit by these four philanthropists illustrates how private capital is an important force not only for sustaining the labor of paleontology but, more fundamentally, for determining which pieces of rock undergo the arduous process of becoming fossils, and which do not. Excavating the Trike was not essential for answering Darrell's research questions, but it was crucial for sustaining the Badlands Station.

The collection of fossils from the rural stretches of the Western United States for museums, universities, and private collections elsewhere continues unabated today, only now the popularity of tourist dinosaur digs has reformulated paleontology's labor regime. Ranchers with local knowledge are being supplemented, although not entirely replaced, by middle-class professionals from outside the region looking for unforgettable vacations. Retirees and other volunteers with sufficient finances and free time do most of the fossil preparation work needed for paleontologists to study the specimens. The stories of a few of the Triceratops that have been unearthed in the United States since the end of the nineteenth century demonstrate that philanthropic capital and volunteer labor are not only essential for the practice of science, they are crucial for the very coming into being of paleontology's object of study: fossils.

As this chapter has shown, the transformation of a prehistoric animal into a rock and then a fossil is an uncertain process, more likely *not* to occur than to occur. When it does, it is neither because of an accident of natural history nor because of human ingenuity; rather, it takes place thanks to a fortuitous collaboration of disparate and unequal forces. Rock, brushes, awls, and excavators' bodies form the outline of the bone. Rock, glues, plasters, putties, and preparators' bodies make it a solid object. Yet we must recognize that some forces are especially powerful. These include biological and geological processes such as the growth and decay of bone, the sedimentation and mineralization of rock, and its erosion by wind and rain. The powerful forces also include the capital of wealthy patrons of science and the labor of middle-class volunteers. Both these groups' high valuation of scientific discovery make the work of paleontology possible, but they also shape its direction. The next chapter extends this analysis of how individual fossils come into being by examining the more-than-human process that simultaneously creates the matter and meaning of whole skeletons, and even of entire prehistoric species.

2

———

Making Charismatic Violence

As visitors enter the Denver Museum of Nature and Science, they are greeted by the gigantic skeleton of a *Tyrannosaurus rex* standing upright on one foot, with the other extended in front of its body, its head twisted to the side, and its mouth wide open with teeth prominently showing (see fig. 6). This *Tyrannosaurus rex* has been an icon of the museum since it was installed in 1987. It has been described as dancing, playing soccer, or, for those with a more juvenile sense of humor, peeing on a fire hydrant.[1] However, most visitors see it as poised for attack. Many children recoil or cry upon seeing it. Countless adults stand under its perpetually gaping mouth, pretending that it is reaching down to bite them, as a friend or relative takes a photo. Parents hold up their children as though they are going to be carried off by the *Tyrannosaurus rex*. Four flights above this scene in the museum lobby, visitors "ooh" and "ah" over giant *Tyrannosaurus rex* teeth in the exhibition *Prehistoric Journey*. Many speculate about what the animal would eat if it were alive today. Would it be hamburgers or humans? Behind these jokes, comments, and photo ops, there is a common recognition of *Tyrannosaurus rex* as an enormous, masculine, and vicious killer with an unusual charisma that attracts people. This chapter explores the origins of this charisma.

Tyrannosaurus rex is the best-known prehistoric animal in the United States, possibly in the world. In fact, more people know this Latin binomial than know homo sapiens, their own taxonomic classification.[2] Theropods—the group of carnivorous bipedal saurischian dinosaurs to which Tyrannosaurs belong—is the most extensively studied and debated among vertebrate paleontologists.[3] The extent of this popular and scientific attention is surprising given how few *Tyrannosaurus rex* skeletons have existed until recently. The first *Tyrannosaurus rex* were unearthed and named in the first decade of the twentieth century, three decades

FIGURE 6. Robert T. Bakker mounted this cast of AMNH 5027 in the lobby of the Denver Museum in 1987. It has remained there since, despite no longer representing current knowledge of *Tyrannosaurus rex*'s physiology. Photo by author.

after the Great Dinosaur Rush (1877–92), when Triceratops and other gigantic dinosaurs were named. No more *Tyrannosaurus rex* skeletons were found until 1987, when an almost two-thirds complete dinosaur known as "Stan" (BHI 3033) was uncovered. The most famous *Tyrannosaurus rex* today is a nearly complete skeleton called "Sue" (FMNH PR 2081), which I discuss in chapter 4. Before this, a partial skeleton known simply by its accession label, AMNH 5027, was the most popular *Tyrannosaurus rex* in the world. The *Tyrannosaurus rex* in the lobby of the Denver Museum is a cast of AMNH 5027 (see fig. 6).

Unlike other famous prehistoric and modern animals, AMNH 5027 was never given a nickname.[4] Yet its charisma has long been evident in the reactions of scientists and nonscientists alike. Millions of people have gone to the American Museum of Natural History in New York City to see AMNH 5027. Millions more have sought out the casts of AMNH 5027 which have circulated to museums and other places around the world. Charles R. Knight's illustrations of AMNH 5027 have enthralled even larger audiences.[5] This specimen has been models for dinosaurs in movies from the silent short *Ghost of Slumber Mountain* in 1919, to *King Kong* in 1933, to *Jurassic Park* in 1993.[6] Remembering when he visited the American Museum as a five-year-old, the eminent paleontologist Stephen Jay Gould recalls that AMNH 5027 both "terrified" and "inspired him to become a paleontologist."[7] The scientists with whom I conducted fieldwork told similar tales of being captivated by dinosaurs' violence as children. While many paleontologists turn to research tiny organisms with far more extensive fossil records, those who study dinosaurs joke that they "never grew up." But dinosaur paleontology is a serious science. Despite the scant fossil record, these prehistoric megafauna, like modern megafauna, have an outsize influence. A graduate student reported that his advisor joked that if he wanted to get published in *Science* or *Nature*, he needed to have "dinosaur" in the article's title. Why? Because they are charismatic. And AMNH is the archetypical prehistoric charismatic megafauna.

This chapter traces the history of the *Tyrannosaurus rex* displayed in the lobby of the Denver Museum—from excavations in the badlands of Montana, to fossil preparation at the American Museum, to descriptions in academic and popular texts, and then, finally, to casting, mounting, and display at the Denver Museum—to examine how *Tyrannosaurus rex* came into being as a species of vicious and masculine charismatic megafauna.[8] Building on the previous chapter's examination of how fossils are made, in this one I explore how a whole species of dinosaurs has come into being in material and symbolic form. This chapter thus demonstrates how the meaning and matter of charismatic prehistoric megafauna emerge together. To do this, I extend the concept of *charisma*—long used to explain how extraordinary human individuals draw large numbers of dedicated followers—to understand how some scientific specimens, and the species they represent, become popular figures that inspire awe and devotion. In this analysis I develop the concept of more-than-human charisma.

PALEONTOLOGY'S CHARISMATICS

What exactly is charisma? It is usually defined as the exceptional magnetic appeal of, say, a preacher who attracts adherents and inspires extraordinary dedication in a way that cannot be rationally explained. The charismatic's unique spark leads large numbers of people to commit themselves to that leader and, often, to freely give them their labor, their money, and even their lives. Charismatics use the awe they incite to persuade people not only to understand the world as they do but also to act as they wish. Charisma began as a Christian concept that named a person who was divinely bestowed a "gift of grace," that is a special talent such as divination or healing, that drew a mass of followers. Max Weber broadened the Christian notion of *charismata* to designate "the personal quality that makes an individual seem extraordinary," thus "rendering that individual a 'leader' [*Führer*]."[9] Charisma, according to Weber's formulation, is an exceptional form of modern rule because its authority resides in a unique person, rather than in the established patterns of society (tradition) or formal rules (law) that ordinarily define authority within states. "The concept of charisma," anthropologist Peter Worsley wrote in the late 1960s, was "once confined to narrow scholarly circles" but had become "one of the most popular pieces of coinage in the intellectual exchange market."[10] By the second edition of his study of millenarian movements, charisma was being used in public discourse to explain the rise of John F. Kennedy to the presidency of the United States, as well as in scholarly studies of religion. Today, charisma is used to describe the leaders of political and social movements, as much as religious leaders, who inspire both awe and loyalty.[11] This term applies equally well to pop culture icons who draw large numbers of devoted fans. Scientists sometimes become charismatic figures, too.

While exhibiting charisma can advance a scientist's career in even the most arcane quantitative fields, vertebrate paleontology is unusual in the importance of charisma for professional success. Paleontology might well be the scientific field that has the largest number of charismatics. From Othniel C. Marsh and Edward D. Cope in the nineteenth century, to Stephen Jay Gould, Robert T. Bakker, and Scott Sampson at the turn of the twenty-first century, a few North American paleontologists have become "rock stars" who draw large numbers of adherents. Dinosaur paleontologists regularly appear in the news, in movies, and on television. Their books become bestsellers. They have millions of social media followers. They draw people—children in particular—to pursue scientific careers and funds to scientific institutions, especially museums. Few other scientists have achieved the level of public standing that they have. They have used their charisma to persuade scientists and nonscientists alike to adopt their views of prehistory and evolution.

The famous charismatic paleontologists have all been white men. This is not surprising, as both charismatic leadership and scientific discovery have been characterized by normative white masculine qualities. As anthropologist of science Sarah Franklin points out, "scientific pursuit is often described in terms of masculinity and adventure—as a domain of seminal breakthroughs, trail-blazing pioneers and unchartered territories."[12] This is especially true of paleontology in the United States. With westward expansion in the nineteenth century, earlier versions of white masculinity were "increasingly superseded by visions of masculinity predicated on physical hardiness . . . These intrepid, muscular virtues found their most profound expression in the figures of the explorer and frontiersman."[13] They "became new icons of 'the strenuous life'" by "liberat[ing] the overcivilized man trapped beneath" urban refinement.[14] It is thus no coincidence that prospecting and collecting fossils is called fossil *hunting*. "Hunter" marks the masculine ideal of physical strength, fearlessness, persistence, visual acuity, anatomical knowledge, and surgical dexterity. It extends the "colonialist sporting spirit of conquest of the unknown . . . frontier."[15] Charismatic scientists today draw on these meanings to attract large numbers of followers. Take, for instance, the vertebrate paleontologist Robert T. Bakker who mounted AMNH 5027 at the Denver Museum.

Robert T. Bakker is one of the most charismatic paleontologists in the United States today. Sarah Franklin's statement describes him well. Bakker first experienced the "adventure" of hunting fossils in Montana as an undergraduate and went on to pursue a PhD in paleontology and conduct fieldwork in the "unchartered territories" of the Western United States, Mongolia, and South Africa, among other places. However, Bakker is unusual in paleontology because he rose to fame owing to his "trail-blazing" work on the theory of dinosaur behavior more than on having made novel discoveries in the field. Almost exactly one hundred years after the Great Dinosaur Rush led by rival paleontologists Marsh and Cope, Bakker helped initiate another dinosaur renaissance in the 1970s that returned scientific and popular attention to dinosaurs. Bakker is widely credited with tearing down the long dominant theory that dinosaurs were cold-blooded, slow, dimwitted, reptile-like creatures. Following his advisor, John Ostrom, Bakker argued that dinosaurs were warm-blooded, acrobatic, swift, and intelligent animals. His mount of AMNH 5027 at the Denver Museum embodies this theory. Throughout his long and storied career, Bakker has proposed radical ideas about dinosaurs, a significant number of which later became accepted theory.[16] Although he has numerous critics and detractors, Bakker has a loyal following of dinosaur aficionados, amateur paleontologists, and professional scientists who eagerly await his latest theory.

I met Bakker at the Morrison Museum of Natural History in the Rocky Mountain foothills outside Denver. Although he was the curator of paleontology at the Houston Museum of Natural Science in Texas, he spent a great deal of time at this small museum dedicated to the paleontology of Colorado. Bakker had been involved with the Morrison Museum for decades and he has continued coming back to its scrappy laboratory to prepare fossils. That is where I found him one spring day when I arrived at the museum for some early morning fieldwork. Nick, a college student who was working at the Morrison Museum during the time of my fieldwork, had a similar experience. He met Bakker for the first time on the day he interviewed for an intern position at the Morrison Museum. He had read Bakker's articles and books in high school and immediately became "in awe of him," but did not expect to meet him in person, certainly not that day.

Nick had arrived early and knocked on the locked door. He was turning to walk back to his car to wait when "the door opens and this bearded guy in a cowboy hat pops his head out and says 'Hello?'" Nick imitated the big-eyed face of surprise and awe he himself had in that moment. He continued:

> I'm thinking "holy crap, that is—" but I didn't actually put it together because my brain recognized him, but my mouth didn't really register it. So I just say, "I'm here for the interview with Matt." He said something like, "Oh, follow me." So I walked, and I didn't even notice [the exhibits] because I was just staring at his back. We go to the back of the museum and start walking up the stairs. I couldn't take it. I [blurted out,] "Are you the paleontologist Bob Bakker?" Immediately in my head I'm thinking, "Oh my god, [I said] Bob? I'm being really informal with him. I need to call him 'doctor.'" He answers, "I am." I'm thinking, "Okay, I've got to redeem myself." I say, "That is so cool!" And he replies, "It is." I'm thinking to myself that that was the most embarrassing thing that has ever happened in my entire life.

Bakker led Nick into Matt's office. Taking one look at Nick's face, Matt asked him, "Did you just meet Dr. Bob?" Nick only eked out a "yeah," to which Matt responded, "He tends to have that effect on people." When Nick recounted this story to me, it sounded more like a teenager meeting the lead singer of his favorite band than a nerdy kid meeting a scientist. By the time I got to know both of them several years later, Bakker had become Nick's informal mentor, but I could see that Nick was still very much in awe of him. Bakker is iconoclastic in his thinking, assertive in his opinions, encyclopedic in his knowledge, clever in his language, and energetic in his style. At events large and small, I saw the throngs who came to hear his newest ideas, shake his hand, get his autograph, or simply be in his presence.

Bakker not only talks like a "scientific prophet"; he also looks like the stereotypical American paleontologist presented in the media.[17] In fact, he is recognized as the model for the paleontologist Robert Burke in the *Jurassic Park* series, for which Bakker was a scientific consultant. The resemblance in their names is matched by their bodies: both have searing blue eyes, long wild hair, and beards; both wear flannel shirts, cargo vests, and, most importantly, white cowboy hats. Bakker has appeared in more than a dozen documentaries in this attire. On the day I first met him at the Morrison Museum, he looked much like his movie doppelgänger. The only difference was that his signature cowboy hat rested on the microscope he was using to examine a fossil rather than on his head. Although Bakker maintained he wore it to avoid sunburns, and not as "a political statement," I saw him wearing it indoors—while teaching school groups, speaking at a fundraiser, giving a talk at the Denver Comic Convention, and working in the fossil prep lab—as well as outdoors in the sun. In fact, after my initial encounter, I can hardly remember a time I saw him without a cowboy hat.

Bakker's cowboy hat is more than the pinnacle of his signature style. Wherever he was, and whatever he was doing, his cowboy hat indicated the rough frontier masculinity of the white men who went on expeditions across the "Wild West" to "discover" riches, be they agricultural lands, minerals, or fossils. As I elaborate in chapter 4, paleontology emerged as a scientific discipline in the context of the colonization of western North America and the dispossession of its native inhabitants. Films, such as Disney's *Ten Who Dared* (1960), use cowboy hats to symbolize the independent and unfettered life of the intrepid rugged white men who endured adversity to locate and extract fossils from the Western United States. Bakker's cowboy hat follows in this tradition. It captures the charisma of the "scientific pursuit" and its "seminal breakthroughs, trail-blazing pioneers and unchartered territories," to quote Franklin once more. It expresses the masculine individualism and physical strength of the charismatic paleontologists in the United States.

The classic scholarship on charisma extended a notion of personhood—the individual—that is specific to the European philosophical tradition.[18] It ignores the social forces that disallow many people from becoming charismatics. The fact that charismatic paleontologists in the United States have all been white men is not because more white men are inherently charismatic, or because they work harder to develop their charisma. It is because they are more often encouraged to, recognized as, and rewarded for being charismatic. Charisma is instilled in them as much as it emanates from them. But is this unique to people? Isn't the cast of AMNH 5027 in the Denver Museum at least as charismatic as Bakker, the man who mounted it? Like human movie stars, some paleontological specimens draw large audiences of awe-struck and devoted fans. The charisma of certain fossils and certain paleontologists coproduce and reinforce each other.

CHARISMATIC PREHISTORIC MEGAFAUNA

Mounted dinosaur skeletons, casts, and reconstructions have demonstrated at least as much magnetic appeal as the charismatic scientists who study them. The American Museum of Natural History unveiled AMNH 5027, the world's first mounted *Tyrannosaurus rex*, in 1915. The anthropologist Franz Boas—who worked at the museum at the time—remarked that "mounted dinosaurs had an uncanny power to capture people's imaginations, and thus ensured a steady stream of visitors to the museum."[19] The charismatic appeal of *Tyrannosaurus rex* has only grown since then. The day the exhibit opened to the public in May 2000, ten thousand people came to the Field Museum of Natural History in Chicago to see "Sue" (FMNH PR 2081), which has supplanted AMNH 5027 as the most famous *Tyrannosaurus rex*. Sixteen million people viewed Sue over the next ten years, while another 6.5 million saw its cast as it traveled the world.[20]

In describing *Tyrannosaurus rex* as a charismatic figure, I draw on the use of *charismatic megafauna* to designate animals that have inspired enthusiasm for environmental movements. This term was first coined in the 1980s to describe the individual animals that environmental conservation organizations were utilizing in their effort to turn abstract and overwhelming environmental problems into something that potential supporters could relate to personally. Charismatic megafauna are not necessarily the species that scientists identify as most threatened, but those that best "serve as symbols and rallying points to stimulate conservation awareness and action."[21] Like charismatic religious leaders, charismatic megafauna strongly draw some people, while they appeal not at all to others. Yet these animals' magnetism has captivated a wide range of scientists and nonexperts. Environmental organizations' promotional materials featured images of popular large animals to "give a face" to habitat destruction and the threat of species extinction.[22] Their awesome bodies and sympathetic gaze appeal directly to the human viewer, saying, "Save us!" Like the *Tyrannosaurus rex* Stan and Sue, modern charismatic megafauna are often given personal names. These individual animals have been effective in convincing large numbers of people to sign petitions, lobby their representatives, volunteer for conservation projects, and the like. Zoos, natural history museums, and environmental conservation organizations have long relied on charismatic animals for their economic survival.[23] In recent years, bears, elephants, and primates have become their "breadwinners."[24]

While charming large mammals have been effective in raising capital for conservation organizations, critics have shown that their use by environmental organizations has led to "saving" individual animals or species but neglecting less charismatic ones and the larger environmental webs on which their lives rely.[25] This occurs, I believe, because the concept of charismatic megafauna rests on the same notion of the individual as the traditional conception of the charismatic leader does. Charismatic megafauna species have been given "singular, self-contained value" that is "a product and reflection of bourgeois individualism, or of the atomization . . . of social relations" in the dominant US way of thinking.[26] Like charismatic people, charismatic animals are seen as "heroes of their own lives."[27] That is, their actions—often their very survival—arouses awe and inspires devotion to them as individual animals or species. This erases the larger webs of social and material relations on which they depend.

Tyrannosaurus rex and other dinosaurs have played a similar role as modern charismatic megafauna in generating support for paleontological research and institutions. Not only have AMNH 5027 and its casts drawn paying customers to museums but, as I showed in the previous chapter, they have also attracted philanthropists and volunteers to paleontological digs and laboratories. *Tyrannosaurus rex* fans have included twentieth-century capitalist tycoons, such as Andrew Carnegie and John D. Rockefeller, as well as twenty-first-century ones, such as David H. Koch. Much as modern charismatic megafauna have inspired volunteers to dedicate huge amounts of time and resources to their care, prehistoric charismatic megafauna have inspired volunteers to give thousands of hours of labor to excavating and cleaning fossils.[28] However, unlike seemingly cuddly pandas and placid elephants, *Tyrannosaurus rex* attracts people because of their ferocious reputation. Although many modern charismatic megafauna fight and kill, this aspect of their lives is not the focal point of their charisma. The charisma of most modern charismatic megafauna lies in the combination of their giant size with their sympathetic, even cute, faces. AMNH 5027 does not seem as sympathetic as even the fiercest megafauna at zoos and amusement parks. Its "face" is comprised only of a skull, albeit an unusually well-preserved one. In the place of seemingly pleading eyes, it has only holes. Instead of offering a friendly smile, it bares its enormous teeth. Yet this iconic visage has attracted unprecedented attention, enthusiasm, and even love, as well as fear. How did this come to be? Is the vicious charisma of this *Tyrannosaurus rex* a cultural construction that scientists and movie-makers have invented? Is it inherent in its anatomy or the nature of its species? Is there a better way to explain *Tyrannosaurus rex*'s long-lasting charisma?

In this chapter I develop an analysis of more-than-human charisma that builds on the anthropological scholarship showing that charisma does not reside in extraordinary individuals but emerges through relationships among leaders, disciples, and followers. Charisma does not work like gravity; it is not universal. Decades of anthropological work has illustrated that a charismatic's power depends on a particular assemblage of people, cultural forms, and social forces within a specific context.[29] This is as true of scientists and scientific specimens as of religious leaders.[30] Charisma can emerge through relationships among humans and nonhumans too, including stones. In the practice of Candomblé in Bahia, Brazil, for instance, charismatic priests and sacred stones come into being together. The power of the stone emanates not from human belief but from its "physical presence and its dialectical relation with the human body."[31] Yet the charisma of stones and people is different, in part, because their materiality is distinct. Charismatic people speak to their devotees and lead them to a promised land, while charismatic stones act through their physical features.

While anthropologists have studied more-than-human charisma within religious contexts, science and technology studies scholars have explored "nonhuman charisma" beyond the charismatic megafauna that draw publics to science, including the charisma of computers and scientific data.[32] In these contexts, like in religious ones, charisma is not a characteristic of a specific individual but a powerful affective force generated through human-nonhuman interactions at a historically and geographically specific site. I build on this work by showing how scientific specimens, paleontological knowledge, and charisma are produced together in a more-than-human process. My ethnographic research with fossils and fossil-loving people reveals that charismatic prehistoric megafauna are mutually generated by animals, people, and objects, and through biological, geological, and political-economic processes that extend over millions of years. In the remainder of this chapter, I use this conception of more-than-human charisma to examine how *Tyrannosaurus rex*, particularly its most famous representatives, came into being as vicious and masculine prehistoric charismatic megafauna. I now turn to one of the important forces in doing this, scientific taxonomy.

THE NAMING OF THE BEAST

A species is a group of organisms that share common biological attributes and are recognized by a common scientific name. Taxonomists separate organisms into species based on one or more distinctive morphological (structural), physiological (behavioral), or biochemical features. Distinguishing species of prehistoric vertebrates is complicated by the fact that there are so few specimens and none are complete. Moreover, many of the criteria scientists use to divide modern organisms into species cannot be used to sort prehistorical ones. They cannot test which can interbreed or compare their genes, for instance. In some fields, taxonomy has been superseded by phylogenetics (see conclusion). Yet, when the fossilized remains of an animal is entered into the scientific record, it is still identified with a species name.[33] This taxonomic identification is often based on commonalities among a handful of incomplete skeletons; it is sometimes based on the difference between a single bone and existing ones. As more fossils are identified, a specimen's classification is revised. Because of this, vertebrate paleontology—possibly more than other scientific disciplines—recognizes taxonomy as an ongoing and contested process.[34] What is not often acknowledged is how the European roots of the ostensibly universal scientific system for categorizing organisms has left its mark on dinosaurs.

As I mention in the introduction, it is likely that the large fossilized bones that people have long encountered in their environment inspired the beasts in folktales.[35] Whatever their origins, beasts have been used—in religion, literature, science, and politics—to label beings that fascinate people at the same time that they frighten them.[36] In the eighteenth century, the Swedish botanist Carl Linnaeus began a process of taming monsters and beasts with an orderly and hierarchical schema for naming and categorizing all life on earth. Linnaeus divided the beasts into different groups of fauna, his Latin invention for animals. For instance, he called four-footed beasts mammals, and quite radically, placed humans within this class.[37] Ever since the codification of the Linnaean system of taxonomy in the second half of the nineteenth century, each organism newly identified by science is labeled with a two-part Latin or Greek name that places it on the ostensibly universal "tree of life" on earth. The first part of the name specifies the organism's genus and the second its species.[38] This naming convention emphasizes common ancestors above shared characteristics in the present, and it implies a linear progression of evolution. It is also an exclusive system that allows an animal to exist in one or another group, but not in more than one. While Linnaean taxonomy introduced a rigid and hierarchical ordering of animals, it did not entirely replace older ideas about monsters and beasts. Instead, these became embedded in dinosaurs' scientific names.

Regardless of who locates, extracts, or prepares a specimen, the scientist who publishes the first scholarly description of a new taxonomic group gets the privilege of choosing its name. This individualization of knowledge production parallels the individualization of charisma in religious and political contexts. Richard Owen exemplified this when he coined the term *dinosauria* in the early 1840s. Owen's new taxa of prehistoric fauna brought together ancient terrestrial animals that he saw as sharing essential anatomical features.[39] He grouped together prehistoric creatures publicized by the European scientists—both professional and amateur—who were competing to discover the largest and most bestial animal to have lived on earth. The English reverend William Buckland asserted that Megalosaurus ("large lizard") was the most bestial, while the surgeon-scientist Gideon Mantell claimed that distinction for Iguanodon ("iguana tooth").[40] The following year, the French anatomist Georges Cuvier argued that Plesiosaurus ("almost lizard") was "the most monstrous [animal] that had yet been found amid the ruins of the former world."[41] As they and their colleagues amassed their descriptions of extinct species, they began to conjure a dark and deadly "antediluvian world" full of enormous beasts.[42] When Owen invented the new taxa *dinosauria*—meaning "fearsome lizards"—he incorporated this concept of beastliness into the Linnaean classification scheme, thereby giving it scientific legitimacy.

The quest to identify the beastliest of the prehistoric creatures, which began in Europe, grew with the hunt for fossils in North America. As new species were recognized, their Latin binomials continued to incorporate notions of bestiality. This was particularly true for the carnivorous dinosaurs that were found beginning in the 1870s. In 1881, Marsh combined the Greek words for "wild beast" and "foot" into *theropoda* to name a new order encompassing all the carnivorous terrestrial dinosaurs that were being excavated from the Mesozoic rocks of the Western United States. It made sense—to both scientists and a lay public with growing interest in dinosaurs—to classify the enormous extinct carnivores into a taxonomic group that highlighted their common violence above all other shared features. But this was far from inevitable. Compare Owen's and Marsh's choices of names with those of the British entomologist George Willis Kirkaldy, who apparently named insects after his sexual conquests, including a bedbug he called Dolychisme (pronounced "Dolly-kiss-me"). Kirkaldy's names "bent the starched collars of British entomology" and were officially censured by the Zoological Society of London.[43] In contrast, the names scientists chose for dinosaurs received no such condemnation on either side of the Atlantic. The historian of science Stephen T. Asma has argued that the nineteenth-century naturalists who codified the new scientific approach to animals discredited the monsters and beasts of European tradition.[44] Yet the taxa *dinosauria* and *theropoda* indicate that the concept of the beast became embedded within the emerging science of paleontology. As the Linnaean system of Latin or Greek binomials replaced older terms, the understanding of prehistoric creatures as beasts did not disappear; it just became subtler. *Tyrannosaurus rex* exemplifies this.

A KING IS NAMED

In October 1905, Richard Fairfield Osborn introduced into the scientific record a new genus and species of extinct animals (Tyrannosaurus and *Tyrannosaurus rex*, respectively). Osborn, the American Museum of Natural History's leading paleontologist, narrowly beat the Carnegie Museum's paleontologists, who had some analogous fossils, in naming the species. Osborn invented the new taxa based on his study of one lower jaw, a few other bones, and a couple of isolated teeth. He also reclassified some fossils that he earlier called *Dynamosaurus imperiosus*, along with a few vertebrae Cope had named in 1892 and isolated teeth others had collected, as belonging to this genus and species. This is not remarkable. During the "bone wars," Marsh and Cope had made it common practice to name new dinosaur species based on only a handful of fossils. What is remarkable is not only how Osborn defined the species' distinctive anatomical features but how extensively he elaborated its distinctive character traits, based on a small number of bones.

Tyrannosaurus rex, the binomial that Osborn invented for his new species, means "tyrant lizard king." Osborn thus joined his predecessors' competition to name the "most bestial" of animals that ever lived on earth. He initially stated that he named it this because of "its size, which far exceeds that of any carnivorous land animal hitherto described."[45] "This animal is in fact the ne plus ultra [zenith] of the evolution of the large carnivorous dinosaurs," he explained, and that is why "it is entitled to the royal and high sounding group name which I have applied to it."[46] Osborn thus used the animal's enormity as evidence that it ruled over other prehistoric animals. In other words, its size indicated tyrannical behavior toward other animals and dominance over its environment. In a later publication, Osborn explained his reasons for labeling the new species a "king" and a "tyrant" somewhat differently:

> *Tyrannosaurus* is the most superb carnivorous mechanism among the terrestrial Vertebrata, in which raptorial [predaceous] destructive power and speed are combined; it represents the climax in the evolution of a series which began with the relatively small and slender Triassic carnivore *Anchisaurus*.[47]

Osborn consistently describes Tyrannosaurs as the evolutionary apex of the carnivorous land animals of the Mesozoic era. In this second statement, however, Osborn places more emphasis on the species' strength and swiftness than on its size alone. Brian Noble points out that in a period in which scientists, including Osborn, were describing white men as the pinnacle of human evolution, Osborn declared Tyrannosaurus the "ultimate [being] in the evolutionary order of the Mesozoic."[48] Noble identifies correspondence between Osborn's ideas about human races and dinosaur species. I want to add that Osborn not only identified the anatomical distinctions of the species but also behavioral and temperamental ones. He used the specimens' dentition, above other features, to justify calling the species the "tyrant kings" of the dinosaurs. For instance, he notes that Tyrannosaurs' dentition was "powerful" and "of extreme raptorial carnivorous type."[49] They had the typical shape of carnivores: they were curved, serrated, and mostly conically shaped. This made them well-suited for grasping and slicing flesh, but not grinding food.[50] Osborn noted that *Tyrannosaurus rex* had a shorter jaw and fewer, but larger, teeth than Allosaurus, a similar dinosaur found in the early Jurassic Period. Tyrannosaurus teeth are not as sharp as Allosaurus teeth, and their serrations are smaller. Paleontologists have interpreted this as indicating that *Tyrannosaurus rex* bit through, rather than sliced off, the flesh of its prey. Yet Osborn went further than identifying these dental features as evidence of the species' diet. He labeled *Tyrannosaurus rex* not only as a carnivore but also as a predator. In recent years, this has become a point of dispute as some paleontologists have argued that

Tyrannosaurus rex were scavengers instead of, or in addition to, predators. For Osborn, predation mirrored the species' name. At the same time that Osborn and his fellow eugenicist scientists described white men as the pinnacle of human evolution, he declared *Tyrannosaurus rex* the "ultimate [being] in the evolutionary order of the Mesozoic."[51] Although they conceived their own "rational" progress through eugenics as the antithesis of dinosaurs' brute struggle for existence, violence underpinned both claims to superiority.[52]

The media coverage of the time amplified the scientific designation of *Tyrannosaurus rex* as the vicious rulers of the prehistoric world. The *New York Times* and *Scientific American* both carried frequent stories about paleontology at the American Museum, to the point that a historian has called the latter "a vehicle for promoting" the department.[53] These news sources featured articles about Osborn's 1905 announcement of the new species *Tyrannosaurus rex*, the display of a pair of composite hind limbs the following year, and the opening of the fully mounted skeleton of AMNH 2057 four years later. In their stories, they frequently used "beast," "monster," and "war lord" to describe the new specimens, as well as the genus and species they represented. One *New York Times Magazine* piece, for instance, described *Tyrannosaurus rex* as "the most formidable fighting animal of which there is any record." It further called it the "most remarkable prehistoric beast," adding that "the newly discovered monster was the absolute war lord of the earth in his day."[54] These references to the European tradition of monsters and beasts imply that *Tyrannosaurus rex* acted with gratuitous violence that exceeded what they needed to survive. Together with the notions of "tyrant" and "king" embedded in the species' scientific name, they transformed the animal's diet of meat into the violent practices of a bloodthirsty ruler.[55] The notions of "tyrant" and "king" also masculinized the whole species. Furthermore, they helped make these traits charismatic. While tyranny is usually the antithesis of charismatic leadership, monsters and beasts have long engendered "the simultaneous lure and repulsion of the abnormal or extraordinary being" that unites charisma and cruelty.[56] As Boas noted, *Tyrannosaurus rex* drew large numbers of people to the American Museum. Osborn's taxonomic invention and the extensive media coverage of his museum's work played important parts in making *Tyrannosaurus rex* into a species of ferocious, masculine, and charismatic prehistoric megafauna.

An account of the "discovery" of AMNH 5027 published in *Scientific American* a few years later further elaborated this characterization of the species. It was written by Barnum Brown, the fossil-hunter who led the American Museum of Natural History expedition that uncovered AMNH 5027. Brown opened his article this way:

> Rarely is it safe to speak of anything as ultimate in prehistoric life, but there is little doubt that the American Museum now exhibits a skeleton of the largest flesh-eating animal that has ever lived.
>
> This is *Tyrannosaurus*, the tyrant lizard, a dinosaur that lived during the close of the Cretaceous period. It was one of the very last expressions of its race and, judged

by size and structure, was king of its kind. An idea of its immense size can be formed from measurements of the skeleton, 47 feet in length, and, as mounted, 18 ½ feet in height. When fully erect this animal would have reached a height of 20 feet.[57]

In recounting his story, Brown implies that he immediately recognized the exceptionalism of the dinosaur he claimed to be the first to discover. As I discuss elsewhere in the book, "discovery" implies the unique accomplishment of an individual, usually an educated white man, who claims intellectual mastery and ownership over nature. Like the concept of charisma, the notion of discovery disregards the other people, institutions, and forces that made that person's achievement possible. Brown claims the new dinosaur as his discovery, despite not having the academic qualification or position to enter it into the scientific record. This ignores the many people who made his "discovery" possible, as well as their claims to the fossils.

In his writing, Brown, like Osborn, highlighted AMNH 5027's teeth as prime evidence for the excessively violent character of this prehistoric animal, and the species for which it stood. Brown stated that the skeleton's "massive head was armed with 13 dagger-like sawbladed teeth in each jaw, the largest 5 inches long." From this, he quickly concluded that "*Tyrannosaurus* was capable of destroying any of the contemporary creatures and was easily king of the period and monarch of its race."[58] This mirrored the *New York Times*'s earlier assertion that "most of his attacking and rending were done with his teeth," describing them as "razor-edged" with "many [that were] a foot long."[59] These examples show that scientists and journalists singled out AMNH 2057's teeth as signs of its incredible violence. Yet note a discrepancy: Brown's description in *Scientific American* states the longest teeth were five inches, while the *New York Times* asserts they were more than twice that length. What accounts for this? Is it simply a case of exaggeration by a publication that needed to sell stories? The answer, I argue below, lies not in scientific or journalistic texts but in the materiality of AMNH 5027, as well as the processes of fossilization, excavation, preparation, and mounting that created it.

UNEARTHING AMNH 5027

Barnum Brown led the American Museum of Natural History expedition to the Montana badlands where AMNH 5027 was unearthed. This fairly complete skeleton with a well-preserved skull was not their first or only find. Brown and his team had been dispatched to the Western frontier of the United States by Osborn, who was then the director of the American Museum's Department of Vertebrate Paleontology and later the museum's president. Following the competition between Marsh, with his Peabody Museum at Yale University, and Cope, at the Academy of Natural Sciences in Philadelphia, Osborn aimed to amass the best fossil collection and develop the best paleontology program in the world at the American Museum in New York City.[60] Brown did a great deal to make the American Museum the world's premier institution for vertebrate paleontology. Although Osborn sent him out West to build a comprehensive collection of Jurassic mammal fossils, Brown was more interested in gigantic dinosaurs than tiny mammals. Brown used Osborn's ambitions for the American Museum to convince Osborn to let him pursue dinosaurs, and even to provide the additional funds needed to extract these much heavier fossils.[61] In the process, he unearthed the museum's most famous prehistoric specimen.

Barnum Brown and his team found the fossils that would become *Tyrannosaurus rex* in an outcrop of Cretaceous rock along the Big Dry Creek near Jordan, Montana, that is part of the Hell Creek Formation. Brown had the premonition that this part of Montana would be a good place to look for fossils several years earlier, when a friend showed him photographs of a Triceratops horn in situ there. Although the geology of the region had not been thoroughly mapped, Brown recognized its similarity to an area in Wyoming where Cretaceous dinosaurs had already been found. He was right. Two summers after he located a few fossils of an enormous animal in Wyoming, Brown sighted similar fossils "before the cook's call for dinner" on his first day of prospecting near Jordan.[62] However, these fossils were encased in extremely hard sandstone, and it took Brown and his crew two field seasons to excavate and transport them to the railroad. Worse yet, the skeleton was far from complete. Most of the animal had not fossilized, or had fossilized but then disintegrated. The American Museum's fossil hunters worked at this site for another six years before Brown found a similar skeleton with nearly half its fossilized bones intact and a well-preserved skull. It was tagged AMNH 5027 to indicate that it was the 5,027th specimen to enter the American Museum's collection.

When Brown and the American Museum team excavated these enormous creatures, they found their bones partially disarticulated, with their teeth scattered near the skull and the jaw flattened. These are common outcomes of the fossilization process that occurred millions of years prior, as I recounted in the previous chapter. The geological process by which soft sediments turned into increasingly hard rock permitted the bone's structure to last for millions of years. However, during the fossilization process, globular bone structures like skulls often become flattened by the pressure of rock layers above them. Teeth are composed of denser material that tends to maintain its shape through fossilization. Therefore, as AMNH 5027's skull slowly became compressed, its teeth remained rounded and dislodged from the jaw. That is how the American Museum expedition found them millions of years later.

At the end of the 1908 field season, Brown's team packed the fossils they had excavated in crates and sent them by train to New York. When the boxes arrived at the American Museum, the staff quickly got to work preparing to mount AMNH 5027 for display in its new Hall of Fossil Reptiles.[63] Adam Hermann, the American Museum's chief fossil preparator, worked with Barnum Brown to prepare the fossils for study and exhibit.[64] After they and their assistants cleaned and repaired the skull, they tried to reinsert the teeth into the jaw. They could figure out which tooth belonged in which socket based on their distinct sizes and shapes, but the teeth did not fit as far into their sockets as they had when the animal was alive. While at least a half of each tooth would have been hidden below the living animal's gums, much less of the tooth fit inside the jaw once it had fossilized. Because of idiosyncrasies in the fossilization process, some of the tooth cavities had flattened more than others. Therefore, the preparators were able to insert different teeth at different depths. This gave AMNH 5027's mouth a jagged look. Furthermore, with some of the teeth protruding considerably, the upper and lower jaws did not fit together and could not be closed. Fossilization had thus created a skeleton that was different from the animal's living form. The distinction between the fossil skull and its living form explains the discrepancy between the *New York Times* and *Scientific American* accounts. The former describes the length of the fossil from crown to root, while the latter measures the tooth exposed above the gum.

The preparators at the American Museum accentuated the differences between the animal's living form and its fossilized one when they assembled AMNH 5027's skull with its jaw open wide and its teeth protruding. Since skulls are too heavy to mount, the preparators made a plaster cast to use on the skeleton destined for the centerpiece of the museum's dinosaur exhibition. They created this cast with the same open jaw and jagged, protruding tooth positions as the skull fossil, which was displayed nearby. The cast and the fossil skull matched as exactly as possible. As a result, the throngs of people who came to the museum to see the enormous new dinosaur viewed AMNH 5027 with its mouth open and its larger-than-life teeth prominently visible. This shows that Osborn, Brown, and the journalists who publicized their work did not simply impose familiar ideas about vicious beasts onto long-extinct animals. This history begins to demonstrate how, instead, *Tyrannosaurus rex*'s vicious charisma emerged, in both material and discursive form, out of the confluence of multiple factors: the fossilization of bone, the history of European folktales, the development of scientific taxonomy, and the practice of fossil preparation and casting. Several paleontologists and preparators helped me to better understand this, despite disagreeing with my conclusion that *Tyrannosaurus rex* has come into being through human and nonhuman forces.

CHEATING TYRANNOSAURUS REX?

I first heard that *Tyrannosaurus rex* skulls were consistently reconstructed with larger-than-life teeth and overly gaping jaws during the question-and-answer period at the end of a training for museum volunteers. The paleontologist answering the questions said that preparators routinely "cheat" in mounting *Tyrannosaurus rex* to "make him look scarier."[65] He further asserted that they mounted *Tyrannosaurus rex* with extra space at the spot where the upper and lower jaw would have met so they could open its mouth wider than they should. I was intrigued but also a bit skeptical. So I asked a few other paleontologists and fossil preparators about it.

A few days later, I found myself standing in front of a cast of the skull of the *Tyrannosaurus rex* called Stan (BHI 3033) with Matthew Mossbrucker, the director and curator of the Morrison Natural History Museum. Stan is the second most reproduced *Tyrannosaurus rex* in the world after AMNH 5027. It was found following eight decades of numerous unsuccessful expeditions to obtain a more complete and articulated skeleton than that of AMNH 5027. A team of collectors from the commercial Black Hills Institute of Geological Research unearthed Stan in South Dakota in 1987. Like AMNH 5027, this Tyrannosaurus was embedded in an outcrop of the Hell Creek Formation in the Badlands of the Northwestern United States. The team called the skeleton Stan, after Stan Sacrison, the amateur paleontologist who first spotted it. Casts of Stan soon joined those of AMNH 5027 on display at major museums across the Global North. Smaller institutions that cannot afford replicas of AMNH 5027, such as the Morrison Museum, often purchase casts of Stan, which come with relatively low price tags, ranging from $100,000 for a complete skeleton to just forty dollars for a 4.5 inch tooth.[66] One remarkable feature of Stan is that its skull has puncture holes in its sides that are the shape of Tyrannosaurus teeth. These holes healed around their edges, so they are not related to the dinosaur's death. Paleontologists have interpreted them as signs that *Tyrannosaurus rex* fought each other for dominance.

Matthew Mossbrucker concurred with the paleontologist who first told me that the preparators at the Black Hills Institute left extra space between the upper and lower jaws when they mounted Stan's skull. Matthew showed me the gap on the cast in front of us. Then, he pointed to the visible line on each tooth where the crown enamel once met the gum. On this cast of Stan, the teeth are not a uniform black, as they are on the casts of AMNH 5027 I have encountered. While the crowns of Stan's teeth had been painted a shiny black, the roots were made a matte dark gray. The coloration of this cast matched the fossils of Stan on display at the Black Hills Institute's museum in Hill City, South Dakota, where I later visited. At the Morrison Museum, I could see for myself that the line between black enamel and gray root was uneven from one tooth to the next. The roots stuck out more than an inch on many of Stan's teeth, and almost not at all on others (see fig. 7). This, Matthew explained to me, was because the preparators at the Black Hills Institute, like those at the American Museum nearly a century earlier, inserted the teeth as far as they would go without modifying the cavities in the jaw cast. Fossilization had not flattened Stan's skull as much as the skull of AMNH 5027, yet Stan's teeth were found disarticulated from its jaw. In this case, too, the preparators were able to identify where in the jaw the found teeth belonged but were unable to fully fit them into their sockets. At the Black Hills Institute, like at the American Museum, the skull was assembled with its teeth protruding unevenly from the jaw and its mouth wide open, making it look especially vicious. The subtle line where

FIGURE 7. On the cast of the *Tyrannosaurus rex* "Stan" (BHI 3033) at the Morrison Museum, the tooth crowns are a shiny black while the roots are a matte dark gray. Photo by author.

the tooth enamel ended, and the root began, indicated the difference between the position of the fossilized teeth and their living form. Yet, in my many months of fieldwork with multiple casts of Stan, I had not previously noticed this detail. I suspect that the vast majority of museumgoers did not either.

I encountered another cast of Stan at the facilities of a small commercial collecting, casting, and molding company. There I met the company's president, Steve, who is among the most respected fossil preparators in the United States. His company's fossils and casts can be found at museums, universities, and private collectors around the world. Steve is a friend, and competitor, of the preparators at the more well-known Black Hills Institute. As Steve described what had occurred with Stan, he brought my attention to one tooth in particular, saying, "There is no way that tooth was sticking out that way." He asserted that the Black Hills

Institute's preparators "left it that way" because "it looked cool." Steve added, with a slightly sarcastic tone, "You do not have gigantic teeth sticking out ten inches from [the jaw] with half of it being root! Hello!" Steve told me that his company sold "a fairly significant number of specimens as research specimens, unmodified and as preserved" to scientists, while they sold specimens they restored to the "way it looked when it was alive" to exhibit designers. He explained:

> If it's a really important specimen, we almost always mold it as is. And then we take a copy and we reconstruct the copy to take the distortion out, or to put the teeth back where they belong, or whatever. Then we have a restored copy. We want to preserve the fossils, but we also know that that is not the way it looked when it was alive . . . So we recreate them using multiple casts, cutting, bending, shaping, and reworking things to recreate the skull back to the shape it was when it was alive. We sell both.

A preparator at Steve's company told me that the people at the American Museum and the Black Hills Institute could have molded the teeth separately from the jaw, reconstructed the jaw cast with wider tooth cavities, and then fit the teeth casts into them so only the crown extended above the jaw line. However, he explained, buyers did not want such a cast of *Tyrannosaurus rex*, unlike in the case of other prehistoric animals. He stated that, for *Tyrannosaurus rex*, "those big teeth are kind of a selling point and they want to show them off." In other words, *Tyrannosaurus rex*'s extra-large teeth are crucial to its charismatic appeal. They attract viewers, ticket and merchandise sales, media coverage, and philanthropic donations. Yet *Tyrannosaurus rex*'s vicious charisma cannot simply be explained by financial calculations.

Paleontology's scientific standards dictate that preparators should restore fossils to the form in which they fossilized, or, in Steve's words, "as preserved." Like gem preparators, fossil preparators regard their work as returning fossils to their original condition before they were damaged by forces such as compression, erosion, or excavation. Legitimate preparation "restores the [stone] to its natural state, what it was 'supposed to' look like" and leaves no "indication of the human labor that went into its creation as such."[67] But to which "natural state" are preparators supposed to restore fossils? Fossils are not static objects, but organic-turned-inorganic matter in a continual process of change. As discussed in the last chapter, rock hardens, compresses, breaks, and crumbles. Vertebrate paleontologists and fossil preparators try to recreate a moment in this process after a bone has turned to stone but before it has become broken or weathered. In the case of AMNH 5027 and Stan, preparators at the American Museum and the Black Hills Institute tried to return the skulls to the moment just before their teeth were dislodged from their jaws, but they could not fully do so. The fossilized teeth would not fit all the way back into the stone jaw. If they pushed further, the fossils would break. Some of the teeth went several inches into the jaw cavities and others, only one or two inches. The rocks were inflexible, and the preparators did not modify them to force them to fit. To do that would have violated paleontology's standards of scientific accuracy. Therefore, it was in part through people's pursuit of scientific accuracy, and the fossils' obstruction of it, that *Tyrannosaurus rex* came into being as the vicious tyrant king of the dinosaurs.

The same rules apply to research-grade casts, which are supposed to be as faithful to the stone as possible. Efforts to conceal preparators' work have only increased since AMNH 5027 was mounted at the American Museum. At the Denver Museum, casts and reconstructions are painstakingly textured and painted to make them indistinguishable from lithic fossils, at least to the untrained eye. Modifying *Tyrannosaurus rex*'s jaw or teeth to fit them together violates scientific protocol, despite reflecting scientific knowledge of the species. The fossil preparators at the American Museum and the Black Hills Institute followed a similar protocol for scientific accuracy in casting AMNH 5027 and Stan, despite the century between them. They both cast these skulls as single objects in their reconstructed fossil form. They could have cast the teeth and jaw separately, in the process eliminating the flattening in the tooth sockets so that the teeth would fit into them as they did in life. But they did not.

It is important to recognize that one casting technique, or one preparation technique, is not inherently more scientific than the other, and neither involves more or less human agency. They are accurate by different standards: one to scientists' best understanding of the live animal, the other to their best understanding of the fossilized one. Furthermore, they both involve human and nonhuman forces: the preparators' dexterous hands and eyes, and the rocks' rigidity and fragility. To further explore the confluence of forces involved in making *Tyrannosaurus rex* the zenith of prehistoric charismatic megafauna, I turn back in time to the mounting of AMNH 5027 at the American Museum in the 1910s.

MOUNTING THE BEAST

Mounted dinosaur skeletons are "complex assemblages" that "closely resemble mixed media sculptures."[68] Even a single fossil is an assemblage of rock, putties, glues, and metals; the labor of scientists and volunteers; tools and hands; knowledge, meaning, and value.[69] They not only coalesce materials, but also ideas. In the beginning of the twentieth century, Osborn wanted to mount AMNH 5027 in the American Museum's Hall of Fossil Reptiles in a way that would make clear that they were once living beings engaged in a fierce "struggle for existence." Osborn therefore favored displays that positioned them fighting or eating each other.[70] For instance, the year before the American Museum preparators started working on 5027, they mounted an Allosaurus standing over a prostrate Apatosaurus. As William Diller Matthews, a paleontological illustrator at the museum, stated, this display painted "a most vivid picture of a characteristic scene of that bygone age, millions of years ago, when reptiles were the lords of creation." In that era, Matthews continued, quoting the British poet Alfred Lord Tennyson, "'Nature red in tooth and claw' had lost none of her primitive savagery."[71] Following the success of this exhibit, Osborn asked his staff to design a display for 5027 that would emphasize "the raw power and ferocity of nature's creations" that *Tyrannosaurus rex* embodied even more fully than other dinosaurs.[72] He believed that AMNH 5027's mount should surpass the museum's previous ones, much like Tyrannosaurus surpassed Allosaurus in size, aggression, and tyrannical masculinity. Osborn wanted the mount to demonstrate in three dimensions that *Tyrannosaurus rex* lived up to its regal name. He also hoped it would draw crowds of awestruck visitors to the museum.

The proposed display for AMNH 5027 portrayed it attacking both its prey and another *Tyrannosaurus rex*. Osborn described the scene he wanted it to convey in these terms: "One [Tyrannosaurus] is crouching over its prey (which is represented by a portion of a [Trachodon, now called Edmontosaurus] skeleton) . . . The second reptile is advancing, and attains very nearly the full height of the animal." This pose, he explained, "represents the animals just prior to the convulsive single spring and tooth grip which distinguishes the combat of reptile[s] from that of all mammals."[73] For his part, Brown imagined the same "scene of daily occurrence" in this way:

> It is early morning along the shore of a Cretaceous lake three [*sic*] million years ago. A herbivorous dinosaur *Trachodon* venturing from the water for a breakfast of succulent vegetation has been caught and partly devoured by a giant flesh-eating *Tyrannosaurus*. As this monster crouches over the carcass, busy dismembering it, another *Tyranno-saurus* is attracted to the scene. Approaching, it rises nearly to its full height to grapple the more fortunate hunter and dispute the prey. The crouching figure reluctantly stops eating and accepts the challenge, partly rising to spring on its adversary.
>
> The psychological moment of tense inertia before the combat was chosen to best show positions of the limbs and bodies, as well as to picture an incident in the life history of these giant reptiles.[74]

Here, as elsewhere, Brown portrays *Tyrannosaurus rex* as a hunter like himself. The use of this term both for the men who searched for and excavated fossils in the Western United States and for the carnivorous dinosaurs they found is telling. As I mention earlier in the chapter, "hunter" signifies a particular masculine ideal that includes physical attributes (strength, ruggedness), personality traits (fearlessness, persistence), knowledge (morphology, physiology), and skill (dexterity, aim). This term builds on the associations of "king" and "tyrant" with masculinity. The *Tyrannosaurus rex*'s masculinity mirrors the rugged masculinity of the paleontologists who hunted for dinosaur fossils on the Western frontier, and then named them at museums on the East Coast. It naturalizes this form of masculinity by making it seem prehistoric, even timeless, and therefore not generated within the particular historical and social context of the violent colonization of the Western United States. The mount of AMNH 5027 would make the species' character evident through its substantial materiality.

The hunting scene that Osborn and Brown devised would highlight the virile charisma of AMNH 5027. In the tripartite interaction that Osborn and Brown devised, AMNH 5027 is the apex predator, challenging another member of its species who has killed a herbivorous animal. Like the beasts of European folktales, this engenders a "simultaneous lure and repulsion of the abnormal or extraordinary being."[75] Although this elaborate scene was never brought to fruition because it was too large to fit within the American Museum's hall, the position of the second *Tyrannosaurus rex* described above was used to mount AMNH 5027 alone.[76] The dinosaur's reason for rearing up was left to the viewers' imagination, but it was clearly poised for attack. Was it obtaining its dinner, fighting for dominance over another *Tyrannosaurus rex*, or both simultaneously? Whatever interpretation one gives it, the skeleton has stood tall on its hind legs, with its mouth wide open, baring its oversized teeth and showing off its ferocity, for over a hundred years.

Eight decades after the first mounting of *Tyrannosaurus rex* at the American Museum in New York, Bakker mounted a cast of AMNH 5027 in the Denver Museum's lobby. As mentioned in the earlier discussion of his charisma, Bakker first became famous for promoting the hypothesis that dinosaurs were nimble, even athletic, warm-blooded animals, in contrast to the prevailing view of them as obtuse and lethargic reptilian swamp-dwellers. He captured this theory in the pose he devised for the cast at the Denver Museum, which emphasizes the dinosaur's agility and agency in addition to highlighting *Tyrannosaurus rex*'s famous size, aggression, and vicious nature. With one leg in the air, the pose was more energetic than the one at the American Museum. Like it, the dinosaur's open jaw showed off its oversized and jagged teeth. Later anatomical and physiological studies led vertebrate paleontologists to conclude that *Tyrannosaurus rex* did not stand upright and could not have balanced on one foot. In the 1990s, the American Museum remounted its skeleton of AMNH 5027 with its spine horizontal, its four legs planted on the ground, and its tail extending backward.[77] The *Tyrannosaurus rex* Sue, and its replica at Disney World, were assembled in similarly horizontal positions. Yet Sue's mouth was not modified, and its teeth jaggedly protrude from its jawbone just as much as the two earlier *Tyrannosaurus rex*. The staff of the Denver Museum debated remounting their cast of AMNH 5027 too, but they decided against it for several reasons: Bakker's mount was historically important; it continued to have great popular appeal; and remounting is difficult, risky, and expensive. So the Tyrannosaurus on one foot has continued to greet thousands of visitors to the Denver Museum each day. They bring to their encounters with this dinosaur their ideas about Tyrannosaurus—based on their education, media consumption, cultural background, and imagination—just as scientists do. Yet these encounters have been molded by the history recounted in this chapter. Despite research indicating that *Tyrannosaurus rex* scavenged for food in addition to, or possibly more often than, hunting, countless visitors continue to pose for pictures pretending to be prey for the "tyrant king" of the prehistoric world.[78]

It is no accident that paleontologists and fossil preparators have exaggerated *Tyrannosaurus rex*'s violence in both textual and solid form. But they did not invent it out of whole cloth. Biological processes formed the living animals' bones and teeth in certain shapes, and geological processes turned them into fossils with somewhat different shapes. Moreover, it is important to recognize that paleontologists have conducted extensive, painstaking, and rigorous research about these specimens. As their publications reveal, they examine, measure, and compare every millimeter of each fossil and combine multiple lines of evidence to make a case for their interpretations.[79] This history of *Tyrannosaurus rex*'s formation has made a difference in the interactions of people with the fossils and casts at my field sites, all of which had at least one representative of the species. To more closely examine these interactions, I shift my attention from the towering *Tyrannosaurus rex* in the Denver Museum's lobby to its exhibition *Prehistoric Journey*, where visitors could get much closer to, and even touch, its teeth and bones.

ADDING A PERSONAL TOUCH

Visitors to the Denver Museum interact most tactilely with *Tyrannosaurus rex* at the "Bone Bar," the most popular of the several exploration stations in the Denver Museum's exhibition *Prehistoric Journey*, where the museum's trained volunteer docents encourage visitors to engage with fossils and casts. The Bone Bar is a cart stocked with trays organized around different themes, which docents take out to enable visitors to touch what the trays contain (see figs. 8 and 9). Because the objects on the Bone Bar get handled, and dropped, frequently, the museum mostly stocks it with fossil casts. While most of the trays are organized around biological processes, such as "Predation" or "Eating and Digestion" rather than by species, there is one tray that is focused solely on *Tyrannosaurus rex*. The "T. Rex Encounter" tray contains a brain endocast, a cast of a furcula (a fused clavicle or "wishbone"), and a cast of a jaw fragment with four teeth and a new tooth growing beneath them. The "Eating and Digestion" tray also has a *Tyrannosaurus rex* tooth cast. The roots of its teeth have single apexes that vertebrate paleontologists believe were held in their sockets by soft tissue when the animal was alive; indeed, while Tyrannosaurus teeth would not have broken easily, they would have fallen out frequently as the animal ate. This explains why far more Tyrannosaurus teeth have been collected than any other Tyrannosaurus remains. However, only the longest ones are routinely cast and sold, so it is not surprising that the tooth cast on the "Eating and Digestion" tray is almost a foot long.

Perched on a stool next to the Bone Bar, I saw the same scenes repeated innumerable times: As docents and visitors held the *Tyrannosaurus rex* tooth casts in their hands, they discussed their size and power, calling them "giant," "huge," "incredible," and the like. They repeatedly joked about *Tyrannosaurs* eating people for lunch or dinner. Not surprisingly, the vast majority of volunteers and visitors referred to *Tyrannosaurus rex* using masculine pronouns. After showing visitors the teeth, many docents liked to point out the hole in the rib of the Edmontosaurus skeleton that stands next to the Bone Bar (see fig. 9). They told them that a *Tyrannosaurus rex* tooth fit perfectly into it, then pointed out

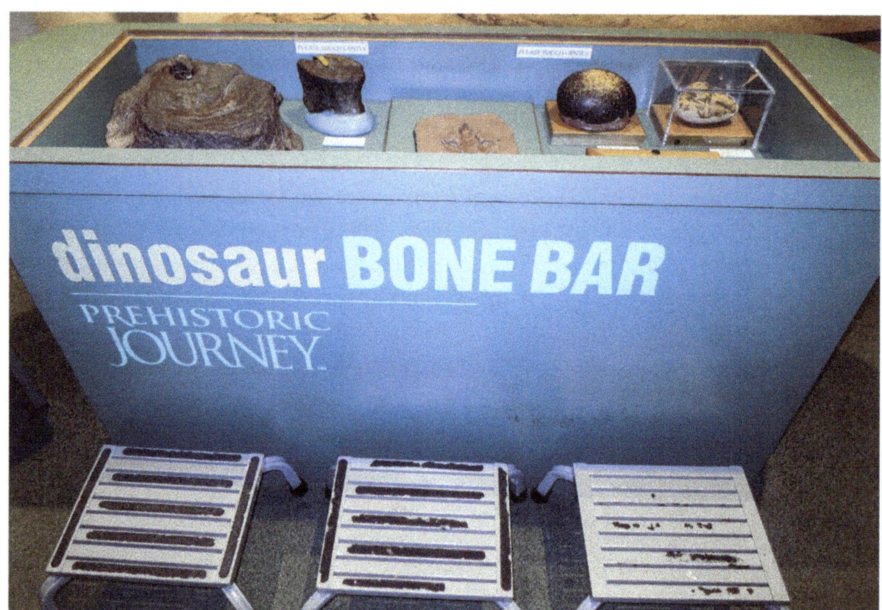

FIGURE 8. The exploration cart stationed in the dinosaur hall of the Denver Museum's *Prehistoric Journey* exhibit, known as the Bone Bar. Photo by author.

that the bone around the edge of the wound hole showed signs of healing. Therefore, they explained, a *Tyrannosaurus rex* bit this large herbivore, but it escaped before the *Tyrannosaurus rex* killed it. This and other punctures have been used to argue that *Tyrannosaurus rex* was a predator. The depths of the holes have been used to calculate the bone-crushing force of its bite.[80] Yet academic and amateur discussions of *Tyrannosaurus rex* dentition go beyond this evidence of the species' diet to understand its character, and to enhance its charisma.

I witnessed the charisma of *Tyrannosaurus rex*'s violence when I joined a "backstage" tour of the Denver Museum that a paleontologist gave to a school group. As we stood amid the rows of long metal shelving holding jacketed fossils in the Big Bones Room, Seth, the paleontologist, called our attention to a series of *Tyrannosaurus rex* fossils encased in their lumpy field jackets. "Everyone knows the T. rex, right?" he asked. As the students affirmed this, Seth continued, "He is the biggest, biggest predator of them all, we think." One particularly knowledgeable boy picked up on Seth's slight caveat to probe whether recent excavations might have revealed a carnivore bigger than *Tyrannosaurus rex*. Seth evidently enjoyed being challenged and answered with a chuckle, "Well that's a good question. But it turns out that no other theropod has bigger bones than the T. rex. And no other theropod has more bone density than a T. rex. So, there are other big predators—" The student cut him off to ask "longer?" Seth conceded, "They can be longer and things

FIGURE 9. Docents and museum-goers gather around the Bone Bar to examine fossils, casts, and models. Photo by author.

like that, but their bones are much more fragile. But their heads are smaller and their teeth are much more narrow. *Tyrannosaurus rex* teeth look like giant bananas and their heads are bigger, and their bones are bigger. They were probably the most fearsome of all. They were probably crushing bones." The student interjected, "instead of slicing," and Seth affirmed, "Yes, exactly. So we think *Tyrannosaurus rex* might have been the biggest, the biggest of all."

Unlike the famous paleontologists discussed earlier in the chapter, museum volunteers play a supporting role and allow the specimens, casts, and models to take center stage in captivating audiences. One June day, I shadowed a longtime museum volunteer named Herbert at the Bone Bar. Herbert started volunteering at the Denver Museum when he retired, fourteen years before I met him. He told me that he always loved working with children and was a classroom volunteer as well as a Boy Scout leader for many years. By the time I knew him, he was quite frail and was having difficulty hearing and walking. Yet he still came into the museum for his docent shifts three days a week without fail, and recently he had received the museum's "10,000 Hour Award." When interviewed for a profile in the museum's newsletter at the time of this award, Herbert said the Bone Bar was his favorite place to work. Herbert told the writer, "I like to put real fossils in [visitors] hands." As I perched on a stool next to him at the Bone Bar that June day, I watched Herbert do that repeatedly, as I had seen him do on many other days. From this experience, I knew that he was including high-quality casts when he said "real fossils" in his interview.

As a white family of five approached the Bone Bar that June day, Herbert asked the boy if he had a favourite dinosaur, as he usually did with elementary school-age children, especially boys. This particular boy did not answer his question but pointed to the model of an *Oviraptor* egg sitting on top of the cart. After they discussed it, Herbert took out the Bone Bar's "Eating and Digestion" tray for them to examine. The boy's father pointed to one of the teeth and asked, "What's that tooth from?" It was a *Tyrannosaurus rex* tooth cast that immediately caught his attention. Herbert drew the father and son into examining it closely, while the mother and daughters stood behind them. "Notice, they've colored it so you can see," Herbert said. He then touched the crown and continued, "This is what sticks out to do the biting and tearing." Fingering the other end, he continued, "All of this is a root." Herbert asked the boy, "How big is the root on your tooth?" The boy answered "small" as his father responded, "Not *that* big!" while he fingered the tooth. The boy's mother and older sister laughed at the father's exclamation. Herbert used his fingers to measure the size of the child's tooth against the dinosaur's, saying "about that big." He then asked rhetorically, "That's because you don't go out, find a cow, and tear pieces right out of it while it is still alive, do you?" He said, "If you did, you'd have to have a tooth like this," as he picked up the *Tyrannosaurus rex* tooth again, and added, "and your teeth would all fall out." Herbert explained that *Tyrannosaurus rex* regularly lost teeth while eating, and new ones grew throughout "his" lifetime. But Herbert did not explain how *Tyrannosaurus rex* teeth are mounted in a way that makes them seem scarier than they would have been in life. Instead, Herbert led the family to compare the *Tyrannosaurus rex* tooth to several other dinosaur teeth on the tray. Unlike their discussion of the extraordinary size and strength of *Tyrannosaurus rex* teeth, the discussion of these teeth stressed their similarities to those of modern animals, such as elephants and parakeets, and to human utensils, such as spoons and salad tongs. *Tyrannosaurus rex* teeth were thereby made exceptionally violent. The boy left the Bone Bar not only more informed about, but also enchanted by, *Tyrannosaurus rex* than when he arrived.

Seemingly unremarkable moments like this one illustrate how interactions among visitors, docents, and museum objects help generate *Tyrannosaurus rex*'s virile charisma. This interaction shows how a boy initially interested in an Oviraptor egg, an object gendered female, is guided instead to engage with a *Tyrannosaurus rex* tooth, an object gendered male. While his father was immediately excited by the tooth, the boy needed to be taught that it should fascinate him. Herbert and the father did this by emphasizing the tooth's enormity and dangerousness. Other guides, educators, and paleontologists compared *Tyrannosaurus rex* teeth to daggers, knives, and other tools of destruction. The paleontologist Seth and many others likened them to seemingly benign bananas, only to underscore their size and blunt force. These analogies recall the discussion of *Tyrannosaurus rex*'s aggressive nature and superfluous violence when AMNH 5027 was first unearthed. Herbert emphasized the tooth's utility not merely for eating meat but for ripping off pieces of flesh while the prey was still alive. He reinforced the appeal of *Tyrannosaurus rex* by noting the differences between its behaviors and those of modern animals, including humans. In moments like this, museum guides do not merely attempt to interest museumgoers in "dry bones" who had been raised on *Jurassic Park*; they sustain the gendered and racialized legacy of the excavations, casting, mounting, and study of *Tyrannosaurus rex* by intensively engaging white boys using feral teeth, while largely ignoring the other people and objects in front of them. In intimate instances like this one, violent colonial masculinity is embodied not by the scientific experts but by the specimens. The combination of sight, touch, and talk that occurred in places like the Denver Museum's Bone Bar continually brings *Tyrannosaurus rex* into being as a masculine, vicious, and charismatic species.

MORE-THAN-HUMAN MAKING

This chapter has traced the history of one species of dinosaur, its most famous representatives and casts, and the humans who have worked with them—from the field to the laboratory to the museum. Exploring how *Tyrannosaurus rex* came into being as a species of violent killers shows how the charisma of this species and the scientists who have studied it emerged together. *Tyrannosaurus rex* is more than a *symbol* of masculine viciousness because its violence is not a cultural veneer placed on top of a presocial material reality. The attributes that have endowed *Tyrannosaurus rex* specimens with their ability to awe people through their ferocity are physically embedded in their dentition, bone size and shape, and posture. *Tyrannosaurus rex*'s violence is also in the fossilized bones of the dinosaurs they bit. Yet this does not mean that *Tyrannosaurus rex*'s viciousness is innate. It is true that biological processes meant that *Tyrannosaurus rex* teeth had long roots, grew up to a foot in length, were able to puncture bone, and fell out easily. Yet the species' charismatic violence was not determined solely by its genes but by the aggregation of disparate forces. Geological processes meant that teeth usually fossilized separately from the animal's jawbone and retained their organic shape while the rest of the skull was compressed. Scientific processes meant that *Tyrannosaurus rex* dental fossils were prepared, cast, assembled, and sometimes painted in ways that exaggerated their size and prominence. These practices were carried out in the pursuit, not in the neglect, of scientific accuracy. Cultural processes have meant that carnivores have been associated with a masculine ideal of physical strength and aggressiveness. Recognizing all these forces leads us to question whether the species was, to quote Osborn once more, "the most superb carnivorous mechanism among the terrestrial Vertebrata, in which raptorial destructive power and speed are combined."[81] However, it does not mean that we should conclude that Tyrannosaurus is a "schizosaur"—that is, "a shape-shifting, transitional figure that can seem to mean almost anything one minute and almost nothing the next."[82] *Tyrannosaurus rex*'s charisma is not the creation of individual scientists, fossil preparators, artists, or media producers any more than dinosaur paleontologists are themselves "self-made" heroes. The materiality of *Tyrannosaurus rex*'s charismatic violence was cogenerated by biological, geological, and cultural forces that came together without anyone or anything alone driving it. In short, its charisma is a more-than-human creation.

Understanding the diffuse forces that brought *Tyrannosaurus rex* into being as an enormous, vicious, masculine, and charismatic killer helps us to question the widespread acceptance of aggression and violence as natural in the United States today. Our social/material world is so saturated with violent competition that it is easy to take it for granted as an inevitable "fact of life." I, for one, did not appreciate that dinosaurs are consistently posed in aggressive positions in the United States until I visited natural history museums in Argentina, where they are not. There, they are often posed with their feet firmly on the ground. In Argentina, dinosaurs are not symbols of violence. In a country where military dictatorship is a living memory, violence is widely understood as political, not natural. Despite the petroleum industry's support for paleontology in both countries, paleontology has not been used to justify other forms of extraction in Argentina as it has in the United States. In the contemporary United States, it is too easy to blame the celebration of violence on the entertainment and media industries. Yet the *Jurassic Park* movies have been transnational favorites. Americans' attraction to violence comes from many places, including scientific practices, material forces, and national historiography.

I do not question whether *Tyrannosaurus rex* had some of the longest teeth of any animal, prehistoric or modern. However, I do want to challenge the too-easy move from the physical reality of *Tyrannosaurus rex* teeth to the understanding of the character and behavior of the species as a whole. Having some giant teeth does not necessarily mean they were used viciously on others. No animal spends its whole life fighting. Yet based on *Tyrannosaurus rex*'s material and textual elaboration, one might reasonably think that this species did. Scott Sampson, another charismatic paleontologist who was the chief curator at the Denver Museum while I was conducting fieldwork there, told me:

> If you look at dinosaur books and dinosaur art, they are capturing some of the most violent moments. If you've ever actually spent time in an ecosystem with big carnivores and watched them, you would know they spend most of their time just sleeping. I've spent years and years going to Africa and sitting around water holes and watching lions and leopards. There's no violence involved at all. So we pull these little moments and we highlight them. That is true in virtually all of the media related to dinosaurs. It gets a little annoying after a while because it misses the point about how the world of dinosaurs really worked.

Sampson asserts that dinosaurs were violent, but that was only one, and probably a minor, part of their lives. He includes himself in the indictment that "we" exaggerate dinosaur violence, as he is part of this, particularly through his roles as a museum director, a television star, and an author of scholarly and popular texts. While lions are charismatic despite their violent practices, AMNH 5027, Stan, and other *Tyrannosaurus rex* are charismatic because of them.

This chapter has charted how the meaning and materials of charismatic prehistoric megafauna come into being together. The next chapter further complicates the story of charismatic violence by showing how people's physical interactions with fossils and casts lead them to make connections between themselves and prehistoric animals that challenge the recognition of these animals as simply violent beasts.

3

—

Forging Connection Across Millions of Years

Joanne began to cry. She wiped away the tear rolling down her cheek with the back of her dirty hand and kept working, delicately brushing away the earth from around the fossilized rib embedded at the bottom of a butte. She cried because this rib reminded her of the horse she had back at home in rural New England. Joanne was not particularly interested in paleontology or prehistory. She loved living animals. She ran a horse farm and spent her time caring for her own and others' horses, while also homeschooling her teenage daughter Megan. It was Megan who loved paleontology. Megan had been interested in dinosaurs since she was a child, and her fascination had only expanded as she grew older. Joanne nourished Megan's curiosity, taking her to the library to find books about paleontology, on trips to science museums and, once Megan was old enough, on paleontological digs at the Badlands Paleontology Station in the Dakotas.

In chapter 1 I recounted how Darrell, the vertebrate paleontologist who runs the Badlands Paleontological Station, identified a well-preserved Triceratops skull sticking out from the rock of a butte. While I was there, he led the team of students excavating "the Trike," and he sent another team of volunteers to scour the surrounding area for any pieces of the Trike skeleton that had weathered out of the rockface and fallen downhill. They did not find any, but they did find evidence of another prehistoric creature at the base of the butte. One of Darrell's colleagues led Joanne, Megan, and a few other volunteers in excavating it. As they removed the overburden, they began to suspect that they might have found a remarkably complete, if weathered and broken, skeleton. This made several of them quite excited, but not Joanne. Frankly, she told me, she found paleontology tedious and boring. That was, until the morning when she uncovered the rib bone that had the same curve as the ribs of her horses. Now she saw the emerging skeleton as the remains of an animal that was similar to the ones she cared for on her farm. That made her cry. She told me that the death of an animal always made her cry, but her emotions caught her off guard this time. She had not seen her reaction coming until it overwhelmed her because she had not previously thought of dinosaurs as animals. Working intimately with fossils led her to understand paleontology as an expression of her love for animals, and excavation as a form of caring for, even mourning, long dead ones.

This story is probably not what you would expect after the last chapter, which examines the coming into being of dinosaurs as violent and vicious beasts. Most of the time, people distance themselves from prehistoric animals, ignoring their similarities with humans and other modern animals. In such moments, dinosaurs are everything that humans are not. They are enormous, vicious, and exotic beasts. Their death definitely does not induce people's sorrow. But in the course of my fieldwork, I noticed another dynamic emerging when people came into personal tactile contact with the material remains of prehistoric creatures. Looking closely at and touching fossils led them to find anatomical commonalities and even shared experiences and emotions that existed between themselves and prehistoric animals. In this chapter I explore how people identify with Mesozoic animals in different ways. The stories I recount complicate the discussion of charismatic violence in the previous chapter. They show how physically interacting with fossils, rather than distantly viewing them, led experts and nonexperts alike to forge affective connections between life in the Anthropocene and the Mesozoic Eras. Not only did these nonexperts reconsider the pervasive conception of dinosaurs as beasts wholly unlike themselves but, at least in some instances, they also perceived prehistoric animals as sharing common features, experiences, and emotions with themselves. One paleontologist has written, "There are times while doing paleontological research when one is suddenly reminded that we are dealing with what used to be living animals in all their rather unusual glory and that what we hold in our hands is concrete evidence of a world so uniquely different from our own that it almost seems imaginary. And yet it *was* real, and that is what makes paleontology so enjoyable."[1] While Joanne's realization that she was handling the bones of a dead animal led her to grief, the experience of touching fossils led to joy for most excavators. In both cases, prehistoric animals moved from seeming strange to being familiar.

Comparisons of dinosaurs with humans is controversial because it raises the question of anthropomorphism, or the potential projection of ostensibly exclusive human qualities onto nonhumans. The guide in the scene from Dinosaur Ridge that I recounted in the book's opening pages could easily be criticized for imbuing prehistoric animals with uniquely human desires and emotions. Remember how he told the tale of a mother and child dinosaur strolling along the riverbank millions of years ago? He explained that "babies are curious" to justify the circuitous path of the juvenile footprints next to the adult's straighter track. The guide pointed out that after the tracks of the mother and the baby diverge, the mother's front footprints disappear, indicating that she stood up on her hind legs. He explained that this was because "she's looking around for her baby." Turning to the children in the group, he analogized: "it's just like if you guys go to the mall with your parents. You always stay right with them?" Dinosaurs clearly did not go to the mall. But did adults protect their progeny from predators? Worry about them? Do such interpretations impose people's modern American experience of kinship onto prehistory? Examining a series of experiences in which people—scientists and nonscientists, adults and children—found commonalities with prehistoric animals has led me to rethink anthropomorphism in vertebrate paleontology.

ANATOMY LESSONS

Two enormous vertebrae of sauropod dinosaurs—one from an Apatosaurus and the other from an Allosaurus—are affixed atop the Bone Bar, the exploration station in the Denver Museum's *Prehistoric Journey* exhibition discussed in the previous chapter. I spent countless hours around this spot, observing the interactions among docents, museumgoers, fossils, casts, and models. When there is a docent on duty, the two huge sauropod vertebrae are usually the first thing visitors ask about. Even when there is no docent at this station, these vertebrae draw a great deal of attention. Some feel a magnetic attraction to them as soon as they see them, walking quickly past giant mounted dinosaurs to reach the vertebrae. Museum visitors come over not merely to look at them but also to touch them. I found a remarkable consistency in visitors' and docents' engagements with these fossils—engagements that equally emphasized the difference in size and the similarity in shape between human and dinosaur bones.

I spent one Saturday morning at the Bone Bar with Robert (Rob), a retired geologist and volunteer docent. Rob had been volunteering in *Prehistoric Journey* since it opened in 1995. When I met him two decades later, he had limited mobility and a soft voice, so he tended to sit on the stool behind the Bone Bar waiting for people to approach rather than move about the hall to engage people as more agile docents did. Early in his shift that day, Rob had taken out the "T. Rex Encounter" tray and placed it on top of the cart, partially obscuring the sauropod vertebrae and other objects affixed there. When a woman and her grandson came to the Bone Bar, they first examined the casts of *Tyrannosaurus rex* teeth and furcula in the tray. Then the woman asked about the vertebrae sticking out from under it. As Rob moved the tray aside, he told her, "This is a real fossil of a backbone." The woman exclaimed, "Look at *that!*" Turning to her grandson, she added, "That's a little bit bigger than your backbone, isn't it?" Rob used his hands to measure a human's vertebra on top of the Apatosaurus one, showing them how much smaller theirs was. They laughed as the woman again exclaimed, "My goodness, that's huge!" A little while later, another woman came over to the Bone Bar and asked her son, "Do you know what a vertebra is?" He did not. She explained to him, "Those are like your back bones. Do you feel those bones back here? Let me feel your bones, do you have bones?" The woman moved her hands up and down her son's back. "Yeah, you have back bones," she joked. "I feel your backbone!" Turning to the vertebrae on the Bone Bar, she told the boy, "This is the dinosaur's back bone. This is much bigger, huh? This dude is *big*." Rob agreed. Both he and these women highlighted the difference in size of dinosaur and human vertebrae, yet the sauropod vertebrae did not incite the same emotions that *Tyrannosaurus rex* teeth did. Instead of focusing solely on their differences, docents and visitors identified a common anatomical feature: a spine. By touching the fossilized vertebrae and the living vertebrae of the kin beside them, they recognized both difference and commonality between themselves and sauropods. Although not as strongly affective as Joanne's grief, these moments reveal an incipient comprehension of a shared animality and a positive connection with prehistoric animals that complicate the more pervasive unsympathetic response.

The Denver Museum was hardly the only place that portrayed the spine as an essential evolutionary link between humans and dinosaurs. The subphylum vertebrate, to which both mammals and dinosaurs belong, is defined by its backbone, although it is not the only anatomical feature they share. In fact, their similarities are so pronounced that vertebrate paleontologists often teach human anatomy. Several graduate students told me that, although it would be extremely difficult for them to find a university or museum position in paleontology, it would not be difficult for them to get a job teaching anatomy in a medical school. Many eminent vertebrate paleontologists began their careers this way. Among them is Scott Sampson, who was the chief curator at the Denver Museum when I was conducting fieldwork there. Sampson is a tall white man with a charming smile, a knack for well-turned phrases, and lots of charisma.

At the Morrison Museum, I observed scientists, educators, and volunteers help visitors who initially regarded dinosaurs as fantastic beasts to reconceive them as animals that they had a great deal in common with. Unlike at many museums, visitors here are encouraged to touch most of the fossils. One docent explained that "you can see Stegosaurus fossils in tons of museums, but you can't see the original Stegosaurus bones, and touch them as well, in any other museum that I'm aware of." Matthew Mossbrucker, the museum's director, is as charismatic as Scott Sampson, but less famous, at least for now. As the curator of a small-town museum, he is also far more accessible to the staff, volunteers, visitors, and one curious anthropologist. I developed a closer relationship with him than with Sampson, who left the Denver Museum before I got a chance to talk to him again one on one. Matthew became the person I went to first with my questions, both technical and philosophical. In noting just how nearby the fossils had been located, he elaborated on the question of proximity and touch:

> You can see the outcrops where a lot of these bones were found from the museum's parking lot, or going up behind the museum. There's a connection that people have when they're onsite, and they see fossils that were found nearby. They can see where the dig sites were. There's the connection that people get with natural history [when they see for themselves] this is where Stegosaurus was found, and the [ancient] seaway was over here. You can see lightbulbs turn on. When you allow people to touch the fossils or casts, it really seems to deepen the connection.

With a subtle gibe at the Denver Museum, Matthew concluded, "I think there's a deeper understanding to be had [at the Morrison Museum] than in big box museums."

A docent-led tour was included with admission to the Morrison Museum. At the beginning of each tour, the docents explained the proper way to touch fossils without damaging them. Addressing children, they usually instructed them to "gently pet the dinosaurs" with just two fingers. Most adults and children were immediately drawn to the *Tyrannosaurus rex* skull near the entrance. As I discussed in the previous chapter, the adults and children commented on the species' seemingly gargantuan teeth and its reputation for biting through bones. But something else occurred when they encountered the delicate Pteranodon on the second floor (see fig. 10). Take, for instance, a tour for an afterschool program led by a docent named Weldon. As the children sat on the floor, looking up at this thin-boned skeleton on its pedestal, Weldon began, "This is the Pteranodon. His name means 'flying no tooth,' so he doesn't have teeth, kind of like a pelican." He asked the children to guess what the animal ate, and they answered fish. Weldon affirmed, "This guy would go flying around and look over the ocean. Then he'd see a fish, dive down, pick it up, and he'd eat that." A boy asked, "He just opens his mouth and swallows?" "Yeah, pretty much," Weldon answered. In this initial encounter with Pteranodon, the animal's lack of teeth underscored its difference both from the *Tyrannosaurus rex* at the museum's entrance and from these children's experiences of eating. Yet this shifted somewhat in the next moment. When Weldon asked the children, "How many fingers do you think he has?" One kid shouted out "two," while another one answered "four." Weldon held up his hand and put his fingers on each of the Pteranodon's digits and counted "one, two, three, four." A kid offered, "I see four, too." Weldon asked, "Do you guys see any more?" The group responded with a chorus of "no." "He actually has five," Weldon told them. "What?" a boy exclaimed. Many of his classmates also seemed surprised. Weldon slowly traced his finger along a long, slender bone. "This is a really long finger. Do you know what finger it is?" A girl called out "a pinky." "You're right!" Weldon replied with enthusiasm. He then explained that Pteranodon "flew on his pinkies." "Can you fly on your pinkies? I can't. My pinkies aren't that cool." The children oohed and aahed with admiration over how the animal flew.

FIGURE 10. A docent at the Morrison Museum shows students Pteranodon's five fingers. Photo by author.

On a tour for an older student group, the docent, Lukas, placed greater emphasis on humans' and Pteranodon's shared anatomy. He asserted that the Pteranodon's wing "looks pretty much like our arm. [They have the] same bones." Lukas fingered the cast as he explained, its "got the humerus, ulna, radius, wrist," as well as "claws and feathers." While most docents ask visitors to guess what the extremely long and thin bone is, as Weldon did, Lukas got right to the point: "This thing over here is its pinky." He told the group that humans, bats, and Pteranodon have "the same bones, just some bones are modified." However, Lukas told the group, what made Pteranodon "completely weird" was the enormity of its pinkies. In these instances, the recognition of a common anatomical feature was coupled with a focus on its different functionality. Noting the evolutionary linkages and resulting anatomical similarities between prehistoric and modern animals, including humans, docents challenged museum visitors to look past the differences in size and function to find commonalities between their own bones and those of the skeletons on display. Notice that Lukas used the anatomical terms, *humerus* and *ulna*, yet both he and Weldon called the Pteranodon's fifth phalange "his pinky." The docents and visitors at the Denver Museum similarly interchanged the words *vertebrae* and *back bones* in conversation. When using these common words, rather than the technical terms, they did not merely make a detached assessment of the morphological differences between species; they equated the prehistoric animal's physical form and structure with their own body's morphology through both talk and touch.

Encounters like these led visitors to question the assumed radical alterity of prehistoric creatures. Although this realization was often subtle, schoolchildren more frequently made it explicit. After touring the Morrison Museum with her elementary school class, a teacher told me that the museum taught her students to realize that "we have the same vertebrae." She instructed them, "Look at their rib cage. We have the same rib cage. Look at their two bones and look at your bones. What's similar? What's the difference?" She recounted the "ah-ha" moments in which her students came to appreciate that both humans and dinosaurs were animals. Another teacher explained that she brought her students to the Morrison Museum because she "wanted them to touch it, feel it, participate, be part of it. I wanted them to have that direct interface" with the prehistory of the place where they lived. She explained, "When you look around, you can connect with these different parts of our global history. That's fantastic!" Yet, she asserted, it is more "juicy and delicious if you can *touch* those guys" and know they once lived in this part of Colorado. She particularly noted that her students loved to "fit their hands in those [Stegosaurus] tracks," putting their own bodies "where the baby Apatosaurus walked." She told me that this was more than a fun activity and a break from classroom learning. "All of those things were really rich and meaningful for my little people."

I asked Matthew if transmitting humans' and dinosaurs' common animality was part of the museum's mission. He replied that he hoped the museum would lead people to understand that "dinosaurs are not dragons. They are not fantasy beasts. These are not monsters that were bloodthirsty that would walk out of a *King Kong* movie." Matthew maintained that people, scientists and educators included, overemphasized the exotic aspects of dinosaur anatomy. He told me, "I try to help folks to understand that dinosaurs were animals, and that they would live and die just as modern animals would do. They behaved in similar ways. Even dinosaurs like T. rex were not always bloodthirsty killers." Matthew argued that most common notions about dinosaurs, such as the view that *Tyrannosaurus rex* was a vicious beast, are not supported by current science. They, like us, were not innately cruel or indifferent to others' pain. His comment illustrates how a recognition of shared animality that starts with anatomy might extend to emotion as well. In the remainder of this chapter, I explore the affective connections between humans and prehistoric animals forged across millions of years.

STEGOSAURUS'S HEEL

Stegosaurus was among the most popular dinosaurs at my Denver-area field sites because it was first found in this area. In 1877, an itinerant teacher and clergyman named Arthur Lakes found the remains of Stegosaurus, among other gigantic dinosaurs, in the eastern foothills of the Rocky Mountains. Lakes is credited with initiating the Great Dinosaur Rush (1877–92), when many iconic species were "discovered" in the Western United States, but his story is less well-known than Barnum Brown's role in US paleontology.[2]

Arthur Lakes was born and educated in England. Despite his keen interest in natural science, he followed his father into the Anglican Church and came to Colorado in 1867 at the behest of a missionary bishop who was desperate for clergy to minister to the colonists flooding Colorado and adjacent territories. Between preaching in mining camps and rural communities on Sundays, and teaching drawing and writing the rest of the week, Lakes pursued his passion for natural history and the new science of geology.[3] He collected flora, fauna, minerals, and fossils for the United States Geological Survey and for the paleobotanist Leo Lesquereux at Harvard University. Lesquereux named several plant species for Lakes. But Lakes's greatest fame came from the dinosaur fossils he unearthed from land that is now part of Dinosaur Ridge park. This occurred three decades before Barnum Brown found AMNH 5027, the first famed *Tyrannosaurus rex* skeleton, which I discussed in the previous chapter.

In March of 1877, the thirty-three-year-old Lakes tarried in Morrison for a few days on his return from his Sunday preaching duties so he and a friend could work on their study of Morrison's rocks. Lakes wrote in his journal:

> At Morrison is a remarkably fine development of the stratified rocks of the Triassic, Jurassic, and Cretaceous periods. The sandstone lying uplifted along the base of the mountains forming long parallel ridges or "hogbacks" . . . Bear Creek Canyon cuts through the whole series of rocks and gives an admirable section.[4]

Lakes had already found some isolated teeth and fossil bones in this area, so it was not a huge surprise that they found another one on a ridge a few miles north of the town of Morrison that Tuesday.[5] While his friend thought it was a fossilized tree trunk, Lakes realized it was "a very large bone belonging to some gigantic animal," as he recounted in his journal. "The question was where was the rest of him and where did his majesty's remains repose?"[6] Lakes's question was soon answered as he and his friend found more fossils nearby. By sunset, they had loaded many into the bed of a borrowed wagon and had driven them to town. As soon as his teaching was completed for the week, Lakes returned to the ridge to unearth more of these giant body fossils. He wrote first to Lesquereux's friend, the geologist Ferdinand V. Hayden, about his finds, and then to rival paleontologists, Edward D. Cope and Othniel C. Marsh. After entering into negotiations about the fossils with Marsh, Lakes eventually developed a fraught but fruitful relationship with him that lasted several decades. Marsh paid Lakes to excavate dinosaur fossils and send them to him at his new museum, the Peabody at Yale University. Lakes's excavations on the ridge above Morrison enabled Marsh to name several new dinosaur species, including Apatosaurus (formerly *Atlantosaurus giganticus*) and Diplodocus.[7] Stegosaurus was the first, and it has remained the most famous of them.

Marsh identified the fossils that Lakes unearthed from the ridge above Morrison in the spring of 1877 as belonging to a new species of Jurassic dinosaur. He named it Stegosaurus, meaning "roof lizard," because he thought its plates covered its back like a turtle's shell. Lakes's fossils were entered into the Peabody Museum as the specimen YPM 1850, and they became the holotype (the specimen that establishes a new taxonomic group) for *Stegosaurus armatus*. They are now displayed on the first floor of the Morrison Museum. A century after Lakes worked in the Denver area, fourth graders began a letter-writing campaign to the governor asking him to make Stegosaurus the Colorado state fossil. It was made official in 1982. Stegosaurus is an especially popular dinosaur in large part because it has three anatomical features that make it easily recognizable: square-toed feet, a four-pronged tail spike, called a thagomizer, and, most famously, the plates that run down its neck and back. These features not only differentiate Stegosaurus from other dinosaurs; they also seem to define its alterity from modern animals. Despite this emphasis on difference, I saw that children and adults forge corporeal, experiential, and affective connections between themselves and this prehistoric species.

The Morrison Museum's Stegosaurus display is its most prized. This display features a few of the fossils that Lakes excavated in the nineteenth century, fossils that sat unappreciated in the basement of Yale's Peabody Museum for decades before the Morrison Museum acquired them. The display also contains boulders with well-preserved Stegosaurus tracks. While most prehistoric footprints cannot be linked to a particular species, and are therefore given their own names, Stegosaurus's relatively small, square-toed feet are sufficiently distinctive that their tracks can be identified. Robert T. Bakker recovered Stegosaurus tracks from Dinosaur Ridge, less than a mile from Lakes's first quarries, during road construction in the park. Although Bakker worked at the Houston Museum of Natural Science, he spent a great deal of time at the Morrison Museum. He was Matthew's mentor, and known by the staff and volunteers at the museum as "Dr. Bob." Following his lead, Matthew found more boulders with Stegosaurus tracks at Dinosaur Ridge to add to the museum's display. They revealed the first baby Stegosaurus footprints ever found.

Matthew did not hide his pride in the Morrison Museum's Stegosaurus holdings. On one of his tours that I shadowed, Matthew told an elementary school class that the footprints found by Dr. Bob were "the best example of Stegosaurus tracks in the whole world." They were "the first example of Stegosaurus's front and back claws together." Matthew continued: "You might be wondering 'how do you know that they're Stegosaurus tracks?' Well, easy. Matching. You see, Stegosaurus hind paws have three toes, they're short and thick, and they have squared-up claws. They're the only Jurassic dinosaur that lived here that had feet that shape." In contrast with its hind feet, Stegosaurus's front feet were not unusual. He joked: "If you have five fingers, and most of you do," you have the same number of fingers as a Stegosaurus. He touched the indents in the rock with each of his own fingers as he narrated, "thumb, pointer finger, middle finger." However, Stegosaurus's "ring and pinky fingers were probably stuck together." Just as in the instances I recounted earlier in this chapter, Matthew gave equal weight to the anatomical similarities and differences between dinosaurs and humans. He also used the digits' common names to help the students associate the fossils in front of them with their own bodies. He then made a further connection: "Between these two footprints there's something really cool going on. Look at this. There's another bonus footprint. There's a short toe there and a short toe there." He explained this as he touched each one. "Most of this track has been erased by this adult stepping on the footprint." A boy shouted, "A baby Stegosaurus!" Matthew affirmed, "He's right," before clarifying, "It's a kid Stegosaurus, like a fourth-grade Stegosaurus." Everyone laughed. Matthew told them that the trackway at another site confirmed "that Stegosaurus traveled in groups called herds." He added, "Fossils like these are indications that Stegosaurus grown-ups and Stegosaurus kids hung out together in the same place at the same time."

Matthew took this lesson further as he showed the class a boulder with the footprints that he identified. Matthew held up a small model of a Stegosaurus and told the class that it was the "actual size of the baby that made the track." One child asked if this one was also a "fourth grader." "This would be like a preschool Stegosaurus. He'd be quite a little one." The animal that made these tracks was probably only about six pounds, the same size as a newborn human, Matthew told them. He added, "This little baby is so small that it could curl up and sleep in the footprint of its mom or its dad." Confirming the evidence of herding, Matthew affirmed that this "tiny little Stegosaurus, cute little thing, wasn't traveling by itself." He pointed out the adult tracks next to the baby ones. "We have at least three individuals making tracks heading in the same direction, so [there was] a little group of baby Stegos. Stegosaurus didn't like to be alone, they liked to be in groups, kind of like people." To conclude his story, Matthew called the Morrison area "a Jurassic playground."

Matthew encouraged the students to touch the footprints in the rock before moving to the next room. They eagerly rose from their spots on the floor and crowded around the boulders, jostling to get a turn to put their fingers in the foot impressions. When it was time to continue their tour, the teachers had to repeat their instructions several times before they pulled the last of the students away from the tracks. Not only did Stegosaurus have the same digits as they did, but they had the same experience of being "herded" around by adults. Mary Leakey, who discovered the tracks of early hominins, described their footprints as "a kind of poignant time wrench."[8] For many, the footprints humanize early hominins.[9] I witnessed a similar dynamic with the Stegosaurus footprints at the Morrison Museum. Despite the more tenuous evolutionary link between the individuals who made these traces and those who touched them, the footprints nonetheless indicate that they were like us. In the previous chapter, touching *Tyrannosaurus rex* teeth reinforced their viciousness and exoticism. Here, touch made Stegosaurus more sympathetic and familiar.[10]

The docents at the Morrison Museum interpreted the Stegosaurus trackways in a similar manner to Matthew, who trained them. Yet some of them introduced variations. As I stood with a group of mostly adults on another day, Lukas picked up a plastic model of an adult Stegosaurus foot and held it up to the boulder. He told the group, "It's a perfect fit here in this footprint, you see?" They nodded in agreement. He then pointed to another footprint, this one missing its heel. He asked, "What do you think has happened?" One man suggested, "Somebody got him." Lukas countered, "The ground was very muddy around the lake. When you're as big as this Stegosaurus, with tiny feet, you are not very stable walking through the mud . . . it just slipped in the mud." Lukas used the plastic foot to make himself into the Stegosaurus and mime how it was walking along and then suddenly slipped. "It's not very elegant. It's pretty funny," Lukas remarked, as the group laughed.[11] A different docent gave the story a slightly different spin, commenting to a group of first graders that this must have been "an embarrassing moment in the life of this Stegosaurus." He asked the students if they had ever slipped in the mud, and most knowingly nodded or said yes. In these instances, the museum staff likened the experiences of the dinosaur to that of the people in front of them. While visitors often suggested violent scenarios to explain the missing heelprint, as the man quoted above did, docents offered explanations that related the Stegosaurus's experience to one that museumgoers had had themselves. In interpreting the tracks in these ways, the experts helped visitors, young and old, to identify not only a shared experience but also a shared feeling, whether amusement or embarrassment. Instead of focusing on the differences between people and Stegosaurus, here they stressed their common physical and emotional reactions. It thus seems they identified much more than the dinosaur's Achilles's heel. This raises the thorny question of whether these trans-species connections across millions of years are a form of anthropomorphism, the interpretation of nonhumans in human terms. Are museum docents and science educators imposing uniquely human thoughts and feelings onto nonhuman beings and things? Or are they identifying commonalities and generating emotional connections between humans and prehistoric animals?

THE PROBLEM OF ANTHROPOMORPHISM

Before exploring anthropomorphism in vertebrate paleontology, it is important to situate it within its larger history in the Euro-American scientific tradition. Prior to the emergence of modern science, anthropomorphism was used to denote the attribution of human form or qualities to deities, particularly the Christian god. In the nineteenth century, the concept expanded to include attributing essential human thoughts and feelings to nonhuman beings or objects as well.[12] Across this shift from the sacred to the profane, anthropomorphism retained an insinuation of immorality as well as inaccuracy, at least in some contexts. On the one hand, it has long been a standard device in arts from children's stories to highbrow literature, and from fine painting to documentary films and animated movies. On the other hand, anthropomorphism has been highly controversial in scientific discourse and practice, in part because of its use in fiction. For many scientists, to call something anthropomorphic is to deride it for raising "the possibility that humans and nonhumans share perceptions, behaviors, and responses." This pejorative use of the term reinforces a "binary distinction between humans and other animals" in which humans are assumed to be inherently superior.[13]

Despite the prevalence of opposition to anthropomorphism in science for more than a century, prominent scientists have deployed anthropomorphism in both their research and scientific writing. Charles Darwin is famous, or infamous, for making "generous, unabashed use of the commonplace terms of (human) mind and action as resources through which to witness and understand animal life."[14] Beginning in the 1930s, the naturalist tradition of animal observation, to which Darwin had contributed earlier, was transformed into the "proper science" of animal behavior, called ethology. With this shift, scientists denounced descriptions of purposeful actions by individual animals, and any deduction of humanlike thinking or feeling from the behavior of other animals.[15] In its place, ethologists developed a mechanistic vocabulary that explicitly dissociated animal behaviors from human endeavors. These ethologists portrayed the actions of nonhuman animals as dictated by stimuli they could neither comprehend nor control, and ultimately as determined by natural selection.

Charged debate over scientific anthropomorphism was reignited at the end of twentieth century with the emergence of cognitive ethology and comparative psychology, as well as growth of the animal rights movement.[16] Although many scientists continue to maintain that there is no place for anthropomorphism in scientific practice, some have distinguished valid forms of anthropocentrism from "Bambification," the uncritical projection of human characteristics onto other animals.[17] The ethologist Frans de Waal, for one, has developed an "animalcentric anthropomorphism" to investigate why nonhumans behave as they do.[18] Rather than assuming the impossibility of trans-species commonality, these scientists investigate the possibility of common mental and physical processes among humans and other animals. Contemporary ethologists go further to assert that a wide range of species exhibit culture, a social process that is usually understood as exclusively human.[19] Because of anthropomorphism's long-standing moral taint, these theories carry more "social and moral baggage" and are often condemned for not meeting rigorous scientific research standards.[20]

I follow Lorraine Daston and Gregg Mitman in finding it more useful to approach anthropomorphism as a way of thinking and acting that many people—including erudite scientists—engage in, rather than as a creed that should be either supported or prohibited, whether in science or elsewhere.[21] Anthropomorphism has been a common practice in scientific research for a long time. Today, it is engaged in to different extents and in a range of different ways. For instance, behavioral ecologists studying meerkats in the Kalahari discriminate between "useful analogies" and "deceptive metaphors."[22] Yet even scientists who have condemned anthropomorphism nonetheless demonstrate identification with the organisms or molecules they study.[23] In fact, the ethologists who have denounced anthropomorphism could not entirely rid themselves of its practice. While claiming that nonhuman animals could only react and not think or feel, the ethologist Konrad Lorenz spoke with birds and adopted several of his research subjects.[24] Some anthropomorphism is inescapable because it is impossible for scientists to describe animals, or anything else they study, in purely objective language free of human reference because language itself is a human creation. The Latin binomials used in taxonomy are colorful examples of this. As I mention in chapter 2, names of scientific species are full of references to particular humans and human ideas about nonhumans that, in turn, shape how species are represented.

I expand Daston and Mitman's analysis of anthropomorphic discourse and vision by highlighting the importance of touch. When discussing Pteranodon pinkies, Apatosaurus vertebrae, Tyrannosaurus teeth, or Stegosaurus footprints, people initially stressed the difference in the size of the prehistoric animals compared to themselves. This was true of experts and novices alike. Yet, several moments recounted earlier in this chapter illustrate how people realized their commonalities with prehistoric animals by touching fossils along with talking about and looking at them. It was the experience of handling Apatosaurus and Allosaurus vertebrae and then rubbing their own backs that led people at the Denver Museum's Bone Bar to look beyond the enormity of these dinosaurs and recognize their shared anatomy. Tracing the juvenile Stegosaurus footprints at the Morrison Museum similarly reinforced children's sense that both they and young dinosaurs were herded around by adults. Physical contact between people and fossils forges a stronger connection across millions of years than discourse and vision alone.

BECOMING STEGOSAURUS

The practice of anthropomorphism is clearly apparent in "Dinosaur Detective," a program at the Denver Museum for early elementary school groups created to accompany class visits to *Prehistoric Journey*. It was redesigned in the early 2010s both to align with the latest state standards for earth science education and to integrate new research on child learning through dramatic play. The museum staff also wanted the program to take advantage of its new high-tech classrooms called "exploration studios." The redesign team framed "Dinosaur Detective" as an "enactment." It had a script but left a lot of room for the teacher and the students to improvise. The child psychologist who led the team explained:

> I made sure that there was a key dramatic play element because that's actually what synthesis looks like in early childhood. They start with some of those key [science] concepts and then they take off with their imagination which is great. I also especially attended to having them understand how it is that we use these things [fossils] as clues. And then, based on the evidence, artists and scientists work together. Artists use their imaginations, and scientists use their imaginations but it's based on evidence. I did not want it to be inhibiting their play, but [I still wanted to make it] an actual enactment with the correct science . . . You can have both . . . and it makes it more fun.

The program's official description advertised that "students become paleontologists," yet when I observed the program I discovered that, first, they became dinosaurs.

One of the times I observed "Dinosaur Detective" was with a class of first graders from a dual-emersion Spanish-English elementary school. I was sitting with them along the wall outside one of the exploration studios when a seasoned museum educator named Ace came out wearing a full-body dinosaur costume and announced, "My name is Ace, and I'm going to be your head Stegosaur today." A minute later, Ace told them, "You're now going to learn about the Stegosaurus by becoming the Stegosaur." Ace asked them to "picture a Stegosaurus." "What does that Stegosaurus look like?" The children yelled out different colors. Ace next asked how Stegosaurus moved and what they ate, supplementing and correcting their answers. Once they established that Stegosaurus were herbivores and their main predator was the Allosaurus, Ace told them, "We're going to try not to get bitten by Allosaurus." "What might the Stegosaurus use to protect him or herself?" Ace asked. The children yelled out "tail!" and Ace informed them that "the scientific word" for a Stegosaurus tail is a "thagomizer." Ace instructed them to repeat the term, first "like a scientist" and then "like a dinosaur." The students enthusiastically complied with gruff, low tones.

With this introduction, the group entered the exploration studio. It was a large open classroom, set up with tables along the walls, a rug in one corner and flowerpots of synthetic ferns scattered around the floor. The most notable feature was that the floor-to-ceiling windows on two sides were nearly covered with screens showing a dense green "jungle" scene with a Stegosaurus in the middle. As the Stegosaurus—made using computer-generated imagery (CGI)—began to move, speakers around the room filled the space with sounds of thumping, panting, buzzing bugs and breaking sticks. With help from the class chaperones, each student put on a vest that had plates running down the back and extended into a tail with a thagomizer. As the redesign team leader explained, each one was different because scientists do not know exactly what Stegosaurus looked like. Once they were "dressed" as Stegosaurus, Ace instructed the students to "move around the room like a Stegosaurus" and "get some food!" As the children wandered around on four legs, stomping and roaring, Ace had to remind them, "Pretend to eat the plants. Don't actually eat them. They're plastic, guys." They were so engaged that they did not notice an Allosaurus emerge on one of the screens. Ace called their attention to it, saying, "Uh oh! Here comes the Allosaurus! How are you guys going to protect yourselves?" The children roared and screamed. Some hid under a table. When the Allosaurus appeared on the screen a second time, it approached a Stegosaurus that was quietly eating some plants. Ace asked, "What do you think is going to happen?" One student answered, "He's going to hit him with the tail!" and another offered, "Eat him!" Ace observed, "I don't think the Stegosaurus sees him!" Ace told them that they would "act out the end of this to see what happens." Once Ace had gathered the students in a circle on the rug, Ace asked, "Who wants to raise their hand and tell me how you think this is going to happen now?" One boy suggested, "The Stegosaurus will hit it with his tail," and others agreed. Another boy said, "There's going to be a big war between the Stegosaurus and the Allosaurus." Ace asked them which one would win this "war," and the children were divided. Ace said, "Well, let's find out."

Ace asked for a volunteer "who has a pretty scary roar" to become the Allosaurus. Many children started roaring to demonstrate their fitness for the job, and Ace selected a boy called JJ who roared extremely enthusiastically. Ace then chose a girl named Bianca to be the Stegosaurus. While Bianca mimicked eating a "delicious plant" nearby, Ace helped JJ change into the Allosaurus costume. Ace commented that he was a "terrifying monster," and the rest of the class made noises to show they were scared. Next, Ace orchestrated the "big war" between the two dinosaurs, telling JJ to "get your three claws out" and "sneak up on" Bianca, then "let out a really big roar." Ace narrated the scene as the two kids enacted it: "She gets a good hit from the Allosaurus. The Allosaurus takes a good bite out of her back!" The students joined in, one girl shouting (in Spanish), "He bit her tail!" Ace then asked Bianca to "lie down and die." With that, Ace dismissed JJ, saying, "After a really big meal, the Allosaurus likes to go back, hang out and watch cartoons and let that sink in." But Bianca was not done. Ace told the class, "Some other dinosaurs might come up and take some gobbles out of her muscles." "What would be left of Bianca?" The students shouted "bones!" Ace then explained that "fossils aren't made overnight," that it can take millions of years for bones to turn into fossils. Ace dramatized this process by covering Bianca in a series of sheets, each representing a different stage in the fossilization process. Ace had the students join in by making the sound effects, such as wind, a volcanic eruption, and rain, while the screens illustrated each stage and Ace explained what was occurring. "A hundred and fifty million years" after Bianca the Stegosaurus had turned to stone, Ace explained to the students, we had finally arrived back in today's Colorado. Ace asked another girl to become a paleontologist. She put on a pith hat and vest and helped to uncover Bianca sheet by sheet. After this, the rest of the students transformed themselves from dinosaurs into paleontologists and enacted the excavation process, digging casts of the museum's juvenile Stegosaurus out of bins of sand and reassembling them into a skeleton in the center of the floor.

In all the times I observed "Dinosaur Detective," no one believed that they were really dinosaurs, yet they took their roles in the enactment seriously. They remained on all fours for a long time, moving around as much like a Stegosaurus as they could. They added their own vocalizations to their movements. More than one child actually tried to bite into a plant. Many jumped back in fright at the sight of the Allosaurus on the screen. Ace and the other museum educators who ran this program reinforced this performance by impersonating a Stegosaurus with their whole body. Ace, for one, liked to crawl around on the floor shaking their thagomizer and warning the students not to get too close so they did not get hurt by it. Throughout the dramatization, Ace and the other educators spoke to the students in the second person as dinosaurs, albeit as dinosaurs that understood English. This impersonation led to some difficulties. Because the participants "become Stegosaurus" and "identify" with the animal, the leader of the program's redesign team told me that they had to be extremely careful to avoid traumatizing the students. The video does not show any bloodshed or violence. In creating the video, she explained, "We made sure that the Stegosauruses were all coming from a distance, so you have a little bit of time to adjust to the fact that there's dinosaurs coming towards you." Moreover, "when we had the Allosaurus run by, I made sure that the Allosaurus never looked at the audience. So it was never 'he saw us.'" Even with these precautions, after piloting the video, the team had to readjust the size of the dinosaurs on the screen because they scared young children. "The kids kept asking me, over and over, 'Do you think he [the Allosaurus] is going to find and eat our Stegosaurus?'" The team decided the educator needed to instruct the students to "wag your thagomizer to protect yourself" so that they would feel safe.

"Dinosaur Detective" seems an obvious case of anthropomorphism. Having the students become dinosaurs implies that dinosaurs and humans share the same fears and desires, thoughts and feelings. Critics of anthropomorphism argue that this "is a form of self-centered narcissism."[25] They would likely assert that the scientists, science educators, students, and other visitors described in this chapter failed to recognize prehistoric beings for what they really were, and instead envisioned them as reflections of themselves. While many scientists assert that anthropomorphism distorts scientific knowledge, not all agree.

WALKING WITH DINOSAURS

Several of the paleontologists, curators, and science educators I have worked with conceived of anthropomorphism in other ways. Matthew Mossbrucker often commented to me that too many adults underestimate children's ability to recognize fiction for what it is and to separate it from science. In his view, telling anthropomorphic tales about dinosaurs makes learning about paleontology enticing to children, leading them to want to know more, and inspiring some to later pursue careers in science. In fact, many paleontologists credit anthropomorphic dinosaur tales in books or movies for first piquing their interest in paleontology and setting them on the path to becoming scientists. When speaking to students, from preschoolers through graduate students, they often spoke to them as scientists in the making. I observed Scott Sampson doing this numerous times, both in person and on screen.

Scott Sampson has shifted his perspective on dinosaur anthropomorphism over the course of a career that has involved simultaneously conducting paleontological research and working on television shows and films for the likes of PBS, Nature, Nova, and the BBC. Before coming to Denver, he was concurrently a museum curator, research scientist, and "Dr. Scott," the host of the PBS Kids television series *Dinosaur Train*. He is among the most charismatic public science communicators today; at the same time, he has continued to be a vertebrate paleontologist with an impressive research and publication portfolio. A tall and broad-shouldered white man, he could have cultivated paleontology's stereotypical Indiana Jones look, but instead he comes off as charmingly nerdy. It is this nerdiness that makes him seem genuine, despite having the polish of a public speaker.

Sampson told me that early in his career he conceived of anthropomorphizing dinosaurs as irresponsible, but he later came to believe that "it's actually kind of tough to go too far." He did not object to the puppeteer who sometimes cruised the museum wearing a handmade dinosaur skeleton. He never spoke, but greeted adults with a friendly nod of the head and bent down to let children feel his teeth (see fig. 11). "Most adults know immediately that this is not serious, that this didn't happen, that dinosaurs did not do this." Although he had once worried that children could not separate fact from fiction, his experience working as the science advisor and on-screen paleontologist "Dr. Scott" changed his view. The "producers pushed to anthropomorphize everything, all the characters and their interactions." He opposed it at first, but explained:

> I slowly came to the conclusion that for preschoolers, anthropomorphizing makes them feel comfortable, and it allows them to engage with the content in a different way, in the same way that they anthropomorphize their teddy bears. There's no reason why we can't do the same thing with science.

Like Matthew Mossbrucker, Scott Sampson came to recognize anthropomorphism as assisting children in gaining scientific knowledge, rather than as undermining it. He argued that they better understand and trust science lessons conveyed through an animated dinosaur, with which they could identify, than through a stone fossil, with which they could not. Anthropomorphism is a powerful pedagogical tool. Yet *Dinosaur Train* explicitly separates pretend and reality, science and fiction. The majority of each episode is a cartoon story about the lives of a Pteranodon family and their adopted *Tyrannosaurus rex* son. Then, Dr. Scott comes on screen to teach

FIGURE 11. A puppeteer in a dinosaur skeleton wanders the Denver Museum on Free Days, greeting visitors such as this girl. Photo by author.

a science lesson. Anthropomorphic cartoon dinosaurs signal fiction; Sampson's human body signals "real science." In chapter 5, I explore how Matthew blurred the line between science and fiction when he participated in making a television special for National Geographic. On *Dinosaur Train*, Sampson reinforces this line.

When I met Sampson at the Denver Museum, he had recently finished working as the lead science consultant for the animated feature film *Walking with Dinosaurs* (2013). This movie had pushed anthropomorphism beyond his limit. In a public lecture about his work on the film, Sampson recounted this experience. The producers "wanted to create a love interest" but Sampson told them, "'You can't create a love interest with dinosaurs. That's anthropomorphizing. It doesn't make sense at all to do that.' But they said, 'No, we're going to do it.'" The audience laughed knowingly as the screen behind Sampson began to display clips of the film's hero, a young Pachyrhinosaurus named Patchi, and his love interest, Juniper. Sampson recounted that "the animators, of course, went right to facial expressions" to convey Patchi's emotions because the dinosaurs were to be voiceless in the film. The only speaking character was the narrator, a bird named Alex. Their decision to not give human voices to the dinosaurs made the producers especially adamant that they demonstrate the emotional life of the dinosaurs through facial expressions. But paleobiological research shows that Patchi and Juniper would not have had the muscles to smile at each other. The smooth skull fossils without muscle scars indicate that Pachyrhinosaurus were incapable of facial expressions.

Sampson lost control over the film's storyline but managed to retain influence over the dinosaurs' physical appearance. He and his fellow scientific consultants came up with a clever solution to their concerns about anthropomorphizing the film's dinosaurs by creating an "emotion grid" for the animators. This diagram showed how sixteen emotions could be expressed through a combination of eye, head, and body movements, and vocalization and group behavior, without smiles, frowns, or words. In his office a few weeks after the public lecture, Sampson elaborated that dinosaurs' "eyes can dilate" and "they can move their heads, their bodies, their tails. They get down low when they're being submissive, and they get really big when they're trying to look tough. We created this [rubric] to give to the animators so that they could, literally, dial in any emotions they wanted with body movements." Sampson was proud of how the science team used fossil analysis, modern analogs, and CGI modeling to generate "the most rigorous effort ever to portray dinosaur emotions in a way that they would have actually expressed them." In this production, in which Sampson did not appear at the end of the show to set the scientific record straight as he did in *Dinosaur Train*, he found anthropomorphic portrayals acceptable as long as the animals' morphology (form and structure), physiology (function), and ethology (behavior) reflected current paleontological knowledge. Sampson was willing to consider that dinosaurs could have had emotions, but he was adamant they could not have expressed them in all the same ways humans do. He and other vertebrate paleontologists, like ethologists studying modern species, understand dinosaurs as "social subjects [who] engage in ongoing interpretative work in understanding, reproducing, and contesting their relationships."[26] Based on comparisons with modern analogs, they argue that some dinosaur species, like some modern ones, could have used their eyes to send and receive social information, thereby shaping their relationships with members of their own and other species. While some species can also do this with their mouths, Pachyrhinosaurus could not.

The making of *Walking with Dinosaurs* did not end with this compromise between the filmmakers and the scientists. Shortly before the film was scheduled to be released, Sampson was shocked by a call from a reporter who revealed that Patchi and two other dinosaurs did indeed speak. Sampson soon discovered that, when the project moved from the BBC to 20th Century Fox, voices were added for the lead characters. As Sampson explained during his public lecture, the Fox executives "hired a guy to write a script at the last second and he gave [the animals] surfer dude voices." The audience laughed as Sampson imitated them. "I couldn't believe it. We had worked for years to make this the most accurate depiction of the world of dinosaurs anyone has ever seen. We've gone to all these lengths with emotions and movements . . . and then you give them surfer voices? But that is what happened." In his office, he further lamented to me, "We wasted all our time, trying to make the most accurate dinosaurs ever on the big screen." While Sampson told me that "it's actually tough to go too far" with anthropomorphism, dinosaurs who spoke like "surfer dudes" crossed the line.

Since the emergence of sound cinematography, makers of animal documentaries have worried that adding human speech threatened the scientific accuracy of their films. The early wildlife documentarians Jean Painlevé and Geneviève Hamon argued that human narrative "considerably strengthened the possibilities for, and probability of, anthropomorphism and anthropocentrism."[27] A century later, Sampson concurred that adding speech diminished the visual presentation of prehistoric animals, essentially turning them into "fleshy puppets for human concerns" or worse, pawns for human laughs and ticket sales.[28] Sampson did not entirely condemn anthropomorphism but reframed this black-and-white question into a consideration of how, and for what aims, anthropomorphism should be deployed. Sampson argued that it was a powerful tool to communicate scientific knowledge, but it should not be used to undermine science for entertainment or profit. He was willing to consider that animals past and present had social experiences, desires, and emotions that were familiar to audience members and scientists alike. Dinosaurs too could love and worry, be embarrassed, disappointed, and jealous. Anthropomorphism, in this view, is an important tool not only because it questions the portrayal of dinosaurs as exotic beasts but also because it leads people to recognize themselves as animals.

SCIENTIFIC ANTHROPOMORPHISM AS POSTHUMAN PRACTICE

An even thornier issue than the use of anthropomorphism in scientific communication is its use in scientific research. It is one thing to assert that anthropomorphism helps to teach science, but quite another to argue that it is a valid tool for creating scientific knowledge. In chapter 2, I demonstrated that paleontologists do not simply impose their ideas onto the prehistoric world. Instead, I showed how *Tyrannosaurus rex* became the "tyrant king" through scientific research practices, particularly fossil preparation. In the current chapter, I have extended this to explore how using humans as analogs for prehistoric animals has long been, and continues to be, a crucial part of scientific inquiry. This practice raises a series of much disputed questions: If scientists find, for instance, the fossil remains of an injured animal that could not have fed itself yet lived to an old age, can they conclude that others cared for it as humans do? Can they infer that these prehistoric animals felt filial obligation? Love?

These questions are similar to the ones that Sampson faced while trying to make *Walking with Dinosaurs* the most scientifically accurate portrayal of prehistoric life on the big screen. His work on dinosaur emotion blurred the line between scientific knowledge and media production. In the same year that *Walking with Dinosaurs* premiered, he was the lead author of a scholarly article describing a new species in the same taxonomic subfamily as Pachyrhinosaurus.[29] Sampson drew on this and other paleobiological research to inform the filmmakers that dinosaurs such as Pachyrhinosaurus did not have the musculature to smile, which humans and other primates have. However, Sampson did not conclude that Pachyrhinosaurus were incapable of thoughts and emotions. Instead, he suggested, they could have expressed feelings through eye, head, and body movements. The CGI models and emotional grid that he and the film's science team developed elaborated paleontologists' current knowledge of Pachyrhinosaurus morphology, physiology, and ethology. This emotional grid was a sort of paleontological ethogram. It adopted the standard zoological method for cataloging the complete repertoire of a species' behavior for prehistoric species.[30] Sampson's combined scientific and media work placed him in the middle of the larger debate over scientific anthropomorphism.

Vertebrate paleobiologists routinely use birds and other modern animals as analogues to study the biology of prehistoric animals. Take, for example, the Pteranodon at the Morrison Museum. To figure out how it flew, scientists have examined skeletal remains and trace fossils, such as wing impressions, from this and other specimens. They compare the grooves and ridges where its muscles once connected to its bones with those of modern flighted animals, such as albatrosses, vultures, and bats. Scientists then create biomechanical models to test possible theories of Pteranodon's wing form, strength, motion, and therefore, function. These models are increasingly made using CGI, rather than the plaster, metal, clay, and rubber of earlier research. By combining skeletal and trace fossil analysis with the study of modern analogs and biomechanical modeling, vertebrate paleontologists have concluded that Pteranodon soared through the air far from shore and fed on fish, similarly to how modern albatrosses do.[31] Although dinosaurs are believed to be evolutionarily closer to modern birds and reptiles than to mammals, some dinosaur species' body size and shape make modern mammals useful analogs for research. To understand how Stegosaurus moved and ate, for example, paleobiologists use functional analogs, such as elephants, rather than phylogenetic analogs, like albatrosses for Pteranodon. Both phylogenetic and functional analogs enable scientists to see past the obvious differences between prehistoric and modern animals and recognize homologies.[32] I observed this throughout my fieldwork, but especially when people encountered fossils up close and touched them, and even more so when they spent longer periods working with fossils at paleontology field stations and laboratories. Much as some biology researchers develop personal affective relationships with their nonhuman subjects, numerous paleontological researchers become emotionally attached to the long-dead ones they excavate or prepare.[33] Joanne's grief at the death of the animal she was excavating is a vivid example of this. It is important to recognize that these researchers did not necessarily assume that prehistoric animals shared their emotions, but they were willing to consider the possibility. This is an aspect of the shift from conceiving of dinosaurs as beasts wholly unlike themselves to appreciating them as animals with common evolutionary ancestors.

I consider the work by paleontologists to use anthropomorphism in their research, as well as their teaching and communication, as part of a larger transdisciplinary movement that works to trouble human exceptionalism, if not to reject it outright. As I discussed in the introduction, this emerging movement—often labeled posthumanism—challenges the long-standing assumption in the Euro-American humanist tradition of an absolute and universal gulf that separates the capabilities of humans and other animals. Posthumanist scholars conceive of human exceptionalism as an anthropocentric conceit that has been presumed without question. They ask whether humans are necessarily the only species able to reason, imagine, act with intentional agency, or possess morality, desires, or emotions such as love and sorrow. They thus open up the possibility that humans share far more with other animals than many scholars have imagined possible.

This posthumanist scholarship has led me to see that the censure of anthropomorphism as inherently unscientific is based on the humanist assumption that there is a fundamental difference between humans and other animals. Critics of scientific anthropomorphism rightly contend that it is wrong to impose uniquely human attributes onto nonhumans. But they are unwilling to investigate which attributes are in fact uniquely human. If one does not start from the assumption that only humans can experience love, for example, then one can ask what constitutes convincing evidence of love and where it might be found. That is what I think Sampson is doing. Along with numerous other vertebrate paleontologists, science educators, and students of all ages, he is considering multiple possible commonalities among prehistoric and modern animals, including humans, that move beyond anatomical structures to embrace experiences and emotions. They are neither simplistically reducing humans to other animals nor promoting nonhumans to human status. They are replacing the widespread exoticism of dinosaurs without ignoring the ways in which humans and other animals are different. In the process, they are challenging human exceptionalism and raising the possibility that anthropomorphism is a useful scientific research technique.

In the previous chapter, I noted that modern charismatic megafauna are animals that inspire people to try to preserve their species from extinction and their environments from destruction. Prehistoric charismatic megafauna similarly generate support for paleontological research and scientific institutions. In both cases, the animals are seen as important, but they do not inspire the intimate connection that we see in Joanne's grief over the dinosaur she was excavating. Joanne's realization that a piece of rock was once part of an animal moved her from indifference to sorrow, from boredom to tears. Judith Butler writes that grief is a "slow process by which we develop a point of identification with suffering itself."[34] For Joanne, the process of excavating became an act of mourning. The fossils were part of an animal that had lived, suffered, and died, like the horses she cared for at home. In a strange way, *un*burying a prehistoric animal was an act of bereavement. I see similar emotional connection in the enormous amount of tactile work that scientists, curators, preparators, and volunteers dedicate to protecting the remains of prehistoric animals from disintegration, which I recounted in the previous chapters. The care that fossil preparators take in cleaning, repairing, and assembling skeletons can be seen as equivalent to preparing a human body for burial. The display case becomes a coffin; a museum visit becomes a wake. In the next chapter, I examine how the history of paleontology intersects with the myth of the American West and how entertainment melds with the horrors of colonialism. These are very different ways that paleontology forges connections between the prehistoric world and the contemporary one.

4

Paleontology, Colonialism, and Edutainment

I never expected my research to take me to Disney World. Yet on one of Florida's typically hot and humid days in May, I found myself walking amid the throngs of people heading down a path lined with tall, lush palm trees and giant ferns in Disney's Animal Kingdom. I passed under a dark green archway with painted gold letters proclaiming "DINOLAND U.S.A." and, in smaller letters above it, "Exploration," "Excavation" and "Exultation." Less than a minute later, I found myself surrounded by enormous dinosaurs. There was a larger-than-life skeleton of some sort of Hadrosaur standing on rocks with a rusty trestle bridge between its feet. There were pastel green Triceratops with gold stars on their bellies, and people riding atop their backs, going round and round on an aerial carousel. There was a duck-billed dinosaur, with a ruddy red crest and smooth ridges running down its back and tail, crouching down so children could climb on it. It took me a while to find what I was looking for, but I finally spotted it: the bronze replica of the famed *Tyrannosaurus rex* called "Dino-Sue" (see fig. 12). It was tucked away in a verdant nook on the walkway to DinoLand U.S.A.'s joy ride, "DINOSAUR."

What were dinosaurs doing in Disney World's Animal Kingdom? This theme park is Disney's version of a zoo—it *is*, in fact, an accredited zoo—that features reconstructions of the exotic environments of different continents, known as "lands," in Disney speak. In "Africa," visitors take a safari ride through a savannah filled with charismatic megafauna, such as elephants, lions, giraffes, and zebras. In "Asia," they raft down a rushing river, watch monkeys play, and ascend Mount Everest's snowy peak. These animals, and their habitats, are positioned to exemplify entire continents. Here, Disney has continued the long history of zoos serving as sites to display foreign animals captured from European colonies. Zoos use

FIGURE 12. The bronze replica of the *Tyrannosaurus rex* FMNH PR 2081, named "Dino-Sue" by Disney, is tucked away in a verdant nook on the walkway to DinoLand U.S.A.'s joy ride. Almost no one stopped to look at it. Photo by author.

animals as spectacles that distance visitors from foreign places, making them feel exotic. Yet Animal Kingdom, along with many other zoos, has also created opportunities for visitors to connect emotionally with animals they do not encounter in their daily lives, even large and frightening "beasts."[1] From Animal Kingdom's opening in 1998 to DinoLand U.S.A. closing in 2025, this "land" worked somewhat differently from the park's other "lands." Live animals were scarce. People came instead for animatronic and ridable models of prehistoric ones.

In DinoLand U.S.A., people enjoyed a quintessential town in the American West, but with a twist. Instead of the usual cast of cowboys, Indians, horses, and buffalo or cattle, there were carnivorous Tyrannosaurus and Carnotaurus and herbivorous Triceratops and Iguanodon from the Cretaceous period. While Animal Kingdom's other lands offer adventures in exotic "natural" locales, Dino-Land recreated a dinosaur-themed amusement park, paleontological research institute, and natural history museum. It was an unapologetically human-built environment that did not hide its concrete and plastic behind a lush facade, as "Africa" and "Asia" do. The only animals in DinoLand U.S.A. were a few croco-diles, birds, and of course, the workers ("cast members") and the throngs of tourists ("guests") who visited each day. DinoLand thus seemed out of place in a park where, according to prolific advertising, "the magic of Disney meets the wonders of the natural world." How did dinosaurs come to represent the natural wonders of North America, while modern charismatic megafauna symbolizes Africa and Mount Everest stands for Asia?

Dinosaurs have never been confined to scientific research repositories. They have long been housed in private collections and exhibited in all sorts of parks— nature parks, city parks, zoological parks, and amusement parks—as well as in natural history and science museums. So it is not that unexpected to find dinosaurs in zoos and tourist attractions across the country.[2] Examining what dinosaurs are *doing* there is more surprising. How are they bringing together science and enter-tainment? What role do they play in creating the myth of the American West? To answer these questions, this chapter explores two quite different parks that feature dinosaurs: Animal Kingdom's DinoLand U.S.A. in Florida and Dinosaur Ridge in Colorado.

These two parks differ in several respects. Animal Kingdom is a private amuse-ment park owned and operated by one of the largest entertainment corporations in the world, while Dinosaur Ridge is a small public park with programs run by a nonprofit organization, Friends of Dinosaur Ridge. Yet they both have deployed dinosaurs as "edutainment" that combines science education and outdoor enter-tainment. More specifically, DinoLand U.S.A. and Dinosaur Ridge have provided visitors with immersive experiences that reenact important moments in the prehistory and history of the region mythologized as "the American West." Both parks have reconstructed the Cretaceous period, when gigantic dinosaurs lived along the Western Interior Seaway in the middle of the continent (see map 1). In addition, Dinosaur Ridge revisits the "Great Dinosaur Rush" (1877–92), when many of the iconic dinosaur species were "discovered," while DinoLand recre-ates a mid-twentieth-century paleontological research station and natural history museum. The focus of these two sites' edutainment also differs. Friends of Dino-saur Ridge aims to teach natural history and geology to park visitors, while Animal Kingdom purportedly instructs its visitors in environmental conservation and animal behavior. Yet I found that the dinosaur-themed edutainment at both these sites accomplished something else: it naturalized iconic elements in the mythology of the American West. Dinosaur edutainment makes the violence of the Western frontier seem natural by writing it into the *pre*history of the region. At the same time, it writes violence out of colonialism. These sites substituted fossil hunters for cowboys to reenact heroic tales of scientific discovery on land "thin with population and thick with peril."[3] Together these sites show how dinosaur edutainment has promoted an origin story for the United States that erases the horrors of settler colonialism and replaces it with a heroic narrative—giant beasts, brave hunters, and an adventurous pursuit of science—about which Americans can feel proud.

DISNEY'S SCIENCE EDUTAINMENT

Animal Kingdom and Dinosaur Ridge both have centered dinosaurs in edutainment that promises to enlighten as well as entertain their visitors. The term *edutainment* might not have been coined until the 1980s, but the idea of combining education and entertainment is far older than that.[4] The world fairs and expositions of the nineteenth century linked tourism, science, and education through dinosaur spectacles that drew unprecedented audiences. The English Crystal Palace Exhibition is exemplary of early edutainment, and an influential antecedent for its contemporary form on both sides of the North Atlantic. Beginning in the early 1850s, this exhibition displayed thirty-three life-sized sculptures of prehistoric animals, reconstructed from newly emerging fossil evidence, in order to "instruct as well as delight" the English working classes.[5] Forty thousand people turned out on opening day in 1854, and millions more followed them to ogle at the "gigantic beasts which . . . precede[d] us in possession of this part of the earth."[6] The Crystal Palace exhibited both dinosaurs and Indigenous peoples as beasts to be ogled at by its mostly white visitors.[7] The exhibition suggested that beasts were living in present-day Africa, but they were pre-history in Europe and North America.

The fusion of education and entertainment expanded from world's fairs and expositions to museums and zoos beginning in the early twentieth century in the United States. While carefully distancing themselves from the commercial "dime museums" and traveling exhibits of spectacular specimens described below, the proponents of the New Museum movement urged scientists to combine their research with public education and to make the revered halls of natural history museums appeal to working-class people, whom they had previously spurned.[8] Although museum directors, funders, and staff were initially hesitant, dinosaur exhibits helped convince them of the benefits of making their exhibits alluring to a wider public. When the American Museum in New York unveiled its mounted Brontosaurus in its new dinosaur hall in 1905, hundreds of thousands of people turned out to view it. Other natural history museums followed with their own spectacular dinosaur exhibits, as I showed in chapter 1. In the same period, zoos were founded to provide the growing numbers of working-class urbanites with an escape from the hardships of everyday life and a place for wholesome family entertainment.[9] In the rural stretches of the Western United States, national parks served a similar function as zoos did in urban settings. The National Park Service's carried out a "dual mandate" to protect the country's remaining "wilderness" from further destruction while at the same time creating "national playgrounds" to instill visitors with an appreciation for the country's natural beauty.[10]

American zoos, parks, and museums have continued to develop science edutainment with an environmental focus. Since the end of the twentieth century, many zoos and national parks have taken on more explicitly environmentalist missions. They have aimed to teach visitors about species extinction and habitat destruction and to instruct them in ways they can advance environmental conservation.[11] At the same time, many natural history museums have reimagined themselves as science museums, turning exhibits displaying prehistoric plants and animals into ones narrating the evolution of life on earth. In making these changes, zoos, parks, and museums have increased their deployment of promotional tactics long used in the entertainment industries to make their product—scientific knowledge— attractive to a broader audience. In the next chapter, I examine a recent National Geographic television special, called *T-Rex Autopsy*, which extends this tradition into home media. A century and a half after the Crystal Palace exhibition, National Geographic created a three-dimensional, life-sized dinosaur, based on the available fossil evidence, current theory, and conjecture, to create a show that would wow its audience with the colossal spectacle of a "beast" while teaching them dinosaur anatomy. In the current chapter, I explore how Animal Kingdom transformed dinosaur edutainment from visual spectacle into immersive experience.

Disney's science edutainment has drawn as much from precedents set by private businesses that have sold exotic people and animals as sources of entertainment as it does from public zoos and parks.[12] Since the nineteenth century, commercial "dime" museums and traveling exhibits "catered to the widespread enthusiasm for 'rational recreation' among the middle and working classes." P. T. Barnum was among the most successful at promoting the scientific and educational merits of amusements "designed to educate as well as delight" the public.[13] While museums have refuted the similarities between their scientific exhibits and commercial ones, Walt Disney unapologetically expanded on Barnum's precedent. The corporation he founded has included dinosaurs among its cast of edutainers since imagineers created dinosaur dioramas for Disney's display at the 1967 World's Fair in Montreal, dioramas that were later installed at Disneyland in California. Before this, Disney developed its approach to environmental edutainment in its Academy Award-winning nature film series, *True-Life Adventures*.[14] These films added an explicitly nationalist message to celebrating the "natural wonders" of the United States. Using the narrative strategies and visual techniques that Disney's filmmakers developed for its fairytale dramas, they sought "to wed the excitement and entertainment [of Disney films] to truth" about "nature."[15] As the films were projected in homes and classrooms, they aimed to teach viewers, young and old, to appreciate the "beauty, strength, and natural wealth" of the United States, to use Walt Disney's own words.[16] Yet they portrayed a "nature" as full of violence as of beauty. The series' cinematic style stitched together video recordings of animals in otherworldly environments to portray an exotic, unpeopled "wilderness" that was supposedly hidden within the modern United States.[17] The filmmakers stayed carefully behind the camera, yet staged the animal behaviors and natural environments they wanted to show.[18]

Decades before Disney built its zoo in Florida, the *True-Life Adventures* episode "Prowlers of the Everglades" (1957) presented an ageless and wild natural world rife with violence. In this show, the Florida Everglades is "the swamp that time forgot," where "primitive creatures feed and breed and stalk their prey just as they did millions of years ago." Amid this watery wilderness, alligators are the zenith of a timeless hierarchy of predator and prey in which, each day, "the hunter and the hunted renew the ancient pattern for existence." The show implies that the dynamics of nature have remained the same, although the beasts are now different. In modern times, the bull alligator reins as "king, his right to rule undisputed," just as *Tyrannosaurus rex* did in the Mesozoic era. "Prowlers of the Everglades" even gives the bull alligator a stereotypical dinosaur roar. This cinematic antecedent to Dino-Land U.S.A. thereby naturalizes violence by framing an individual animal's acts of predation as part of a cycle of life that has remained unchanged since prehistory. At the same time, the television series erases humans from the landscape, perpetuating the myth of a timeless and untouched wilderness. What is more, the show hides the irrevocable environmental destruction and social change that swamp drainage projects caused in the Everglades, projects undertaken for the benefit of the Walt Disney Company.[19] While naturalizing violence, Disney films have generated expectations and provided scripts for people's encounters with living and prehistoric animals in Animal Kingdom.[20]

Animal Kingdom is the fourth theme park in Disney's Florida complex. It has combined the company's cinematic tradition with the precedent set by world's fairs, expositions, and zoos in staging nature as live spectacle for middle-class audiences.[21] As the undisputed leader of the entertainment park industry, Disney had already convinced millions of people that its theme parks were "cultural sites that everyone should visit" because they provided "a middle-class form of parenting and family bonding."[22] The company had effectively imparted the message that "good parents" educate their children not only by sending them to school but also by taking them on "quasi-educational vacations" to Disney parks.[23] As an accredited zoological park, Animal Kingdom added affective encounters with charismatic animals and science edutainment to Disney's Florida vacation experience. Like the *True-Life Adventures* films, the park's edutainment has focused on deepening visitors' aesthetic appreciation for and affective connection to flora and fauna, rather than on increasing their scientific knowledge about them. Animal

Kingdom transformed Disney's history of visual spectacles into immersive and tactile experiences.

Long after Walt Disney's death, the Disney corporation's environmentalism still revolved around preserving timeless and exotic "natural beauty," now with a more globalized frame in which visitors are taught to appreciate animals and landscapes the world over.[24] In addition, Animal Kingdom has drawn on the popular notion that children are "the true stewards of embattled nature" and "saviors of vulnerable animals" who stand up to "an unfeeling, cruel adult world."[25] Disney's zoo—like other environmental edutainment sites—has helped adult visitors to feel good about what they were already doing to "save the planet from destruction."[26] For instance, they are repeatedly told that their entrance fees support wildlife conservation and scientific research. At the same time, Disney's name for its zoological park established a parallel with its signature theme park, the Magic Kingdom. The monarchical metaphor of "kingdom" suggests that humans are the "crown" that rule over the natural world today. The park instructs visitors in how to be more benevolent sovereigns.

DinoLand U.S.A., Animal Kingdom's representation of North America differed from Animal Kingdom's other "lands" in its approach to creating memorable experiences of nature for visitors. The "Africa" and "Asia" lands have followed the long tradition of "pair[ing] native people with native animals."[27] In contrast, DinoLand U.S.A. substituted dinosaurs for both "native people" and "native animals," with a few crocodiles on the side. Crocodilians existed alongside dinosaurs in the Mesozoic era, but unlike alligators, crocodiles are not a potent symbol of the Florida Everglades that were drained to create the parkland. DinoLand thus constructed a mythical western plains ecosystem on top of a destroyed wetlands ecosystem. These plains were inhabited by white scientists and prehistoric creatures, not Native Americans, either past or present. Disney replaced Indigenous peoples, and their sovereignty, with dinosaurs who "ruled over" the prehistoric world. While the rest of Animal Kingdom has encouraged its visitors to believe they are valiantly saving modern charismatic species from extinction, DinoLand encouraged them to imagine themselves rescuing dinosaurs for science. These messages extended a long history of Euro-Americans using the geosciences to justify the exploration and colonization of North America.

NOW PRESENTING "DINO-SUE"

On my first day at the park, as I wandered through DinoLand U.S.A., I felt like I had stopped during a cross-country car trip at a roadside attraction sometime during the 1950s. It was as though I had stumbled into a historic, small, Western town containing the seemingly incongruous combination of an amusement park, a natural history museum, and a paleontological research institute. DinoLand had changed little between when Animal Kingdom opened and when I visited, two decades later, because it was intended to feel out of date. In that time, however, there had been an addition to Disney's recreation of the United States' past: Dino-Sue.

The year before Animal Kingdom opened, the Disney corporation helped Chicago's renowned science museum, the Field Museum, purchase Sue—the largest and most complete *Tyrannosaurus rex* yet uncovered. The story of this specimen illustrates the ongoing interconnectedness of paleontology and internal colonialism. In the summer of 1990, an amateur fossil-hunter working for the commercial Black Hills Institute found the skeleton on a ranch that was part of the Cheyenne River Sioux Reservation, which is still legally held "in trust" for the Sioux by the US federal government. The Black Hills Institute paid the ranch owner $5,000, but it is not clear whether the money was given in exchange for the fossils or the right to excavate on his land. Soon after Sue was removed and transported to the Black Hills Institute, there ensued a protracted legal battle over the dinosaur, which garnered significant public attention.[28] The legal saga fostered heated debate over the rise of commercial fossil collecting amid a lucrative global market for dinosaur fossils.[29] It also illuminated the jumble of inconsistent and archaic laws governing the ownership of fossils in the United States. The dispute did not involve the Native American Graves Protection and Repatriation Act (NAGPRA), which passed into law the same year that the Black Hills Institute team excavated Sue. Therefore, the court battles over Sue did not consider whether dinosaur fossils were Native American patrimony and thus subject to NAGPRA's repatriation requirements, but turned on whether dinosaur fossils were an intrinsic part of the land or resources apart from it.[30] After years of wrangling, the courts ruled that Sue was part of the rancher's plot of land that was held in trust for the Cheyenne River Sioux by the federal government.[31] At the rancher's behest, Sue was put up for sale. While some saw the ruling as a victory in the struggle for Native American property rights, many paleontologists saw it as a loss for science. They worried that Sue would be bought by a private collector and not be made available for scientific study or public education.

At the Sotheby's auction house in New York City in October 1997, the Field Museum bought Sue for more than eight million dollars. A sizable chunk of this money came from Disney, another from McDonald's. Funding the Field Museum's acquisition of Sue connected these corporations with the excitement of scientific discovery and the valiant pursuit of knowledge, while also broadcasting that they value education. Some commentators asserted that the partnership to buy Sue brought corporations and a museum closer together than previous edutainment ventures had.[32] Yet this corporate sponsorship followed the long tradition of business tycoons buying public goodwill through their financial support of natural history museums.[33] It also furthered ideas from the New Museum movement described earlier in the chapter.

Tyrannosaurus rex Sue was the object of both fierce fights and tender care along its circuitous journey from South Dakota to Chicago. As part of the partnership among the Field Museum, Disney, and McDonald's, Sue's fossils were prepared by museum staff at two McDonald's-branded fossil preparation laboratories, one set up at the Field Museum and the other inside DinoLand U.S.A. Like the Denver Museum's fossil prep lab described in chapter 1, Animal Kingdom visitors could peer into the lab through glass windows, watch the preparators work, and often talk to one of them. The fossil preparators in DinoLand's lab strangely paralleled Disney's cast members who wove baskets and cooked food in "Africa."[34] Once the prep work was complete, DinoLand's lab closed. A short time later, a bronze replica of Sue's skeleton was erected, standing silent and still along the path to DINOSAUR, not far from the herds of brightly colored plastic dinosaurs in Dino-Rama. Dino-Sue partially imitated the live animals in Animal Kingdom's other "lands" by providing a charismatic megafauna specimen for North America. Dino-Sue was also supposed to enhance Disney's claim that Animal Kingdom offers science edutainment.

Having followed the news coverage of Sue's saga and Disney's financial investment in purchasing and preparing it, I arrived at Animal Kingdom expecting Dino-Sue to be the centerpiece of edutainment in DinoLand. But it turned out to be a side show, if that. On my first few walks through DinoLand, I could not even find Dino-Sue. Once I located it, I quickly realized that its charisma paled in comparison to movie stars like Donald Duck who posed for photographs with tourists on the opposite side of the path. I did not see anyone pose for a photograph as though they, or their children, were about to be eaten by Dino-Sue, as many did with the mounted *Tyrannosaurus rex* in the lobby of the Denver Museum. Nor did Dino-Sue spark the affective attachments that the live charismatic megafauna in "Africa" and "Asia" did. As figure 12 illustrates, no one seemed to pay attention to Dino-Sue. Disney claimed that hundreds of thousands of people have seen Dino-Sue, but I doubt that more than a handful remember it.[35] What most remember from DinoLand was its thrill ride, DINOSAUR, which I discuss later in the chapter.

ROADSIDE DINOSAURS

DinoLand U.S.A. recreated two distinct moments in the North American past that were separated by millions of years. While the thrill ride DINOSAUR represented the Mesozoic era, the rest of DinoLand reinvented a rural town in the American West in the 1950s.[36] Dino-Rama, the amusement park located within DinoLand, was modeled on roadside attractions that dotted highways across the United States for much of the twentieth century. In the 1930s, the Works Project Administration sponsored the creation of several dinosaur-themed attractions in the arid northern plains in an effort to help residents of its small towns use their prime "natural resource"—Mesozoic animal fossils—to capitalize on car tourism to Mount Rushmore.[37] In Disney's nostalgic recreation of this moment, road

FIGURE 13. At Disney's Animal Kingdom, a large wood-plank sign near the entrance announces "Welcome to DINOLAND U.S.A. The Friendliest Fossils in America." The second line is partially hidden behind ferns, one of the few plants that has continued to exist in North America since the Mesozoic era. Photo by author.

signs along the perimeter of Dino-Rama indicated that visitors were on imaginary US Route 498 in fictitious Diggs County. The landscape of DinoLand U.S.A. matched the landscape in classic Western movies, where "the apparent emptiness makes the land desirable not only as a space to be filled but also as a stage on which to perform and as a territory to master."[38] With its flat terrain, reddish-tan color scheme, wide streets, and old "Western" buildings, its aesthetic reminded me of the North Dakota town where the Badlands Paleo Station is located. Of course, DinoLand was far more crowded with people and commerce than the oil-turned-ranching town where I stayed while conducting paleontological fieldwork. But I could imagine the main street of this North Dakota town, a half century earlier, transformed into Dino-Rama's midway by a traveling carnival company.

A large, wooden plank sign near the entrance to the North American "land" welcomed guests to "the Friendliest Fossils in America" (fig. 13). The plastic dinosaurs in Dino-Rama were cartoonish animals that were as gentle and playful as Barney, the purple dinosaur from the popular kids' television show of the 1990s and 2000s. On the gondola ride Triceratops Spin, I sat with other adults and children on the back of a plastic Triceratops that leisurely undulated around

FIGURE 14. The ride "Triceratops Spin" featured cartoonish plastic Triceratops that leisurely undulated around a merry-go-round a dozen or so feet off the ground. Photo by author.

a merry-go-round a dozen or so feet off the ground. My Triceratops had pastel green skin with gold stars on its sides and stubby horns that curved slightly backward from its face in an entirely unthreatening manner. The frills framing its face had scalloped edges and blue diamonds, which made them look more like tiaras than armor (fig. 14). Instead of the charismatic violence of dinosaurs I described in chapter 2, Dino-Rama embodied Disney's signature "wholesome, middle-class, family entertainment" that has made the company famous.[39] Indeed, with the exception of the thrill ride DINOSAUR, all traces of violence had been eliminated from this recreation of the American West. As I explore below, violence was not entirely absent from DinoLand, but it was confined to *pre*history.

Not far from Dino-Rama, the Boneyard Playground recreated a paleontological excavation site frozen in the middle of a field season. On one side, a battered old jeep sat, seemingly stuck, in the center of a sandbox surrounded by plastic walls made to look like rock faces with articulated dinosaur skeletons sticking out of them. None of these dinosaurs showed any indications of violence either. Even the gigantic *Tyrannosaurus rex* looked innocent: its skeleton is positioned lying down with its head near the ground. I watched young children excitedly run and fearlessly climb, using the dinosaur bones as handholds (fig. 15). On the other side of the Boneyard Playground was a sandbox dig pit simulating an active fossil

FIGURE 15. The *Tyrannosaurus rex* skeleton in the Boneyard Playground is positioned lying down with its head near the ground. Children use its rib bones as handholds as they happily climb into it. Photo by author.

quarry. In typical Disney style, this dig pit was far bigger than those I saw in other places. Under an enormous tent offering shelter from the sun, I watched children and adults use yellow plastic shovels and green plastic buckets to move away sand and uncover the dinosaur fossils beneath. Here, visitors enacted fossil excavation as an easy and fun diversion in which they quickly and effortlessly exposed complete, articulated skeletons. The dig pit encouraged them to imagine themselves discovering dinosaurs, or at least of their fossil remains. The edu- part of edutainment was minimal. One could learn about the dinosaur species represented in the Boneyard Playground from the "Site Information" boards positioned next to each dinosaur skeleton. The boards displayed typed sheets with facts about the quarry, colored illustrations of dinosaurs in the style of Charles R. Knight's murals, diagrams of fossils like those found in scientific publications, and excavators' handwritten notes. But I did not see anyone pay attention to these. The children were too busy climbing on the dinosaur skeletons or digging in the sandbox, while their parents rested in the shade. The arduous work of excavation I experienced at the Badlands Station was nowhere to be found.[40]

Here, children and adults played at being fossil hunters, not Indian fighters, in a romantic recreation of a Western town. DinoLand U.S.A. aligned with Disney's other idealized representations of Americana that deploy "nostalgia for . . . a kinder and gentler past in order to promote consumption in the present."[41] Disney's "kinder and gentler past" erase the fact that much of the northern Great Plains is either Sioux Treaty Lands or the Great Sioux Reservation. The treaties between the United States government and the Dakota and Lakota (called Sioux by outsiders) were made in 1851 and 1868, then quickly violated as whites colonized more and more Indigenous land.[42] By substituting dinosaurs and crocodiles for people and bison, DinoLand U.S.A. reinforced the myth that the American West was wide-open, virgin land until scientific explorers arrived, first, to survey its flora, fauna, and fossils and then, to harvest its natural resources. It further suggested that this region is still a no-man's-land today. Although Sue's remains were found on a Native American reservation within the region that DinoLand recreated, there were no signs of Native Americans or other residents who might object to outsiders coming to remove fossils or natural resources. Disney World guests were encouraged to imagine themselves as scientific heroes without facing any of the physical, social, legal, or moral difficulties of paleontological fieldwork in the western United States. DinoLand's Boneyard Playground transformed a site of violent settler colonialism into one of the world's remaining "natural wonders" still awaiting intrepid scientific explorers. It turned a place where the US government and white colonists slaughtered Native Americans into a *dinosaur* cemetery. This begins to show how DinoLand U.S.A. romanticized settler colonialism and reenacted the heroic tales of scientific discovery that have long been key to the American West mythology, while erasing the historical and ongoing processes of violence.

RIDING INTO PREHISTORY

The exception to DinoLand's peaceable kingdom was its signature thrill ride, DINOSAUR, which promised visitors that they would "take off on a scary, prehistoric tour aboard a rip-roaring time rover to save a dinosaur from extinction." The ride was originally called "Dinosaur: Countdown to Extinction," then rebranded after Disney released an animated film—and associated video game and other merchandise—with the name *DINOSAUR*. While the popularity of the film was short-lived, the ride remained a favorite at Animal Kingdom. It used Disney's patented "enhanced motion vehicles" to simulate movement as it took guests on a driverless train ride through a dark tunnel full of coordinated special effects, including animatronic dinosaurs, strobe lighting, and loud sounds. By the time I visited in the 2010s, these special effects paled in comparison to the virtual reality technology deployed on the ride Avatar Flight of Passage in Animal Kingdom's newest land, Pandora. Yet I observed DINOSAUR providing myriad people with the somatic and affective enjoyment of danger—the thrill—they sought. DINOSAUR recreated the land around the Western Interior Seaway in the Mesozoic era as the first "Wild West," a time full of enormous beasts that were even more impressive than the modern charismatic megafauna in Animal Kingdom's "Africa."

I heard the same comments again and again as people exited the ride. One girl delightedly declared, "That was scary!" and her mother affirmed, "That *was* so scary," adding, "It was dark." Many others remarked that the ride was also "rough" and "loud." In another instance, a father turned to his young daughter and exclaimed, "That was awesome! Want to do it again?" The girl immediately replied, "No!" As conquering fear is a stereotypically male-gendered aspiration in the United States (and elsewhere), boys more often claimed they had not been scared and wanted to ride DINOSAUR multiple times. The intensity of the physical experience of the thrill ride reinforced a cowboy masculinity—central to the myth of the American West—that celebrates acting heroically, with no signs of fear, in the face of danger. The premise of DINOSAUR, in which riders rescue an Iguanodon from extinction, further strengthens this message.

The premise of DINOSAUR was that its riders are "guests" of the "Dino Institute" taking a "field trip" to "the age of the dinosaurs." In the pre-ride video that set up the storyline, the Dino Institute's director, a contemporary Black female version of the famed paleontologist Othniel Charles Marsh (played by Phylicia Rashād of *Cosby Show* fame), looked straight into the camera and promised that DINOSAUR riders would experience "a breathtaking journey through a prehistoric world where you will witness the most spectacular creatures to ever walk the Earth." The plot turned frightening when the time rover's controller, a rogue scientist sardonically named Dr. Grant Seeker (played by Wallace Langham from *CSI: Crime Scene Investigation*), takes over. He told riders that he had reprogrammed the rover to go to the final moment of the Cretaceous period, which Dr. Marsh deemed too dangerous to visit, and asked us to bring back an Iguanodon before the asteroid hits and kills them all. With Dr. Marsh out of sight, Dr. Seeker explained, "You follow the homing signal to the Iguanodon, then I'll enlarge the transport fuel and, boom, you're back with one additional passenger, extra-large." He set up the thrill by saying, "Don't worry about that asteroid. You'll be in and out of there before it even breaks the atmosphere. Trust me, what could go wrong?"

At the end of the pre-show, riders were guided to a loading dock and efficiently file into a train of twelve-seater, militaresque jeeps that quickly set off down a track into a dark tunnel. The ride itself lasted fewer than six minutes. As a meteor shower rained down with light and sound, animatronic dinosaurs jumped out at the vehicles from all sides. As the jeeps shook, turned, dropped, stopped, and started abruptly, an audio track announced each dinosaur species we encountered and warned us of impending dangers. A computer-generated voice stated, "Warning. Meteor. Strong," then Dr. Seeker commanded the rover, "Evasion maneuver. Right! Left! Right! Left! That was close." There was an especially dramatic roar as a dinosaur lunged in front of the jeeps, and they came to a sudden stop. "Computer, now what?" Dr. Seeker asked. The computer voice identified the dinosaur as a Carnotaurus, the main predator of the Iguanodon that the riders were trying to capture. As the meteor shower increased in force, Dr. Seeker yelled, "That's it. Abort mission! Abort! Abort!" while the computer announced that its scanner had found the Iguanodon. The voiceover concluded:

> Dr. Seeker: We've got it! Get them out now!
> Computer: Asteroid impact.
> Dr. Seeker: Brace yourselves! We're not going to make it. We're not going to make it.

Finally, the computer announced "mission accomplished" as the train of jeeps arrived back at the loading dock. Riders were quickly unloaded into the gift shop where they could purchase dinosaur-themed merchandise of all sorts. While Animal Kingdom's other "lands" make it fun and easy to save endangered species and their ecosystems through consumption of commodities and experiences, Dino-Land U.S.A. joined the Boneyard Playground in making it entertaining to save prehistoric megafauna—dead or alive—for scientific research.

It comes as no surprise that, despite Disney's claim to provide science education at Animal Kingdom, DinoLand U.S.A. did not teach visitors about paleontology, the prehistoric world, or the history of the United States. Instead, DINOSAUR taught that paleontology is an exciting and daring adventure and suggested that facing and overcoming danger in pursuit of science is a worthy goal. The ride further communicated that immersive experiences with live animals are the most effective form of science education. In Animal Kingdom's "Africa," the safari jeep ride teaches visitors that the best way to learn to protect endangered species, and their habitats, is through immersive experiences that bring them as close as possible to their captive exemplars. In DinoLand U.S.A., the thrill ride taught that dead specimens in natural history museums will one day be replaced by similarly emotive experiences. In the pre-ride show, Dr. Marsh addressed The Dino Institute guests directly: "I hope you enjoyed those quaint exhibits in the Old Wing. That's how dinosaurs have been presented to the public since the study of fossils began over 150 years ago. Today, that bare bones approach is about to become extinct." Guests learned that the Institute's "trans-dimensional joy ride" was the "perfect blending of science and technology."[43] This message was embodied in the architecture of DINOSAUR. The line for the thrill ride snaked through "the Old Wing" of The Dino Institute, a building resembling museum exhibitions like the Denver Museum's *Prehistoric Journey*. In dingy and dimly lit halls, the other riders and I passed by cases of fossils organized around themes, such as "PREDATOR & Prey." (Note the emphasis on violence indicated by the capitalization.) Once inside The Dino Institute's main hall, we circled around a mounted skeleton of a Carnotaurus, the villain of the adventure to come. This late Cretaceous theropod, known from only one specimen found in Argentina, is often described as a small *Tyrannosaurus rex* with horns. Like Dino-Sue, this dinosaur stood frozen in place, posed in a horizontal posture with its mouth wide open and its legs mid-stride. Like

Dino-Sue, it attracted little attention from visitors. It serves merely as the antecedent to the animatronic dinosaurs in the thrill ride we were waiting to take. The ride's plot took Disney's message about how visitors can save endangered species to the maximum: bringing back animals that have been extinct for more than sixty million years. It did this by turning the spectacle of the classic Western film into an immersive prehistoric experience. In DinoLand's Western town, fossil hunters took the role of brave cowboys and dinosaurs replaced "Indians" as "villains . . . threatening the wagon train . . . a particularly dangerous form of local wildlife."[44] The scientific cowboys bravely rode into danger on rovers that move like bucking broncos over rough and uncharted terrain, while being threatened from all sides by predators that jumped out at them unexpectedly and without provocation.

DINOSAUR drew on Disney's nature films' portrayals of the country's violent wilderness and replaced United States' internal colonialism with a whitewashed national storyline similar to those in Disney's western history films. The history films' "unapologetically patriotic" stories ignore inconvenient realities to portray uncomplicated "trail-blazing heroes and brave pioneers" who "created, preserved and expanded the United States," despite being attacked by "savage Indians."[45] In one pertinent example, the film *Ten Who Dared* (1960) depicts geologist John Wesley Powell's 1869 expedition to map the Colorado River. While his crew pans for gold, Powell risks his life to collect fossils. Both this film and the thrill ride present "the Wild West" as an abundant wilderness ripe for extraction, but also full of life-threatening perils, from storms to savages. DINOSAUR replaced the "Indians" in these films with dinosaurs while continuing to portray fossil hunters as valiant heroes who risk their lives in pursuit of vicious beasts that will become scientific specimens. Its riders are the valiant vanguard of science who, like Disney's John Wesley Powell, managed to escape, just in time, with their hard-won scientific prize in hand. In an immersive experience that ostensibly portends the future of science, DINOSAUR reinvigorated the heroic cowboy-scientist narrative that has been central to the historical myth of the American West.

Amid the ongoing conflict over fossil sales, DinoLand's thrill ride also subtly supported commercial fossil collectors, including those at the Black Hills Institute who excavated Sue, in their claims to the scientific value of their operations. While numerous paleontologists have asserted that only scientists with doctoral degrees should be allowed to excavate fossils on public lands, commercial fossil collectors have argued that when their access is restricted, valuable specimens are destroyed before they can be collected and used to advance science.[46] As one of the companies that funded the purchase and exhibit of Sue, Disney encouraged its "guests" to imagine themselves, like the fossil hunters who excavated Sue, as the valiant saviors of prehistoric animals. Disney made itself into a scientific hero, while erasing the claims of Native Americans to the fossils and their potential profits. DINOSAUR celebrated the violent capture of specimens, rather than the gentle care discussed in previous chapters. DinoLand U.S.A. thus joined Animal Kingdom's other lands in communicating that commercial ventures can be engines of scientific discovery and environmental stewardship, divorced from history and politics.

The different parts of DinoLand U.S.A. came together to create a romantic origin story for the American West in which paleontology is a worthy adventure to rescue scientific specimens from an inhospitable wilderness and advance scientific discovery.[47] DinoLand produced an apolitical and morally comfortable past in which neither its visitors nor their ancestors bore responsibility for the slaughter and ongoing dispossession of Indigenous peoples. This is a brazen extension of the doctrine of Manifest Destiny into both the deep past and the future. While Disney's science edutainment contrasts with the view of paleontology offered by the visible fossil preparation laboratories in the Denver and Morrison museums, it shares a great deal with the presentation of the history of paleontology at the "outdoor museum" Dinosaur Ridge.

TO PRESERVE AND TO EDUCATE

As earlier chapters discuss, Dinosaur Ridge is a nature park in the eastern foothills of the Rocky Mountains, in Jefferson County, Colorado. It was designated a National Natural Landmark in 1973 owing to the dinosaur fossils embedded within the exposed layers of Jurassic and Cretaceous rock. While the park land is partially owned by Jefferson County and partially by a family-run ranch, the park's activities are run by the nongovernmental organization Friends of Dinosaur Ridge. The organization's mission is, first, "to preserve the paleontologic, geologic, and historic resources" of Jefferson County and, second, "to educate the public about these resources." It pursues the second goal through educational tours for the hundred thousand visitors who come to the park each year. Friends of Dinosaur Ridge does not need to advertise its programs because, a staff member told me, "Dinosaurs just sell themselves." Unlike Disney World, the organization relies primarily on volunteer labor and struggles to pay its bills. Its small staff and many volunteers offer tours to school classes and other groups during the week and tours to the public on the weekends and throughout the summer. Although Friends of Dinosaur Ridge does not call its tours "edutainment," this term fits them well. As one of Dinosaur Ridge's longtime educators told me, "Once people get here, they realize that it is just like a hiking trail that has cool stuff on it." At this "outdoor museum," unlike at natural history museums, "you can touch everything. You can experience it—rather than just see it—to learn from it." Teaching visitors about geology, paleontology, and the prehistory of the place is the organization's highest priority, while its human history is secondary. Dinosaur Ridge's science edutainment reproduces a historical narrative in which fossil hunters are scientific heroes who rescue specimens for the advancement of knowledge. In doing so it simultaneously erases their participation in white settlement and native dispossession. Dinosaur Ridge thus exhibits a tamer version of the violent scientific adventurism enacted in DinoLand U.S.A. This makes its story of paleontology in the Western United States more convincing than Disney's narrative does.

Let me recount one of the dozens of tours I joined over the course of my field-work at Dinosaur Ridge. It was the Fourth of July weekend and the weather that Sunday morning was beautiful, sunny, and warm with a cool breeze coming off the mountains. The parking lot was full before the Dinosaur Ridge Visitor Center opened. Although people are permitted to hike up and down the ridge whenever they like, most prefer to take a van tour. Visitors had to wait in line quite a while that day, but they seemed less frustrated by the wait than those at Disney World did. Once fourteen people were seated in a van for one of the first tours of the morning, I jumped in and sat down on the floor as I often did. The tour was led by a longtime volunteer-turned-staff member named Jody. She was a fit and ener-getic woman in her sixties who cared deeply about Dinosaur Ridge. As she drove toward the ridge, she declared, "We're going to start out with a little bit of present-day history about this spot." Jody offered this account:

> This valley right here is called Rooney Valley. It was home to the Rooney family starting about the same time as the Colorado Gold Rush. At one point the Rooneys owned about four thousand acres up and down this valley. Today, the family owns about two hundred acres. Many of the Rooney family descendants live right over here, on the left hand side of the road, on the original ranch site. Our Visitor Center and barns were part of the Rooney Ranch at one point in time . . . Back in the 1800s, the Rooneys were friendly with the Ute Indians who also lived out this way. If you look to your right you'll see a very large pine tree just part way up the hillside, just to the right of that rock wall. That tree is known as the "Ute Council Tree." There was a Ute Indian chief by the name of Chief Colorow who used to come out here, hang out with the Rooneys, go up there, and meditate. He called that his "inspirational tree."

Jody's historical narrative draws from the stories of the Rooneys and the Utes in Friends of Dinosaur Ridge's publications. A small group of dedicated volunteers have researched and produced an archeological survey of the area, a field guide to its flora and fauna, and histories of the paleontological quarries and ranch that are now part of Dinosaur Ridge park. While few visitors purchase these publi-cations, they hear their content on tours. Friends of Dinosaur Ridge's dedicated staff and volunteers do not intend to misrepresent the history of the region. Their first goal is to teach science in a fun, hands-on, approachable manner, enriching people's appreciation of the natural beauty of the place and their understanding of its geologic history and scientific importance.

Friends of Dinosaur Ridge's history of the park recounts that Alexander Rooney moved from Iowa to Colorado with a dozen family members in 1860 as part of the Colorado Gold Rush. While he did not get rich on gold, Alexan-der developed several successful livestock ventures. When he found a valley just east of the Rocky Mountain foothills with "wild hay as tall as a horse's belly . . . and water running down the slope," he decided to stay and build a homestead.[48]

Alexander quickly erected a claim shanty to secure a land title from the US government. After he established a cattle and horse ranch there, he brought his second wife, Emeline Littlefield, from Iowa. A few years later, they moved from their claim shanty into a large stone house they had built on the property, now named the Iron Spring Ranch. Alexander and Emeline had six children who "hiked the Ute trail across the [ridge] to attend Harriet Raymond's school" in the 1870s.[49] Dinosaur Ridge's *Field Guide* states that the Ute Council Tree on the ridge, which is estimated to be at least five hundred years old, has been recognized by the National Register of Historic Places and named a Jefferson County Historic Landmark "because it marks the spot where the Ute Indians often held council with the original settler, Alexander Rooney."[50] The Rooneys took full advantage of government land grants and by the end of the nineteenth century they had established the largest cattle and horse ranch in Jefferson County. Their ranch once encompassed more than four thousand acres.[51] In addition to ranching, they farmed; mined coal, clay, and limestone; quarried and cut building stones; and sold mineral water.[52] As Jody noted on her tour, the family sold off most of this land over the course of the twentieth century. While the Rooneys still allow them to use a piece of their extant property as a parking lot, Friends of Dinosaur Ridge worry that this parcel too will eventually be sold to developers who are eager to further expand the Denver suburbs.

Several archeological and historical studies provide accounts that significantly differ from the stories told at Dinosaur Ridge. They document that numerous bands of Utes moved seasonally across an area that extended from the Great Basin to the Rocky Mountains to the western edge of the Great Plains prior to European colonization.[53] With the juridical incorporation of their ancestral lands into the United States territories after the Mexican-American War, many Utes retreated into the Rocky Mountains to avoid conflict with whites and with those native groups pushed onto the Utes' territory by land dispossession further east. This strategy became nearly impossible after 1858, when tens of thousands of whites flooded the Front Range as part of the Colorado Gold Rush.[54] As colonization shifted the regional economy from the fur trade to ranching and mining, the federal government not only violated the treaties it signed with the Utes and enacted laws that allowed whites to claim land; it also turned a blind eye to illegal settlers. As the historian Ned Blackhawk writes, "mineral, ranching, and corporate interests soon combined with sentiment to undermine Indian land claims, while bribery, alcohol, and intimidation facilitated Ute compliance."[55] The Rooneys thus took part in dispossessing the Utes and other native peoples of their land and its resources.

While Friends of Dinosaur Ridge's tours and publications celebrate the Rooney family for having an exceptionally amicable relationship with the Utes, one of its publications hints at a far more pernicious history. This history contrasts the many alleged wrongs that the Utes committed against the Rooneys with the family's "kindly" treatment of "the local Ute Indians."[56] It quotes Alexander Rooney II's assertion that "the Americans broke some 232 treaties with the Indian, but my grandfather was respected by them for his honesty; even the hostile Arapahos became his friends."[57] However, this claim is countered by the family's military service. Alexander Rooney joined the Colorado Cavalry, after "native attacks" in Jefferson County, and fought at Sand Creek in 1864.[58] In other words, at the behest of the Colorado Territory's governor, he took part in slaughtering more than 230 Native Americans who had gathered under a white flag of peace near Fort Lyon, Colorado. The Sand Creek Massacre precipitated more than a decade of extreme violence against Native Americans in the Great Plains.

Alexander Rooney and his family also directly took part in native peoples' dispossession once he returned to the Rocky Mountains foothills from Sand Creek. Those living in the area had long known of the "strong flowing spring that oozed out of the base of the sandstone" at the spot where the mountain met the valley floor.[59] Alexander selected this location for his ranch because of this reliable and plentiful water source. Whites called it "Iron Spring" because of its high iron content, and Alexander adopted this name for his ranch. A Friends of Dinosaur Ridge publication recounts what he did next:

> In 1867, Alex Rooney enclosed the spring in a stone building, complete with sand tanks to filter the water as the Natives had taught him. Tanks of constantly flowing cool water refrigerated crocks of milk, butter, and eggs. Meats hung from hooks on a high beam in the 51° temperature. Eventually, the water, filtered through sand, was piped to all of the ranch buildings . . . Enough water flowed through Rooney Gulch that the family dammed the creek in numerous places to create cattle ponds and to irrigate their fields in the southern part of the Rooney Valley.[60]

Alexander Rooney depended on the Utes for knowledge and skills necessary to succeed in farming and ranching in the semi-arid foothills. While the Utes continued to access the Iron Spring after the Rooney family took over the land, it is also clear that most of the spring's water was diverted to the Rooney's commercial ventures and personal consumption. How friendly Alexander Rooney was with Chief Colorow matters little, while the Rooneys dispossessed the Ute people of this vital resource.[61]

A hundred and fifty years later, Jody told the park visitors a romanticized version of this history to give her tour a "little bit of local color." She then announced, "We are going to go back about a billion years," and the Utes and the Rooneys were not mentioned again for nearly an hour. Traveling the mountain pass where Native Americans once lived, visitors learned only about the Stegosaurus, Apatosaurus, and Iguanodon that roamed the landscape millions of years prior. This is the glorious past of America's charismatic megafauna that DinoLand U.S.A. also recreates. As the van crested the top of the ridge and descended into Jurassic period rocks, Jody returned to human history to tell visitors about the importance of Dinosaur Ridge for paleontology:

Go back to the year 1877. There was a gentleman that lived out this way. His name was Arthur Lakes, and he was a schoolteacher at what's now called the Colorado School of Mines. He called one of his buddies over here. They were just going to dig around and look for plant fossils, which they did not find. Instead, they found something that was a whole lot cooler than that, and that happened to be dinosaur bones! That was the discovery of the very first Stegosaurus, right here at the place we're stopping next.

Told a different way, a decade after Alexander Rooney enclosed the Iron Spring, Arthur Lakes found his first dinosaur fossils on traditional Ute land along the western edge of the Rooney's ranch. Colorado was now a state, and whites had been farming, ranching, and mining there for decades. Lakes described the place he found his most famous dinosaur skeleton as an "old quarry," indicating that it was probably one of the spots where the Rooneys, or other colonists, had mined lime or building stones.[62] Lakes thus benefited from the Rooneys' prior land claims to prospect for and extract fossils from the Ute homeland. These early colonists had built the infrastructure and otherwise paved the way for future generations of whites to hunt for natural resources, fossils, and other scientific specimens in the region. Other fossil hunters used homestead laws to secure ownership of valuable fossil beds in other parts of the Western territories.[63] They did this not only on land designated for white settlement, but also on Indian reservations, which were ostensibly off-limits to whites.

It is within this context that we need to read Lakes's journal entry from the day he stumbled on dinosaur fossils. Lakes writes:

> We had not climbed very high when Captain B . . . called me to come and examine what appeared to be the fossil compression or cast of a branch of a tree upon a loose slab of sandstone . . . But on looking at the impression I saw at once that it was too smooth a cast to be left by any tree, and further on one end there were little patches of a purplish hue which I at once recognized as fragments of bone. Here, then, was a cast of a very large bone belonging to some gigantic animal . . . We soon traced the loose slab to the parent rock of brown sandstone from which it had slipped and, as I jumped on top of the ledge, there at my feet lay a monstrous vertebra . . . It was so monstrous—however, thirty-three inches circumference—so utterly beyond anything I had ever read or conceived possible that I could hardly believe my eyes, and called to my friend Captain B. to confirm the vision. We stood for a moment without speaking gazing in astonishment at this prodigy and threw our hats in the air and hurrahed. And then began to look about us for more. Presently Cap B. cried out "why this beats all!" At his feet lay another huge bone resembling a Herculean warclub ten inches in diameter and about two feet long . . . On digging beneath where this lay under some bushes some smaller vertebrae, apparently concealed, were discovered.[64]

Lakes's diary echoes Barnum Brown's account of finding the *Tyrannosaurus rex* AMNH 5027 chronicled in chapter 2. They both emphasize the prehistoric

FIGURE 16. Arthur Lakes's watercolor *Professor Mudge Contemplating Atlantosaurus Bones*, Morrison, Colorado, 1878. Image courtesy of the Yale Peabody Museum.

creatures' size and capacity for violence. They both naturalize this violence by suggesting that it has existed since prehistoric times. Here, I want to call attention to the location where Lakes found this ostensibly monstrous beast. Dinosaur Ridge asserts that, because Lakes's excavations "produced the first large dinosaur bones in the American West," this site is "probably the most important paleontological site in the country."[65]

Lakes notes in his diary that the ridge where he found these dinosaur fossils was pocked with colonists' quarries and contained caves where Utes and other Native Americans had interred their dead.[66] Yet, he ignores these observations and presents the area as an unpeopled landscape in his writings and paintings (figs. 16 and 17). After establishing a campsite near the old mining quarries, he writes in his journal, "Whilst here and there 'growing where no life is seen' . . . you may see a few pine trees." One of these pines was most likely the Ute Council Tree that Dinosaur Ridge tour guides point out to visitors. Lakes's portrait of the landscape prefigures the one recreated in Animal Kingdom's Boneyard Playground. Much like Disney films and parks create an unpeopled wilderness as the stage for their stories of fossil hunting, Dinosaur Ridge's publications and oral accounts make it seem as if Lakes had been exploring a no-man's-land, not the outskirts of the Rooney's prosperous cattle and horse ranch built on land that had long been inhabited by native peoples.

FIGURE 17. Arthur Lakes's watercolor *Digging Out Bones at Morrison, Quarry No. 10*, Morrison, Colorado, 1879. Image courtesy of the Yale Peabody Museum.

Educated men of European descent, such as Lakes, were not the first to take interest in the fossils embedded in ridge rocks and strewn across the land that became known as the American West. The folklorist and historian Adrienne Mayor has undertaken the difficult task of reconstructing native peoples' knowledge of the prehistoric remains they encountered long before white fossil hunters arrived on the scene and "discovered" them. The significance of dinosaur fossils to preconquest Utes is indicated by burial sites, in which bodies were found wearing small fossils as amulets, and Ute council trees, which were often located close to dinosaur fossil beds.[67] The meaning of these fossils for these people has been lost, yet we do know that these trees held significance as the places where Utes gathered to make important decisions and resolve disputes, first with other native people and later with white colonizers, such as the Rooneys. We also know that white fossil hunters hired Native Americans to guide them to fossil beds, but did not give credit to their guides in their reports or scientific publications. These scientific explorers "seem to have regarded Native Americans as an inconvenient, fairly amusing part of the exotic wildlife of the West."[68] For some, Native American warriors were subhuman beasts who needed to be subdued, if not slaughtered, in order to extract valuable resources, including the remains of prehistoric beasts. Even the most sympathetic fossil hunters took credit for feats they never could have accomplished without the knowledge, skill, and labor of Indigenous and other people living in the area. Marsh may have "paid homage to Sioux origin stories" when he named Brontosaurus, but the names of the Native Americans appear nowhere in his publications.[69]

While Dinosaur Ridge tours do not discuss the Utes, they never fail to tell visitors that Lakes "discovered" Stegosaurus, the Colorado state fossil, in the park. As the notion of "discovery" indicates, the whites who remove fossils and other geological specimens from this ostensible no-man's-land are understood not only as the first to identify them as valuable resources, but also as their rightful owners. The story about the origins of paleontology at Dinosaur Ridge thus dovetails with the grand narrative of Manifest Destiny, in which Euro-Americans were the divinely appointed and rightful owners of the continent and its resources, while the Native Americans living on these lands for centuries were destined for extinction.

ARTHUR LAKES

As the previous chapter recounts, Arthur Lakes was many things: an Anglican minister, a teacher, a fossil hunter, a writer and illustrator, and a geologist. Friends of Dinosaur Ridge's biography of Lakes proclaims that he "was an amazing, productive man who played leading roles in exciting adventures of the settlement of the American West" and was a "pivotal figure in the dinosaur 'bone wars'" between Marsh and Cope. As tours of Dinosaur Ridge recount, he excavated the fossils of several of the most famous dinosaurs, including Allosaurus, Apatosaurus, Stegosaurus, and *Tyrannosaurus rex*. Lakes's biography describes him as both "a Victorian natural scientist" and a geologist "on the cutting edge of the industrial revolution" in Colorado.[70] Lakes is credited with setting off the Great Dinosaur Rush, recognized as one of the great fossil hunters of the United States, and known as the "Father of Colorado Geology."[71] The library at the Colorado School of Mines is named after him. Lakes's obituary proclaims that he "accomplished more for his state in gathering useful geological data and making surveys of the commercial deposits than any other one man."[72] Lakes's fossil hunting did not merely benefit from the Rooney's colonization of the area that is today Dinosaur Ridge; Lakes himself developed a career that combined geological mapping, fossil hunting, and mineral mining to advance settler colonialism.

Lakes was born in rural England in 1844 and died in British Columbia, Canada in 1917, but lived much of his life in Colorado. After studying theology and natural history at Oxford University for a few years, he moved from England to Colorado in response to the Anglican Church's call for more ministers to serve the multitudes of miners arriving there.[73] Lakes also hoped the move would improve his economic situation after he had withdrawn from Oxford without earning a degree. Colorado Territory, it turned out, was not paved with gold, and Lakes struggled to make a living. In addition to serving as a minister in the mining camps and rural communities around Jefferson County, Lakes taught writing, drawing, and geology at Bishop Randall's Collegiate School for Boys (also known as Jarvis Hall), which later became part of the engineering and applied sciences university the Colorado School of Mines. Lakes also collected fossils and mineral specimens for the United States Geological Survey.[74] When these jobs failed to provide him with the income he desired, he wagered hunting dinosaur fossils might prove a lucrative addition.

As Jody and other tour guides conveyed to people visiting Dinosaur Ridge, Lakes found his first dinosaur fossils while conducting a geological survey of the ridge above the Rooney Ranch for the United States Geological Survey. When he initially wrote to Marsh, the paleontologist at Yale University, about his paleontological finds on the ridge, he did not explicitly state he wanted to sell the fossils, but he did make clear that he expected to be paid for his work extracting the rest of the skeleton and future ones. He argued that "whilst I am thoroughly imbued with enthusiasm about such pursuits . . . I have not the pecuniary means to do so." When Marsh did not immediately offer him a job, Lakes unambiguously stated in a subsequent letter, "my circumstances obliged me to sell the specimens."[75] He used an offer from Edward Cope, Marsh's rival at the Academy of Natural Sciences in Philadelphia, to elicit more money from Marsh. Lakes ended up being employed by Marsh for several field seasons, in Wyoming and Colorado, and working one winter in his laboratory at Yale.[76]

Lakes's career in paleontology and geology was enabled by the colonization and resource extraction in the Denver region and the Western United States more broadly. His journals reveal that he "obtained the help" of colonizers including quarrymen, a blacksmith, a stone mason, the superintendent of a local mining company, a mine and hotel owner, and "many [other] hands."[77] After Marsh sent him money, Lakes hired as assistants two of his former students from the School of Mines.[78] Nearly all these men were in Jefferson County, and available to assist Lakes, because of the mining and ranching operations there. So too were the schools where he taught. As Lakes wrote in his journal, he took his students on field trips to the fossil quarries to show them the "workmen in their task of exhuming geological remains for Yale college museum." They also stopped by "Alexander Rooney's fine stock ranch" to enjoy its "excellent mineral spring."[79] Iron Spring ranch was more than a treat after a class trip. Lakes's paleontological discoveries of dinosaurs depended on numerous resources, services, and businesses set up by the Rooneys and other white colonizers.

Lakes's excavations also relied on the railroad lines that were built to facilitate resource extraction and colonization. Lakes sent fossils to Marsh in New Haven, and received back supplies and luxuries, like the canned oysters he liked to eat during fieldwork.[80] These were definitely not the "Rocky Mountain oysters" (fried bull or bison testicles) that ranchers in Colorado ate![81] More important than supplying Lakes with delicacies, the transcontinental railroad "opened a new frontier for the expansion of scientific practice" and enabled the "process of converting nature into knowledge that could travel to metropolitan centers."[82] For example, in 1877, when Lakes found fossils that turned out to be the largest dinosaur known to paleontology at that time, he sent it by train to Marsh's laboratory at Yale. Marsh named the new species *Atlantosaurus giganticus* (meaning "huge Atlas-lizard," after the mythological hero who carried the world on his shoulders, now Apatosaurus). Marsh also named Allosaurus, Diplodocus, and Stegosaurus, Colorado's state dinosaur, based on the tons of fossils that Lakes sent by rail to Yale over the next few years.[83] These celebrated discoveries helped to justify colonization of the region, the enclosure of land, and the massacre and dispossession of the Utes and other Indigenous nations.

I compared Lakes's fossil-hunting prowess to Barnum Brown's earlier in this chapter. They both prospected for fossils across the Western United States and found fossils that would become the holotypes for famous dinosaur species. Yet Lakes did not have the wanderlust, adventurousness, or charisma of Brown.[84] After spending a winter preparing fossils in Marsh's Yale laboratory, Lakes returned to Colorado, where he mostly stayed. He did not work for Marsh again, possibly because Marsh was unwilling to allow Lakes to co-publish articles about the fossils he had found.[85] Lakes became a full-time professor at the School of Mines, and then a mining consultant. He published numerous articles on Colorado's geology, identifying profitable locations and warning potential investors against overvalued ones.[86] Commercial geology eventually provided Lakes with a successful and lucrative career.

Lakes had more formal education than most fossil hunters in the Western United States at the time, but others also moved between paleontology and industrial geology as opportunities grew.[87] As the historian Lukas Rieppel writes, "the almost mythical status of the American West as a land of geological plenty grew out of the development of a booming extractive economy . . . It was the hope of striking it rich" that motivated miners to "scour the landscape" in search of profitable resources.[88] When miners and other colonizers found fossils, they usually attempted to sell them. Yet the work that men like Lakes dedicated to excavating fossils cannot entirely be explained by financial desires. They were also motivated, and rewarded, by the moral values attributed to scientific discovery. As chapter 2 discusses, fossil hunting was a heroic endeavor that proved one's manliness. While this form of frontier masculinity celebrates self-reliance and independence, no one locates or extracts fossils alone. Previous chapters emphasized the distributed agency, involving humans and nonhumans, of fossil making. Here, I emphasize that the fame of Marsh and Lakes depended on colonization of the Rocky Mountain foothills, which made possible paleontologists' access to land, infrastructure, supplies, and labor, while dispossessing the Utes and other Indigenous peoples in the region. Dinosaur Ridge's publications and tours disregard this history when lionizing Lakes as among the greatest fossil hunters of the American West. They recount the perils he faced, the hardships he withstood, and the obstacles he overcame to pursue science for the common good. Dinosaur Ridge's depiction of Lakes pales only in comparison to Disney's portrayal of John Wesley Powell in *The Ten Who Dared*.

PALEONTOLOGY AND COLONIALISM

Colonization, dispossession, genocide, environmental destruction . . . it is unsurprising that none of these themes appear in the edutainment offered at Dino-Land U.S.A. and Dinosaur Ridge. Even if "dinosaurs sell themselves," it is not in the interest of Disney or Friends of Dinosaur Ridge to advertise the dark side of paleontology's history. The whitewashing of the brutality of colonialism at Dinosaur Ridge is not intentional, but neither is it merely the unfortunate result of people trained in geology trying to tell human history.[89] The history told in Disney films and parks is not merely the result of corporate profiteering. The history recounted at both these parks is part of an entrenched narrative about the American West that has been told and retold since the nineteenth century. The Pulitzer Prize-winning historian William H. Goetzmann vividly illustrates how the standard history of the United States paints a romantic picture of geology and paleontology. He writes that the "collection, classification and organization" carried out by scientific explorers, such as Marsh and his fossil hunters, "was the essence of civilization come to the wilderness."[90] In a more recent history, biologist Keith Thomson proclaims that "the doctrine of Manifest Destiny" was "Manifest Opportunity for science":

> The West was . . . a scientific treasure house. Among its most exciting secrets were ancient fossils . . . A small group of scientists discovered and described thousands of previously unknown kinds of fossil animals from the American West . . . These findings depended on exploration and discovery on a greater scale than anything attempted in Europe, carried out during times of adversity and adventure . . . all in the cause of serious science.[91]

Lakes was one among this "small group of scientists" whose story is told through the well-worn narrative of the heroic scientist in the Wild West. Told in a different way, Lakes's life story illustrates how paleontology benefited from white invasion, broken treaties, land and resource grabs, and the dispossession, impoverishment, and decimation of native peoples. The material traces of this human history persist in the landscape of Dinosaur Ridge, alongside the material remains of its prehistory, but they go unrecognized by the vast majority of visitors, guides, and scientists at the park.

The portrayal of paleontology at Dinosaur Ridge and Animal Kingdom reinforces the understanding of this science as a moral good that has advanced knowledge without harming anyone. Yet geology is not a neutral study of inorganic earth that is "immune from the violence and dispossession enacted through extraction of mineral resources" around the world.[92] Paleontology, like the explicitly commercial subfields of geology, both depended on and profited from the colonization of the Western United States. With help from the federal government, scientific expeditions ignored treaties and tribal sovereignty in their pursuit of flora, fauna, and mineral samples, as well as fossils. In the process of mapping the land and locating its natural resources, they provided crucial information that aided further military and colonial conquest. "Like the settlers seeking Indian lands for farming or ranching, and the miners seeking precious metals on Indian lands, [fossil hunters] staked out their own claims to the natural resources in which they were most interested, and 'opened up the West' in their own way."[93] They were hardly alone in this. Every white person who ventured into and beyond the Great Plains in this period was complicit in advancing colonialism, including my anthropological predecessors. Like archeological and ethnographic artifacts, fossil "trophies" from dinosaur graveyards were deployed to justify dispossession as advancing scientific discovery.[94] Geological mapping, paleontological excavations, and the collecting of fauna, flora, and mineral specimens provided scientific cover for economic exploitation, political and social subjugation, and military conquest of Native American homelands. From its emergence as a science alongside European imperialism through its practice today, geology is thus part of the larger story of colonialism and imperialism.[95] The history told in this chapter shows that this is not only true of industrial geology but also extends to paleontology's extraction of scientific specimens. It further shows how edutainment promotes the scientific defense of colonialism to a wide audience.

The dominant history of paleontology—recounted in text, image, speech, and the multisensory stimuli of immersive edutainment—continues to tell tales of heroic pioneers who suffered hardships to advance scientific knowledge for all humankind. Explicating a counternarrative helps to answer the question that opens this chapter: What *are* dinosaurs doing in Disney's Animal Kingdom? Both Animal Kingdom and Dinosaur Ridge carry on a long-standing tradition of using charismatic megafauna to symbolize the immensity and strength of the United States. Even before US independence from England, Thomas Jefferson promoted the American mastodon (*Mammut americanum*, which Jefferson called the mammoth) as proof of America's natural greatness.[96] Jefferson deployed this animal to counter the European impression that North American flora, fauna, and Indigenous peoples were smaller, weaker, and generally inferior to European ones.[97] While Jefferson used fossil evidence to make his case, he and his contemporaries believed that explorers would find mastodons still living in the unmapped expanses of the continent. But they found herds of herbivorous Plains buffalo (*Bison bison*) instead. Although buffalo are native to North America, they never became a national symbol in the way that Jefferson had hoped the mastodon would. Why? The buffalo is disqualified by its close association with the violence of settler colonialism. Whites slaughtered buffalo in an intentional effort to deprive native peoples of subsistence, as well as killing them indirectly through railroad construction, land enclosure, farming, and ranching. Although buffalo were no longer in danger of extinction when Disney was building Animal Kingdom in the 1980s, long-dead dinosaurs offer a more appealing symbol of American greatness because they are not tainted by the ills of colonialism. The Cretaceous dinosaurs that Lakes and his associates extracted have been used to represent the strength, enormity, and grandeur of the North American continent and to glorify the history of its conquest. Stegosaurus and *Tyrannosaurus rex* surpass the megafauna in Disney's African safari, as well as Europe's native fauna. Two centuries after Jefferson's rebuttal of America's natural inferiority, Europe does not even merit a "land" in Disney's zoological world map.

At the world's fairs held in the United States during the twentieth century, dinosaurs were used to demonstrate the country's scientific and pop cultural superiority. Disney, and to a lesser extent Dinosaur Ridge, continue this tradition.[98] Here, dinosaurs are American heroes unscathed by the violence of colonialism or the politics of the contemporary moment. Certain spaces continue to be envisioned as ripe with exotic beasts to be captured, valuable materials to be extracted, and priceless scientific specimens to be removed and studied. Science edutainment's reenactments of pre/history justify both historical and contemporary dispossession in the name of science. While the current chapter has explored how science and entertainment are integrated, the next chapter will examine how they are also divided.

The Paleontological Real

"Is it real?" When people are confronted with a fossil, I found this to be their most common question. However, the paleontologists, fossil preparators, science educators, and museum docents I came to know through my research did not all agree about how to answer it. The real has many others, including the fake, the copy, and, particularly in the case of fossils, the cast, the reconstruction, and the model. What significant differences do scientists and other experts mark between a real fossil and other-than-real ones? How and why do experts differentiate them? Why does it matter? That is, it explores the stakes of asking and answering, "is it real?" when faced with a fossil. An important part of the answer lies in the conflict over evolution and creationism that has raged in the United States for more than a century.

This penultimate chapter of *Making Our Beasts* revisits several of the sites previously explored in the book to look at them anew through the lens of the real. It also introduces one additional site—a paleontology-themed hotel in the Denver suburbs. As I engaged with a wide range of paleontological objects in these places, I found that absolute distinctions between the real and the unreal made less and less sense. Examining the relationship among fossils, casts, and other paleontological objects led me to reconsider not only what is real and what is not, but also the ontology that underlies people's main question about fossils. By *ontology*, I mean an approach to the existence of things and beings, including the concepts used to organize them. The dominant ontology in the United States posits a series of opposite concepts: nature versus culture; growing versus making; science versus fiction; and specimen versus both artwork and artifact. In the previous chapter, I scrutinize the manifestation of the dominant

Euro-American ontology in the practice of "science edutainment" without directly examining ontology. In this chapter, I analyze the ontological conception of the material world that undergirds both the science of paleontology and "dinomania" in the United States. I then describe an alternative ontology that I found emerging at the Morrison Museum.[1] This ontology suggests a multidimensional continuum of paleontological objects, rather than a binary opposition between real fossils and fake ones. First, though, I place the question of whether a fossil is real within the context of the long-standing concern over creationism. This debate demonstrates the real-world implications of different ontologies, which otherwise can seem an esoteric philosophical issue.

THE QUESTION OF CREATIONISM

The specter of creationism hung over my fieldwork. When I attended popular events such as Free Days at the Denver Museum and Scout Days at Dinosaur Ridge, creationists were usually manning a table that looked like a legitimate information station, although theirs was full of creationist texts. They appeared and departed without fanfare. They did not proclaim their beliefs with a megaphone, but patiently waited for someone to approach them. At the Denver Museum's *Prehistoric Journey*, a creationist leader periodically gave unauthorized "Biblically Correct Tours," but I never had a chance to join one. I was not alone in feeling creationism's presence. The majority of the scientists, educators, and volunteers at the places I conducted research had some interactions with them, too. Of the creationists they had met, very few were combative or even tried to convince staff or volunteers of their views. Most were quietly schooling their children on their own tours, pointing out the "errors" in placards and giving alternative interpretations. Despite their limited contact with creationists, museum docents and park guides were painfully aware of the prevalence of creationist views in the United States, and they were enraged about what creationists were doing at museums. Many talked about creationism over lunch, in quiet moments before or after work, and during interviews with me. Creationism is, in classical anthropological terms, a key symbol in the world of US paleontology.[2] It underlies the approaches to the paleontological real and its others I describe in this chapter. By the *paleontological real*, I mean the demarcation of what counts as evidence of prehistory and the evolution of species.

I had my most extended encounters with creationists at "the dinosaur hotel," a chain hotel in the Denver suburbs designed to evoke the history of paleontology in the region, not far from Dinosaur Ridge and the Morrison Museum. Often, creationists booked their stay on the chain's website, unaware of this franchise's distinctive theme. But immediately after entering, they saw the giant cast of an in-situ skeleton filling the wall behind the reception desk. The lobby lounge was decorated to recreate a fossil hunter's salon during the Great Dinosaur Rush. There were period curio cabinets full of fossil casts, and there were mounted dinosaur skulls around the room. The bedrooms, where I stayed a few times, had reproductions of Arthur Lakes's paintings of his field expeditions. Paleo Joe's—the hotel's breakfast room in the morning and pub at night—was adorned with a giant fossil suspended from the ceiling.

The hotel's paleontology theme was not limited to its decor. The owners paid a staff member from the Morrison Museum to set up a "fossil table" next to the coffee urns during the breakfast hour. Hotel guests would stop to look at the display and talk to the person behind the table. When I arrived at the hotel one morning, I found Nick, the docent working there that day, deep in conversation about carbon dating with a middle-aged white man. When I joined them, I noticed that the man asked astute questions about how fossil dating worked. In the course of their conversation, he shyly, almost apologetically, identified himself as a creationist. Despite starting from a different premise, the questions this and other creationists posed were remarkably similar to those asked by the most inquisitive paleontology aficionados I met. After first asking "is it real?" most visitors followed up with "how old is it?" Creationists asked instead, "how do you determine how old a fossil is?" I found that many creationists know a great deal about rock-dating methods, especially their flaws, which have been used to challenge scientific theories about the age of the earth, animal evolution, and extinction. Sophisticated creationists were better-versed in the pitfalls of particular techniques used to date rocks than many museum docents were. For instance, they knew that some dinosaur fossils were too old for radiocarbon dating and too young for argon, uranium, and other isotopes used in geology. In short, creationists wanted to discuss, sometimes sincerely and sometimes disingenuously, how scientists generate knowledge about prehistory and support the theory of evolution. I, for one, learned from these conversations that I could not explain evolution as well as I thought I could. While Nick was not easily flustered, most museum docents and park guides found the experience of having their explanations challenged by creationists quite threatening to their self-confidence and sense of purpose in their work. The experience also threatened the distinction they routinely made between real fossils and other paleontological objects.

Natural history and science museums in the United States have long aimed to promote evolution and discredit creationism.[3] While all the institutions where I did research shared this goal, they took somewhat different approaches to dealing with creationists in person. Docents in the Denver Museum's *Prehistoric Journey* each carried a card stating the museum's official position, tucked into the breast pocket of their green vests. It seemed to work like a talisman that they could take out and use whenever they needed help.

The Denver Museum of Nature & Science promotes the scientific understanding and study of the natural world. The history of life on earth is presented and interpreted so that the public may better understand the origins of different forms of life, the changes in the physical environment that have taken place over time, and the response of life to environmental and ecological change. Evolutionary theory utilizes the scientific method to explain the origins, changes and extinction of all life forms. The Museum presents science-based educational programs and exhibits that explore evolution.

If confronted with creationists, *Prehistoric Journey* docents were instructed to read this card and engage no further. When talking with visitors who accepted evolution and assumed that dinosaurs are extinct, docents often told them that scientists currently think that turkeys are descendants of theropod dinosaurs. But when talking with visitors who claimed that dinosaurs were on Noah's ark, they did not broach the evolution of species.

The Morrison Museum's director, Matthew Mossbrucker, took a somewhat different approach. He trained his staff and volunteers not to debate creationists if "they are proselytizing," but encouraged them to talk with creationists if they asked questions.[4] Matthew instructed that if, after engaging them, docents "cannot find some common ground" with the creationists, they should "bring them to him, or Dr. Bob [Bakker] if he is there." Robert T. Bakker, the charismatic paleontologist discussed in chapters 2 and 3, relished talking with creationists. While Matthew did not exactly enjoy it, he told me that he had "lots of interesting, and mostly positive, encounters with creationists." He told me with a twinkle in his eye that when people mention the possibility of dinosaurs living today, such as the Brontosaurus supposedly in the Congo Basin, "I always look at those people skeptically and say, 'I've devoted my entire life to the study of these animals. If I could see a live one, I'd go do that. But I just don't see any evidence that those animals are real.'" Here, "real" means alive today, but we will see below that it means different things in other contexts.

Mossbrucker and Bakker were the exceptions. Most museum staff and volunteers found interactions with creationists disconcerting because, I think, they challenged their deeply held conceptions about the real versus the unreal, and science versus fiction, as well as their confidence in their grasp of evolution. As I elaborate later, they were steeped in an ontology that posits that only the remains of prehistoric animals that have turned to lithic objects without human involvement are real fossils. This secular Euro-American ontology, which undergirds the science of paleontology, assumes that stones are unique, authentic, and enduring while plastics are generic, fake, and pliant. Moreover, while fossils are the paramount evidence that species evolved over time in the scientific ontology, the creationist ontology claims that the Christian Bible is the primary evidence, and fossils secondary evidence, that species were divinely created.[5] To elucidate the dominant scientific ontology, I now return to the Bone Bar in the Denver Museum's *Prehistoric Journey* exhibition, where volunteer docents and staff educators have been fielding the question "is it real?" on a daily basis since the exhibition opened in the 1990s.

PREHISTORIC JOURNEYS

The Denver Museum is always crowded on Free Days, when there is no charge for admission. On one such day, I was at the Bone Bar shadowing Norman, a gregarious and sprightly retiree in his seventies, when a man came up and asked him, "Are these actual bones of dinosaurs?" We had both heard this question many times before, and Norman was ready with his answer: "Yes and no," he said. He then beckoned the man to follow him over to the mounted skeleton of an Edmontosaurus. He pointed to the display sign and explained, "This guy is 80 percent real." Then, he drew the man's attention to the skull on the mount, saying, "Look at that head. That's not so real, right? It kind of looks like it." Showing him a nearby display, Norman told the visitor, "The real head for that [Edmontosaurus] is right there. Okay?" Then he repeated, "He's 80 percent real." Norman proceeded to take the man on a tour of the other mounted dinosaurs in the hall. As they walked around, Norman reeled off the percentage of "real fossils" in each mount. He began with the largest skeletons in the room: "The Diplodocus is about fifty-fifty. This guy [Stegosaurus] is 75 percent real. The one over here [Allosaurus], he's about 30 [percent]." With each mounted skeleton, Norman elaborated his initial answer by identifying some bones as fossils and some as casts or reconstructions. His answer satisfied this visitor, who moved on to the next exhibit. But this was not Norman's final answer to the question of the paleontological real.

Ten minutes later, a few children accompanied by another man stopped at the Bone Bar. This time, Norman initiated the interaction, asking them, "Have you ever touched a real fossil?" "Not a real one," one of the kids responded. "This," Norman told them, "is a real honest-to-goodness fossil." As he handed them an Edmontosaurus foot bone, he explained, "you have this bone in your body," much like the docents described in chapter 3 had often done. Norman bent down and showed them where the equivalent bone was in his foot. He then reaffirmed, "That's as real as you'll get. That was walking around here." One of the kids turned his attention to the Apatosaurus and Allosaurus vertebrae affixed to the top of the Bone Bar. "Are *these* real?" the boy asked. "Yes," Norman answered without hesitation, reiterating that each was "a real fossil." However, he then complicated his answer in a way he had not done with the man a few minutes earlier. "What does it feel like?" As they touched the objects with their fingers, one of the children responded: "a rock." Norman affirmed this and continued, "Yeah, so why does it feel like a rock and not a bone?" Norman taught the children that a fossil is a bone that has turned into stone through a process of material exchange—fossilization—in which "the living stuff goes out into the soil, and parts of the soil come in." Norman wanted to keep talking with these curious youngsters, but the man accompanying them thanked Norman and ushered them into the next section of the exhibition. Despite their brevity, moments like this one bring into focus the ontological distinction between the real and the unreal.[6]

The division of the real from its others is evident in a few small but telling details in Norman's interactions with these children. When, earlier, a man asked whether he was looking at "actual bones of dinosaurs," Norman did not tell him that none of the objects in the room were bones, or discuss how fossilization turns bones into stone, as he did with the group of children. Rather, Norman interpreted "dinosaur bone" to mean body fossil. More significantly, he did not distinguish between the casts and reconstructions that were not "real fossils" but made up the remainder of the "bones" in the mounted skeletons, which he grouped together as unreal. Norman reinforced this binary even when adding nuance to his discussion of the specimens in *Prehistoric Journey* while talking with the children. This time, Norman differentiated bones, as organic matter, from fossils, as inorganic matter. Yet, while making this distinction, he maintained that both were the "real" remains of prehistoric animals, unmanipulated by people. Norman's description of how fossils come into being left out the human labor involved in this process (described in chapter 1). He implied that bones and fossils are collected and preserved, but not made, by people. This suggests that fossils are natural scientific specimens, while casts, reconstructions, and models are human inventions. He did this out of fear that otherwise he would imply the validity of the creationist assertion that scientists do not have solid indisputable evidence of evolution and so manufacture objects that reflect their ideas.

A senior member of the Denver Museum's education staff told me that she did not try to mandate which words docents and staff used or did not use, but encouraged them "to point out the specimens that are real." While Norman used "real fossils" to describe the specimen in *Prehistoric Journey*, other docents preferred different terms. A docent named Betty told me she favored "actual" over "real" because "everything is real." She reminded me that some of the objects in the exhibition were "not the actual fossil" but a "copy" of one. Another docent, Arnold, routinely told visitors that "real" is "a bad word." Arnold liked to ask museum visitors, "What's the opposite of real?" Their answer, almost invariably, was "fake." Arnold retorted that nothing in the exhibition was "fake." Because of this, he preferred the term "original fossil." He often explained that most of the objects at the Bone Bar were "copies of fossils" because "the originals" were too fragile to be handled. Real, actual, original. Docents and staff disagreed about the best terms for the objects in *Prehistoric Journey*. But only somewhat. Despite the differences in their preferred terminology, the Denver Museum's staff and volunteers differentiated between two discrete categories of objects. The first group was composed of the physical remains of prehistoric animals that had been excavated from the earth. These were either body fossils, the remains of the animal's hard parts, or trace fossils, evidence of an animal's activities, such as footprints. The opposing categories included casts, reconstructions, and models.

At the Denver Museum, like most science museums, the distinction between fossils and other paleontological objects is frequently emphasized, while the differences among the latter group are rarely discussed. Although the scientific ontology groups them together, casts, reconstructions, and models differ in important ways, particularly in how they come into being. Casts are made by molding a fossil and filling it with plaster or plastic, to manufacture a replica. Reconstructions are created by sculpting, say, a right rib based on observations and measurements of the left one. Reconstructions are increasingly made with three-dimensional (3D) scanners and printers, but most of the ones displayed in the museums I have visited were handmade. Every mounted dinosaur skeleton includes some casts and reconstructions, since a complete skeleton is never found. Even when a museum has a fossil skull, they usually mount a cast because of this fossil's weight. Anatomical models are created by adding synthetic flesh, skin, scales, and, recently, feathers, to the cast of a skeleton to recreate what the animal probably looked like when alive. In making these models, paleontologists and fossil lab technicians restore casts to the form the bones would have taken before being flattened and misshaped by the fossilization process (see chapter 1). Many museums are starting to incorporate computer-generated image (CGI) models, which are already familiar to museum visitors from movies. These models explicitly embody a theory about a prehistoric species' anatomy and physiology. Yet, casts, reconstructions, and mounted fossils also materialize theory in physical form. Recall from chapter 2 that Bakker's mount of AMNH 5027 in the Denver Museum's lobby materialized his theory about *Tyrannosaurus rex*'s dynamic physiology. While fossils, casts, reconstructions, and models are conceptually separated, these different kinds of paleontological objects are combined to create a museum exhibition, sometimes even a single display.

THE REAL THING

I wondered what Scott Sampson, the famed paleontologist, public scientist, and then Denver Museum's chief curator, thought about the approach to the real I witnessed on the exhibition floor. Sampson had long confronted questions about the paleontological real as he pursued a career at the intersection of vertebrate paleontology and science education over several decades. Would he confirm my hunch that it was related to the evolution-creationism debate? As we began to talk in his elegant office, Sampson's camera-ready smile faded into a frank look of contemplation when he realized that our discussion was not going to be like the ones he regularly had with media reporters. After a pause, Sampson told me, "I'm a big believer in giving people information so they can tell whether it's real or not." Although Sampson was not involved in training the staff or volunteers in *Prehistoric Journey*, he articulated the ontology that I saw in action in the exhibit halls. Like the museum's docents and staff, Sampson asserted that it is important to distinguish "real fossils" from the other objects in the museum halls and collections because fossils embody prehistoric animals, while the other things merely represent their likeness. However, Sampson qualified this distinction by telling me that he opposed painting the cast fossils and reconstructed pieces a different color from the fossil bones that are part of mounted dinosaurs. In accord with current museum practice, he preferred indicating which objects were "real fossils" and which were not, with a line drawing on a placard below the mount.[7] Sampson told me, "I find it is kind of ugly if the bones are different colors" because "it takes away from the visual impact of the mount." He wanted people to see the animal as a whole before differentiating the real from the rest. For him, the awe-inspiring experience of viewing a dinosaur skeleton took precedence over knowledge about each of its components. Like most paleontologists I have met, Sampson believed that seeing a mounted dinosaur would lead people to understand them as prehistoric animals rather than as beasts or, worse, fictional monsters. He further hoped this experience would persuade the museum's visitors that the animal species in the world today had evolved over millions of years from the species that existed in the prehistoric world. The reality of dinosaurs' existence millions of years before humans, and the validity of the evolution of species, were of greater importance for him than the difference between a fossil and a cast or reconstruction.

I pushed Sampson to elaborate why he advocated distinguishing "real fossils" given this aim. He stood up, walked over to the bookcases lining his office wall, and took an enormous Hadrosaur vertebra fossil, which he had excavated himself, off the shelf. As he handed it to me, he told me how he often takes this fossil with him when he visits schools. When he hands it to students, he tells them:

> You're now holding in your hands the tail vertebrae of a seventy-five-million-year-old duckbill dinosaur. Through this hole went the spinal cords. It went all the way up to the brain—it connected the tail to the brain—to allow this animal to think.

Sampson wanted me to feel the fossil's weight—both literally and metaphorically—as he spoke. He believed that its materiality supported his assertion that the information about this vertebra "means a whole lot more when you are holding it, and you can feel the heft of it." I found that it was remarkably heavy, much heavier than a bone or cast. I also noticed that it appeared exquisitely well-preserved. It showed no sign of having been infused with plastic or otherwise modified. It sure felt real to me, someone raised with the secular Euro-American conception of the real. As I held it in two hands, Sampson continued:

> It is amazing to hold the real thing. If that were a cast, you would say, "that's pretty cool," but it is not nearly as meaningful as actually holding the fossil in your hands. You are making direct contact with the ancient world by holding it in your hands.

Sampson implied that the age, heft, and flawless beauty of this stone give it meaning that a replica could never possess, no matter how accurately it reproduced every curve, bump, and indentation of the bone it came from. For him, as for the docents upstairs in *Prehistoric Journey*, there was an absolute distinction between fossils and everything else in the museum's collection. Only fossils were real paleontological specimens, and only real specimens could embody prehistoric animals. Sampson's act of handing me the Hadrosaur vertebra as he spoke expressed his view that physical contact with "the real thing" would convince me of this more than words alone could do.

Sampson further explained his approach to the paleontological real by analogizing that a dinosaur fossil is like the *Mona Lisa*. "People line up for hours to be able to walk up and look at it, even though they don't even get that close to it." Despite the fact that the painting is behind glass, people "want to feel like they touched" it. "It is that direct connection that I think is important." Few people eagerly wait for hours to see a reproduction of a work of art or a model of a dinosaur. As I pointed out in chapter 4, people did not line up to see the bronze Dino-Sue at Disney World. In contrast, millions of people have waited to see Sue's mounted skeleton at the Field Museum in Chicago because it is 90 percent "real" fossil. For Sampson and many others, what separates a fossil from its others, and what makes it a powerful communicator of science, is a seemingly spiritual "direct connection" to prehistoric life. I could, of course, make direct contact with prehistoric matter by touching any old rock. The pebbles outside the museum would do. But those rocks do not embody the prehistoric world the way the nearly complete skeleton of Sue the *Tyrannosaurus rex* or Sampson's single Hadrosaur vertebra does. As the paleontologist who had excavated the fossil he handed me, Sampson could trace the material flow, over millions of years in one geographic location, from the bones of an animal living in the Late Cretaceous to the stone sitting in his twenty-first-century office. He could not draw such a line back in time for a cast or a reconstruction. These objects have been too greatly transformed by human processes for him to trace a physical path from prehistory to the present, from an ancient animal to a contemporary object. In this ontology, fossils are real natural specimens, while casts are not, no matter how well they reproduce the bones of prehistoric organisms.

The distinction that Sampson and many others make between a fossil and its others is not only a material, temporal, and spatial one; it is also an ontological distinction between living beings and nonliving things, and between science and fiction. A fossil such as Sampson's vertebra embodies an individual, sentient being, an animal both like and unlike ourselves, which lost its life.[8] Despite the transformation of the dinosaur from animate being to inanimate object, fossils are understood as the remains of a singular being that once walked the earth. Sampson sees the moment when a person touches a fossil as a transhistorical connection that convinces them of the evolution of life over millions of years. For him, this simultaneously physical and affective connection between a person and a prehistoric being is more powerful than dispassionate arguments about the isotopic dating of rocks. The power of stone fossils lies not in rational explanations of the mechanisms of evolution but in the act of touching "the real thing" and getting goosebumps from its emotional intensity.[9] Touch, Sampson emphasized, enables people to recognize dinosaurs as moving and thinking animals in many ways like themselves. When a person holds a bone-turned-stone, Sampson claimed, it creates a unique relationship between a human living today and an animal living millions of years ago. Plastic casts, in this view, do not generate the experience of awe, authenticity, and veracity.

Sampson helped me to appreciate why the Denver Museum docents, as well as myriad other experts and nonexperts, vehemently uphold an absolute distinction between fossils and casts, reconstructions, and models: Touching a stone leads people to *believe* in evolution, even if they do not fully understand it. This approach frames evolution as a truth arrived at through an intense somatic and affective experience with a singular and vital object, rather than through argument. As I explain later in the chapter, this way of distinguishing fossils from other paleontological objects confounds the distinction between scientific theory and religious belief, even as it reinforces the separation of real fossils from other paleontological objects. When I contemplated the problems with how paleontological objects are classified, I began to see a novel alternative in Matthew Mossbrucker, the curator and director of the Morrison Museum. He put into words an approach to fossils, casts, and the paleontological real that I had seen, but had not heard, in the course of my fieldwork elsewhere.

BONE-CLONES

When I asked Matthew how he answered the question about the paleontological real, I immediately realized I had struck a nerve. With an exasperated laugh, he blurted out, "'Is it real?' is the most poorly thought-out question ever!" Matthew elaborated that it creates "a false binary, a yes or no answer, when of course the answer is more complicated than that." He told me that after becoming frustrated with repeatedly hearing this question, he began to wonder why people asked it. "As a museum educator, I have to take a step back and try to understand what they are asking." Instead of answering, he turned the question around and asked people to explain what they meant by "real fossil." From this, Matthew figured out what people really wanted to know: "Is it a model or a sculpture of some sort? Is it animatronic?" Was the object in front of them "something human-made, or is this the original specimen?" In other words, was it solid and unmediated evidence of a prehistoric species, or not? In response to the suggestion that the Morrison Museum's dinosaurs were animatronic, Matthew quipped, "Like our tiny museum can afford robots!" Humor aside, he had given great thought to the question of the real and had developed a novel approach as he took over the directorship of the Morrison Museum and expanded its collections.

The paleontological real was not merely a theoretical question for Matthew but also a practical one, as he worked to turn a "homegrown" town museum into a science museum focused on the prehistory of Colorado. The museum had several unique specimens from Arthur Lakes's excavations, and Matthew had borrowed a few more from Marsh's collections at Yale's Peabody Museum. Working with Bob Bakker, Matthew had found some spectacular trace fossils in boulders at Dinosaur Ridge, which the park administration let them put on display at the Morrison Museum. But the museum did not have the funds to purchase more fossils like these, so almost all the objects Matthew bought to fill out its exhibits were casts. Unlike Scott Sampson, however, he conceived of these casts as different from, but no less real, than fossil specimens. Matthew explained:

> I've developed a little bit of canned speech over the years because as soon as you say, "well, that's not the original fossil," people devalue the object as some sort of papier-mâché creation made in ten minutes. Anyone who has done any fossil molding and casting knows that—from the process of exploration and excavation, to the preparation phase, and then to molding and casting—tens of thousands of dollars and years' worth of time came together to make that object. By saying "oh, it's just a fake" it devalues that object. That upsets me.

Talking with visitors, creationists included, had taught Matthew that "original fossil" means different things to different people. He told me that when he previously responded to the question "Is it real?" with the answer "Yeah, that's the original fossil," he found that some people responded, "Oh, so it's just a rock now." This retort bothered Matthew as much as the comments that equated casts with fakes. They both signify the hegemonic binary of the real and all its others, but in different ways.

Matthew's experiences with visitors' questions about what is real led him to begin calling research-grade casts "bone-clones." I had heard the museum staff and volunteers use the term *bone-clone* many times. One could understand it as a reflection of Matthew's desire to defend the Morrison Museum's legitimacy as a science museum following his decision to buy casts. Yet my close examination of museum tours and docents' explanations illustrate that this conceptualization of the paleontological real is not simply an attempt to bolster the museum's legitimacy, finances, or prestige.

Matthew Mossbrucker contested Scott Sampson's claim that bone-clones are not "the real thing" with several arguments. First, research-grade casts enable paleontologists to gather a collection of fossils and compare them side by side. Original specimens are usually not permitted to be taken from their repository, while casts can be sent in the mail and handled repeatedly. Second, Matthew pointed out that even a single bone-clone could provide important scientific information about a prehistoric animal. For example, one can use research-grade casts to investigate how a prehistoric animal's ligaments attached to its bone or whether the bone regrew after a break or puncture wound. Matthew told me that it is sometimes easier to identify sutures (bone junctions) and pathologies (diseases) in casts than in original specimens. Furthermore, casts can more accurately indicate an animal's morphology than original specimens can. As chapter 2 explained, bones often become flattened during fossilization. Casts, unlike original fossils, can be reformed to eliminate the warp caused by geological forces. They can then be assembled and used to study the posture and movement of the once-living animal. Thus, while only stones can be used to figure out *when* an animal lived (through isotopic dating), research-grade casts can be used to study *how* it lived. In chapter 1, I showed that fossil casts and original fossils adhere to different scientific standards. In the current chapter, I contend that they can reveal different, but equally valid and important, scientific data.

Another of Matthew's arguments for why research-grade casts and fossil specimens are equally real rests on the commonalities in *how* they come into being. The process of fossilization replaces animal bone, and sometimes soft tissue, with minerals. Bacteria, many of which were in the animal while it was alive, travel through its bone passages, converting the animal's biochemicals into molecules and atoms that form minerals. Bacteria also enlarge the bone passages, enabling water to enter and deposit minerals.[10] These minerals then harden into rock. Matthew understood casting as a continuation of the fossil exploration, excavation, and preparation processes. Casting replaces the minerals with plaster and plastics. In both the fossilization and casting processes, the overall shape and internal structures remain largely the same while the material composing the bones changes. Returning to Matthew's response to people who brought up the possibility that there is a Brontosaurus living in the Congo, notice that he used the word *real* to mean alive in the world today. By linking *real* with *evidence*, he asserted that the real is not primarily determined by the material composition of the object but by whether it was the result of scientific research, which included the rigorous collection and analysis of data.

The Denver Museum scientists and educators excluded humans from their explanations of how fossils are made, even though many of them were engaged in this work themselves. In comparison, Matthew highlighted people's involvement in fossil excavation, preparation, *and* casting. In this conception, casting is a part, not a perversion, of scientific knowledge creation. In a later discussion, Matthew defined a fossil as a three-dimensional "hypothesis" of a prehistoric animal. As I described in chapter 1, the processes of fossil excavation and preparation not only introduces non-lithic materials into the stone but also form it into a discrete object, distinct from its surrounding sediments. The coming into being of "original" fossils thus involves human actions as well as biological, chemical, and geological forces. Both fossils and casts are real and hypothetical simultaneously. Although Matthew used different words to articulate it, his approach to fossils and casts recognizes that both are created by the conjuncture of human and nonhuman forces. This suggests a dynamic relationship between fossils and casts, rather than a hierarchy of truth and value. Further, this conceptualization better reflects what actually occurs in paleontological quarries and laboratories than the binary of real-unreal does.

It is important not only to recognize that some casts have scientific value, but also to acknowledge that some fossils have none. While United States law requires that all dinosaur fossils be stored in authorized repositories, most will never be used in research. As chapter 1 mentions, the buttes near the Badlands Paleontological Station were so strewn with fossils that it was difficult to walk around without stepping on some. Paleontologists call these "frags," broken pieces of fossilized bone that are of no use for research. But the same was true of many of the fossils we painstakingly dug out of the rock. Some were too weathered to have the features paleontologists need to investigate pathology, morphology, or physiology. Other fossils were well preserved but nonetheless useless because they were "isolated," that is, they could not be linked to other fossils from the same animal or have their species identified. Sampson could tell me no more about the vertebra on his office shelf than that it belonged to some sort of Hadrosaur. It resided on his office shelf because, despite being spectacular, it was not useful for advancing paleontological knowledge. When I was at the Badlands Station, a student unearthed a similarly impeccably preserved but isolated vertebra that was useless to the paleontologists there. A few days later, a volunteer cracked open a broken piece of rock to reveal a plant imprint that was of great interest to a visiting paleobotanist. The distinction between these two objects is not between the real and the unreal, but between scientifically valuable and scientifically meaningless fossils.

THE POWER OF TOUCH

Everyone I spoke to agreed that physical touch is crucial for paleontology, both for conducting research and for transmitting its findings. Yet there are significant differences in how it is carried out at the Denver Museum and the Morrison Museum, which are related to their differing approaches to casts. In Denver's *Prehistoric Journey*, the exploration stations, such as the Bone Bar, are the only places where visitors are allowed to touch paleontological objects. The majority of these objects are casts, although not the vertebrae discussed earlier in the chapter. In contrast, the museum's fossil prep laboratory was enclosed by windows so that museumgoers could observe and talk with fossil preparators but could not touch the fossils. At the Morrison Museum, visitors of all ages were invited to "pet" nearly all the objects on display. Only the fragile fossils that Lakes excavated were kept under glass. Furthermore, if a preparator was working in the museum's lab during a tour, visitors were invited inside to try their hand at fossil preparation. One at a time, they donned safety glasses, sat at a workbench, and attempted to chip away the lithic matrix around a fossil (staying far from the fossil itself). Sometimes it was Dr. Bob who taught visitors how to prepare fossils, often without them knowing their teacher was a famed paleontologist. Many other times I watched Matthew rest his hand on top of the hand of a hesitant young visitor and guide the vibrating airscribe into a groove to help them remove some rock.

The Morrison Museum is, in Matthew Mossbrucker's words, "a site-specific and handcrafted educational experience" that "wouldn't work if [visitors] couldn't touch the fossils." The same could be said about Dinosaur Ridge. Matthew explained the benefits of this approach: "You can see lightbulbs turn on when you allow people to touch the casts or fossils." His equation of fossils and casts counters Scott Sampson's conviction that handling stone has an effect that plastic cannot replicate. In his view, only "real" fossils speak directly to people, even though most of them can only be handled by experts. In comparison, Matthew argued that bone-clones spark people's scientific curiosity just as much as fossils do. In particular, he asserted that touching either one helped people to appreciate "the history of life, its development, and how long it has taken to get to where we are right now." Mossbrucker's conception of the magic of touch depends less on fossils' ability to convert people to believe in evolution than Sampson's does. Matthew's approach relies on paleontological objects' ability both to inspire a desire to understand evolution and to help explain it in an approachable, concrete way.

Tours of the Morrison Museum put Matthew's novel approach to the paleontological real into action. Take, for instance, one morning when the fourth grade from a nearby elementary school visited. It was an all-hands-on-deck moment as there were more than a hundred students in the small museum at the same time. As the children entered in groups, a docent invited them to "pet Stan" with two fingers, as docents did on nearly every tour I attended. I decided to shadow Matthew at the Triceratops skull on the second floor, since he rarely had time to give tours himself. Almost immediately, a girl pointed to the skull and asked with surprise, "Is that the actual—." She had not even finished her question before Matthew told her and the other students:

> What this is, guys, is not the fossil itself, but it's exactly like it. It is called a bone-clone, or a cast. I could take measurements from the cast. I could look at parts of the cast with a microscope and see where the cells were, at least the texture of the cells, and the canals on the bone. Think of an object like this as a clone of the original bone.

Here, Matthew stated that some casts were as real as fossils. He then informed the students that they could "pet" this Triceratops skull—like the *Tyrannosaurus rex* skull at the museum entrance—because they were also "bone-clones," not "original fossils." This implied that these casts were better than fossils. Before he let the children touch the Triceratops, Matthew reinforced this probably unfamiliar argument: "When you look at an object like this cast, it is real." The children clamored to touch the Triceratops skull, undeterred by the knowledge that it was a cast. After joining tours and events at numerous museums, I realized how much touching bone-clones engendered excitement about learning science and communicated knowledge, particularly about evolution, more effectively than viewing fossils alone. At the Morrison Museum, it became clear that this was not because people mistook casts for fossils. This realization led me to question Sampson's claim that touching a cast is "not nearly as meaningful as" touching a fossil.

SECULAR-SCIENTIFIC BINARIES

The Morrison Museum is the exception. The binary division between fossils, on the one hand, and casts, reconstructions, and models, on the other, can be found in most science museums because it reflects the Euro-American ontology that emerged with secular science beginning in the seventeenth century. Among the most fundamental of this ontology's binary pairs is what Philippe Descola calls "the great divide" between nature and culture.[11] In this understanding, nature is the other not merely to culture but to the human broadly construed. Nature is everything in the world that humans ostensibly had no part in making. Rock is a quintessential example of what is natural in this context. In contrast, plastic is understood as unnatural and associated with human creations that are untruthful illusions and unreal fictions. For example, an educator at Dinosaur Ridge told me that when she holds a fossil cast, she "will always get [asked] 'Is that real?'" But when she shows visitors the fossil footprints or a body fossil embedded in the rockface, "they never ask that." Instead, when she tells visitors they can touch "real dinosaur fossils," many remark, "Wow, that's really cool!" The secular Euro-American ontology's bifurcated conception of the world manifests somewhat differently in science museums than in nature parks like Dinosaur Ridge. Here, things that grew or developed on their own are "specimens" while those that were made by people are "artifacts."[12] They are normally housed in separate collections and displayed in separate exhibitions. When a cast or reconstruction is included in a paleontology display, it is marked as different from "real" fossils. Fossils are placed behind glass or otherwise set apart to tell visitors they should not touch them, while visitors are often invited to touch casts and reconstructions. At science museums and myriad other sites across the United States, specimens representing nature are separated, physically and ontologically, from artifacts representing culture. Yet mounted dinosaur skeletons show us that the practice of paleontology muddles this distinction.

An additional binary in the Euro-American ontology separates specimens and artifacts from works of art. Specimens and artifacts are representatives of something greater: specimens, of a natural group, such as a species, and artifacts, of a cultural group, such as a Native American community. Both specimens and artifacts are used to identify the commonalities within a group and the differences between that group and other ones. For example, the fossil remains of a single animal that Lakes sent to the Yale paleontologist Marsh in 1877 became the holotype—the specimen that defines a new species—for *Stegosaurus armatus*. While specimens and artifacts are important because of the commonalities they share with others in their group, artworks are normally conceived as objects that express individual human creativity and are valued for their uniqueness. Fossil specimens inspire awe of evolution, while works of art inspire awe of individual genius. Despite this difference, Sampson analogized a mounted dinosaur skeleton to the *Mona Lisa*. He thereby suggested that it possesses authenticity and "aura" like an original work of art.[13] In Sampson's view, fossils give people an authentic and unmediated connection to the prehistoric world. However, he also asserted that the Hadrosaur fossil I held in his office was ostensibly created without human involvement, while casts and artwork were made by people. In the Euro-American ontology, the paleontological real is thus constituted by the absence of the human, while the artistic real is constituted by its zenith. This inconsistency in the definition of the real demonstrates that paleontological objects do not neatly fit into a universal binary opposition between the real and unreal. Moreover, both oil paintings and fossils are coproduced by wealthy patrons, geological and biological processes, and the unrecognized labor of human and nonhuman assistants.[14] The secular Euro-American conception of the world is so deeply embedded in paleontology that it is maintained not only by docents and educators who work on the museum floor but also by scientists, staff, and volunteers who conduct fossil excavation and preparation. They discounted their own labor and presented fossils as specimens that they found, but did not help make. This included the eminent paleontologist who served as the Denver Museum's chief curator. Why do they uphold this ontology despite their own experiences that countered it? I suspect in part because of their grave concern about another influential Euro-American ontology: creationism.

The secular ontology not only clearly divides nature from culture, and specimens from artifacts and artworks, but it also separates secular scientific theory from religious belief, and thus evolution from creationism. At the heart of Sampson's and others' insistence on an absolute distinction between real fossils and everything else is the worry that to challenge this distinction would undermine the theory of evolution. For many paleontologists, science educators, museum professionals, and volunteers, blurring the line between natural specimens and human creations threatens the status of fossils as the prime evidence for evolution. In this view, acknowledging that people have any role in creating fossils opens the door for evolution to be "just a theory," and thereby one that is akin to "creationist theory."[15] Yet Sampson inadvertently blurred the line between scientific theory and religious belief when he argued that touching lithic fossils had the power to convert skeptics into believers in evolution. This is not simply a mistake on his part. It is a problem with the scientific ontology, which fails to recognize that a definitive line between theory and belief is never possible because science always involves affect as well as reason.

Paleontological objects illustrate a fundamental problem with the dichotomies—real and unreal, nature and culture, growing and making—in the secular Euro-American ontology. While some of the docents in *Prehistoric Journey* were uncomfortable with what this ontology implies about casts, they lacked an alternative way of talking about them. They thus asserted that casts are not fakes, but neither are they "real fossils" or specimens. Yet my experiences conducting paleontological fieldwork and labwork showed me that both the materiality of fossils and the practice of paleontological research challenge the binaries that are used to discuss them.[16] As I have demonstrated throughout *Making Our Beasts*, fossils did not simply grow in the Mesozoic only to be found in the present era. Chapter 1 analyzed the heterogeneous nonhuman and human forces that must come together to turn a prehistoric animal into a precious fossil, or into worthless dust. If a fossil makes it to a preparation laboratory, the lines between finding and making, human and animal, and nature and culture are further blurred. As chapter 1 describes, fossil preparators enlist plastic consolidants to hold pieces of rock together. If the resulting fossil becomes part of a mounted dinosaur, stone, plastic, metal, and other materials are combined to create what Lukas Rieppel calls "mixed-media installations."[17] But they are *not* works of art, equivalent to the sculptures found in art museums. Nor are they artifacts, like the tools, pottery, and cloth found in anthropology exhibitions.[18] They are scientific specimens.

THE PALEONTOLOGICAL CONTINUUM

My experiences participating in tours of the Morrison Museum, conducting paleontological research at the Badlands Station, and talking with Matthew led me to reconsider the question motivating this chapter, "Is it real?" When people ask this, I think they are really posing other questions: Is the object before me the result of rigorous scientific research? Was there once an animal with this morphology and ethology living on earth? In the remainder of this chapter, I articulate an alternative framing of paleontological objects that refuses the real-unreal binary. Neither Matthew nor anyone else explicitly named or described the continuum I describe below. The words are mine, but the ideas are implicit in the practices I saw in action at paleontological quarries, laboratories, and sometimes in museums. While experts and laypeople spoke in binaries, their interactions with paleontological objects revealed a nondualistic, multifaceted continuum of paleontological objects. Conceiving of the paleontological real as a continuum more accurately reflects actual scientific practice than the dichotomy between real and unreal, and it can effectively support evolutionary theory against the challenges posed by creationists.

A continuum approach to paleontological objects implies that they can be arranged along multiple intersecting spectra, each based on a particular set of characteristics, rather than placed into binary categories. The rainbow, for instance,

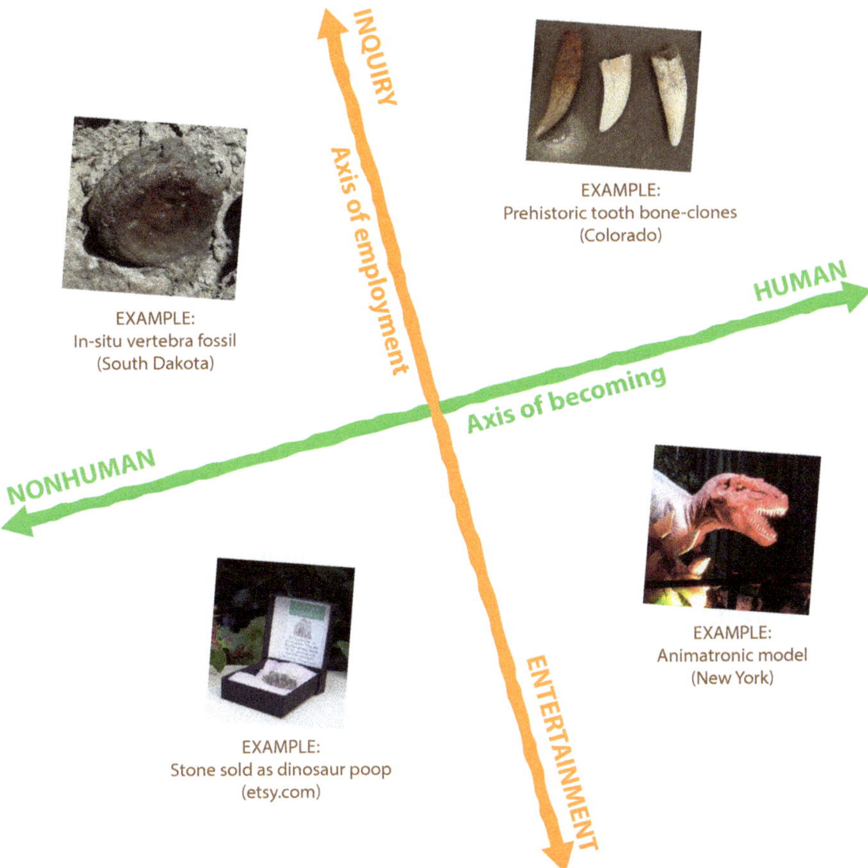

FIGURE 18. The paleontological continuum. Author's sketch of the continuum of paleontological objects.

is the spectrum of wavelengths of light that human eyes can perceive. It is a continuous range of light, in which colors blend into each other without breaks between them. The categories of red, yellow, and blue represent human efforts to divide the light spectrum into discrete units. Although a continuum does not have inherent divisions, it does have a characteristic, or a set of characteristics, that organizes it. For the rainbow, it is wavelength. What would be the principle for organizing a continuum of paleontological objects? I sketch them in figure 18. This continuum has two axes, each with a wide middle ground between its poles. Because an object's placement is based on multiple criteria, the paleontological continuum is neither linear nor stable. It is more like a wavering rainbow in the sky after a rain shower than the color spectrum on a piece of graph paper under a prism.[19]

The first axis represents an object's process of becoming, ranging from human to nonhuman. The second axis represents an object's subsequent employment, extending from research to entertainment. Rather than separate real fossil specimens from all other paleontological objects, this continuum expresses that paleontological objects represent neither proven fact nor pure fiction. My sketch of a continuum of objects in paleontology somewhat resembles the narrative field in primatology that Donna Haraway, following Judith Merril, calls "SF" to include science fiction, science fantasy, speculative fabulation, and science fact.[20] Haraway shows how scientific articles and textbooks, on the one hand, and science fiction stories and films, on the other, "respond to each other across [the] structured space" of SF.[21] The paleontological continuum I describe below extends Haraway's insights about discourse to material objects by identifying the common field—a continuum—shared by dinosaur fossils, casts, models, and even toys, which recognizes that all these paleontological objects are more-than-human creations. Let me elaborate.

The axis of becoming, drawn in green in figure 18, represents the more-than-human process through which paleontological objects come into being. Unlike a binary approach, a continuum reinforces that all paleontological objects emerge from the remains of prehistoric organisms transformed into inorganic matter. Fossils, casts, and models are shaped, first, by biological, chemical, and geological forces and, later, by humans and tools, but to different degrees. At one end of the green axis are objects that have come into being through physical processes that humans had little part in shaping. For example, fossil-rich geological formations, such as Hell Creek, developed over millions of years through physical processes that unfolded before humans entered the scene. Human practices, such as building roads, producing acid rain, and drilling for oil, have also affected these rock formations. But the nonhuman forces play a much larger role in determining the material composition of a Hell Creek fossil than humans do. Nonetheless, the continuum approach acknowledges that even this fossil, which seems unchanged since the Mesozoic era, is actually the result of chance occurrences over millions of years, as well as purposeful actions by people in the present era. A plastic dinosaur femur is no less real than a stone one, but the difference in their materialities is significant.

Objects near the other end of the becoming axis are ones whose coming into being owes a great deal to deliberate human activities combined with nonhuman forces. A plastic model dinosaur, for instance, is made of the remains of prehistoric zooplankton transformed into petroleum by the heat and pressure of geological forces, then, millennia later, extracted, refined, processed, and manufactured into something that bears almost no resemblance to the prehistoric matter it came from. It seems relatively easy to identify where the objects I encountered in my fieldwork fit along the continuum's axis of becoming by determining their material composition. The Hadrosaur vertebra in Sampson's Denver Museum office sits near one end, and the Triceratops skull in the Morrison Museum at the other. Repaired fossils and mounted skeletons fit between them. The green axis of the paleontological continuum recognizes that even the most complete dinosaur skeletons contain some plastic, but casts have significantly more than fossil specimens do. This effectively dissolves the absolute divides between real and fake, as well as between growing and making. But it does so without collapsing the important differences among the processes and actors that form different paleontological objects. While the green axis recognizes that every object in the contemporary world has been shaped by humans to some extent, looking only at materiality would group together the research-grade casts in the Morrison Museum and the dinosaurs in Disney's DinoLand U.S.A. because they are both plastic. This could inadvertently reinforce the misconception that stones are inherently natural, authentic, and enduring objects that embody scientific truths, while plastic things are intrinsically fabricated, malleable, and thus unscientific. Material composition alone is thus insufficient to understand how an object came into being or to place it in the paleontological continuum.

Geography is another key component of a paleontological object's process of becoming, and it thus also defines an object's placement on the green axis of the continuum. Objects on the nonhuman side of the becoming spectrum are site-specific, while objects on the human side are not. This is because a bone only turns into stone under particular conditions in a specific location. A plastic cast, in contrast, is an amalgamation of matter that comes from many locations, none of which determine the cast's form. While a toy cast is not tied to the natural history of any particular place, a research-grade cast most certainly is. The continuum approach accepts Scott Sampson's assertion that a "real fossil" can be traced back from the present to prehistory in a specific location, yet it recognizes that some casts can, too. However, it does so without negating people's involvement in a fossil's coming into being. Although the material composition of a research-grade and a toy cast is the same, the processes of their becoming follow distinct paths, across different spaces, and involve people with distinct expertise.

Casts illustrate that the way an object comes into being matters as much as where and from what it is made. As Matthew explained to me, the coming into being of a research-grade cast is an extension of paleontological excavation and preparation. I experienced this when I helped make a cast of a footprint in the Badlands. It could only come into being after the process of prospection, excavation, and preparation of the rock surface had taken place, but it had to occur before subsequent erosion wiped the trace away. Even though the timing and conditions were satisfactory, it still took several attempts before the rock and silicone rubber, as well as the expert technician's and my hands moved together in the right way to transfer all the ridges and crevices from stone to silicone before it hardened. The coming into being of this cast thus involved a precise combination of physical events and skillful action, each in the right time and place. The process of making this cast for research is also different from the one used for making a cast for play, although both involve a mix of human and nonhuman labor. The making of the cast of the footprint in the Badlands involved much more personal care than the production of a cast in a factory.[22] Even when research-grade casts are produced as commodities by commercial fossil-hunting operations like those described in chapters 2 and 4, the people who excavate, prepare, and cast fossils expend significant time, energy, and money in caring for these objects, despite it not always making financial sense to do so.

The other axis in the paleontological continuum, the axis of employment, extends from inquiry at one pole to entertainment at the other. I have sketched it in orange on the diagram above. I mean *employment* in both its transitive and intransitive senses—that is, how objects are utilized by people to advance their objectives (like a hammer is employed by a construction worker) and how these objects do work (like a saw makes a cut). Near the inquiry pole of this axis are things deeply engaged in generating and transmitting knowledge of the prehistoric world and the evolution of life. These include fossil specimens, research-grade casts, and anatomical models. You are likely to encounter these objects in paleontological research sites like the Badlands Paleontological Station, in laboratories like the Denver Museum's prep lab, and in paleontological exhibits like those at the Morrison Museum. All these objects are engaged in both scientific research and teaching.

Near the entertainment pole are things intimately involved in creating an imaginary world that did not exist in prehistory. This is the world of toys, but it includes much more. I label this end of the axis *entertainment* rather than with a word like *enjoyment* to recognize that scientific inquiry can be as pleasurable to some people as a video game is to others. At Disney's DinoLand U.S.A., pleasure is strongly linked to pretending. Yet my fieldwork at field stations and museums showed me how enjoyable other kinds of engagements with "dry ol' bones" can be. At Dinosaur Ridge, for instance, visitors, young and old, enjoyed searching for fossils still embedded in rock so much that park guides had to repeatedly remind them not to climb up the fragile rockface to touch them. The excitement of fossils can last for decades. Several of the volunteers I worked with at the Badlands Station chose, year after year, to spend their vacations doing the seemingly grueling labor of excavating and cleaning fossils. They did this because they enjoyed the intellectual and physical challenges of paleontological research. Creativity, imagination, and pleasure exist all along the employment axis.

Geography is as important to the axis of employment as it is to the axis of becoming. The orange axis distinguishes two materially identical fossils that are being used for different purposes. For example, a bone-clone is deployed for inquiry at the Morrison Museum, unlike an identical one located in the Creation Museum, the Young Earth Creationists' institution in Kentucky.[23] The axis of employment, likewise, differentiates 3D and CGI models that help scientists to test their theories about the anatomy, physiology, and ethology of prehistoric animals from 3D and CGI models involved in making the *Jurassic Park* movies. At the same time, the axis of becoming separates these CGI dinosaurs from the CGI models that Sampson helped construct for *Walking with Dinosaurs*, described in chapter 3.

I label the upper pole of the employment axis *inquiry*, rather than *research*, to avoid preemptively separating the production, dissemination, and reception of scientific knowledge. In some cases, such as research-grade casts, objects are engaged in both generating and conveying knowledge. However, other objects that are part of paleontological research are not the best communicators of it. The mounted fossils in museums often communicate outdated theories about prehistoric animals. The *Tyrannosaurus rex* standing upright on one foot in the lobby of the Denver Museum is a case in point. In comparison, the CGI dinosaurs in *Walking with Dinosaurs* convey current scientific knowledge of dinosaur posture, movement, and behavior. Edutainment objects, much like the edutainment experiences discussed in the previous chapter, thus lie at different points along the wide middle of the employment axis. As we will see below, some "fake" objects can be surprisingly effective science teachers.

Science and fiction, growing and making, specimen and artwork, nature and culture are converging areas on the paleontological continuum, not opposite categories. The continuum accounts for the materiality, process, and geography of their formation on one axis, and accounts for the context of their subsequent employment on the other. Both axes in the paleontological continuum reveal that fossils, casts, models, and toys all embody creativity, precision, expert labor, and fortuitous occurrences, but in different ways and to different degrees. The continuum similarly recognizes that biological, chemical, geological, and social forces are variably important. Therefore, this continuum approach does not judge objects as scientific or unscientific, because, like *east* and *west*, these terms depend on where you stand. When situated amid the mounted dinosaurs in *Prehistoric Journey*, the objects in the museum's paleontological labs are more scientific because they are currently involved in research projects. Yet, when standing in front of a row of casts and models in the Denver Museum gift shop, the mounted cast of AMNH 5027 in the lobby is the more scientific object. The multifaceted, processual, and contextual character of the paleontological continuum recognizes that a wide range of objects are integral to generating and conveying scientific knowledge.

A SPECTACULAR AUTOPSY

Early in my fieldwork I learned, to my surprise, that many paleontologists love "dino flicks." I witnessed their excitement most vividly when Universal Studios released the fourth film in the *Jurassic Park* franchise, *Jurassic World*. Tickets were sold out before I could get one to join the group of paleontologists, fossil preparators, and other proud "dino nerds" who invited me to see it with them on opening day. But I met them in the lobby when they came out of the theater with big smiles on their faces. A Dinosaur Ridge staff member repeatedly gushed that the movie was "cinematographically amazing" and "so much fun." While many media critics panned *Jurassic World*, sales figures demonstrate that my paleontology friends were hardly alone in enjoying it.[24] Ever since the first *Jurassic Park* film was released in 1993, CGI dinosaurs have excited large and diverse audiences. Why? The critic Alexander Huls suggests that the original *Jurassic Park* was the first film in which "the technology [had advanced] to the point where the seam between illusion and reality completely disappeared . . . Sam Neill and Laura Dern's stunned awe upon seeing a real-looking brachiosaur on its hind legs eating from a tree was a perfect mirror of our own. Audiences believed."[25] What did they believe? That prehistoric animals once looked like they do on the screen today? Or, maybe, for a short time, that dinosaurs were really living on an island that people could visit? Huls's explanation for the success of the *Jurassic Park* franchise returns us to the question about the relationship between awe, belief, and science, and to the dichotomy of the real and the unreal in paleontology.

Some scientists expend significant effort denouncing the inaccuracies of the dinosaurs in popular media, but the paleontologists I worked with did not. They envision dinosaurs as a "hook" to get more people interested in paleontology and in the evolution of life on earth. For them, media dinosaurs are not vehicles for spreading either scientific truth or falsehood, but for stimulating curiosity about long-extinct animals and their prehistoric world. These paleontologists have little interest in spending time correcting the scientific errors in dinosaur movies and television shows. They are also relatively unconcerned with doing the "boundary-work" of delimiting who and what can present paleontological knowledge.[26] As I discussed in chapter 3, they are confident that adults and children understand that animated dinosaurs are pretend. In fact, numerous paleontologists report having been inspired to pursue paleontology by the dinosaur movies of their childhood. They hope that today's media will, likewise, inspire young people to take an interest in paleontology too. In their view, it does not matter that many of the dinosaurs in movies are inaccurate because they encourage people to pursue scientific knowledge and even to contribute financially to supporting it. The extraordinarily oversized and extremely vicious Velociraptors in *Jurassic Park* led people to come see the Velociraptor fossils in science museums. Furthermore, several paleontologists told me that their research had received a funding boost as an indirect result of the *Jurassic Park* movies. This made paleontologists more eager to collaborate with media-makers. While scientists in other fields tend to lose prestige when they work in popular media, paleontologists do not. Both Robert T. Bakker and Scott Sampson have worked on popular dinosaur shows.

Matthew Mossbrucker got his chance to engage in media-making when he was, quite unexpectedly, invited to participate in National Geographic's television special, *T-Rex Autopsy*. He told me, "I literally was sitting in this chair, and I got a random phone call asking if I wanted to participate in dissecting a giant rubber Muppet and talk about anatomy and physiology." The show's premiere was timed to capitalize on the "dinomania" that *Jurassic World* was anticipated to inspire. Matthew's participation in *T-Rex Autopsy* gave me the opportunity to ethnographically explore popular dinosaur media, rather than simply interpreting the television show as a cultural studies scholar would. I have combined my close reading of the show with the perspectives I gained from watching it with others, discussing it with Matthew, and following its news coverage and comments on social media. I joined Matthew and a lively crowd of his family, friends, and colleagues from the Denver area paleontology scene to watch the premiere of *T-Rex Autopsy* on the television in Paleo Joe's, the pub in the paleontology-themed hotel I described earlier in this chapter. Just before it started, Matthew warned us that the show would be a "splatter fest." He later commented to me that he did not anticipate quite how much of a gory "bloodbath" the final cut would turn out to be.

National Geographic's show follows Matthew and three other scientists as they dissect a seven-ton, forty-six-foot-long *Tyrannosaurus rex* cadaver.[27] Drawing from the "drama-doc" tradition, the National Geographic team combined techniques from documentary film and reality television in order to take "the camera to places where the camera, apparently, can't go."[28] Like Disney's *True-Life Adventure* films and its thrill ride DINOSAUR, discussed in chapter 4, *T-Rex Autopsy* made science an exciting, and violent, adventure, yet the blood-soaked drama played out on a set resembling a surgical theater, not the "Wild West." The television special shows four scientists cutting into flesh and getting covered in blood, sitting inside a dinosaur's abdomen and pulling out yards of intestines, reaching up its cloaca and extracting its eggs, and other destructive acts. Interspersed between these scenes are CGI sequences of *Tyrannosaurus rex* hunting in a Cretaceous jungle that highlight *Tyrannosaurus rex*'s ability to rip apart flesh and swallow it whole. The scientists too appear to take pleasure in dismembering a dinosaur. The show's storyline could be considered a necrophiliac rape.

Although the cadaver at the center of *T-Rex Autopsy* can easily be dismissed as a "fake," National Geographic nonetheless pledged it would give viewers a scientifically accurate glimpse "under the skin of a full-size T-Rex for the first time ever" and "reveal how the 65-million-year-old beast may have lived." Like cartoons and many other animal television shows, *T-Rex Autopsy* focused on an individual animal. It also worked to "inspire awe and respect through the pornographic presentation of extremes of difference."[29] However, unlike most animal shows, it did not engender sympathy for an individualized being but curiosity about the cause of its death. While *T-Rex Autopsy* is undoubtedly an edutainment spectacle, it communicates some different messages about science than do Spielberg, Disney, and most other media productions.

Before analyzing *T-Rex Autopsy*, it is important to point out that Matt, the television character, is not the same as Matthew Mossbrucker, the museum director and curator. Matt, the television character, was cocreated by the show's producers through their selective inclusion, exclusion, and framing of his words

FIGURE 19. Matthew Mossbrucker gently guides a child's hands as he learns to use an airscribe to remove sediments from around a fossil without touching it. Photo by author.

and actions.[30] Matthew was not invited to give input into the editing of the show and did not even view it before it aired. The resulting character "Matt" differs from the person "Matthew" in several crucial respects. Matthew, the museum curator and science educator, deploys an inclusive feminist pedagogy as he teaches interested but nervous girls how to prepare fossils in the Morrison Museum's lab. As mentioned above, he gently guided children's hands as they used an airscribe to carefully remove sediments from around a fossil without touching it (fig. 19). Matt,

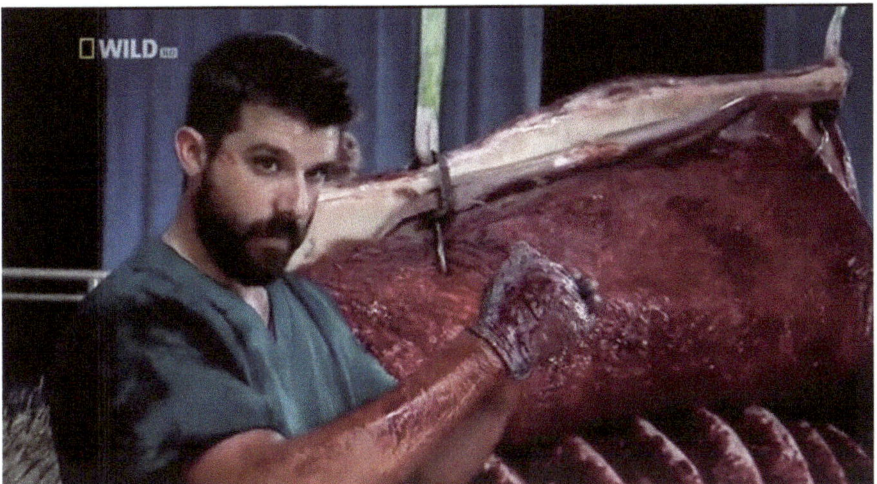

FIGURE 20. Matt in National Geographic's television special, *T-Rex Autopsy*. Photo by author.

the character in the show, embodies the frontier hunter masculinity discussed in chapter 2. Wearing bright green medical scrubs and white rubber boots, he and his costars dismember the dinosaur corpse using surgical implements, giant knives, chainsaws, and their hands (fig. 20). This distinction applies to Matt's human costars as well. Tori, Steve, and Luke are not the same as Dr. Victoria Herridge, an evolutionary biologist at the Natural History Museum in London, Dr. Stephen Louis Brusatte, a professor of paleontology and evolution at the University of Edinburgh, or Luke Gamble, a veterinarian.

Back at the Morrison Museum after filming in London, Matthew told me about his experience making *T-Rex Autopsy*. He only agreed to participate after the show's director assured him that they would not attempt to mislead viewers into believing its superstar was a real dinosaur. Once in the London studio, Matthew and his human costars were not given a script, only an outline of the body parts they should explore. The production crew "never said you can't," with one crucial caveat: "The one thing they did not want us to say is that this is a model because they wanted us to take it seriously." If any of them used words such as *model, fake, reconstruction,* or, conversely, *real, original,* or *actual,* they were edited out later. In the final cut that aired, Matt and his costars refer to the Tyrannosaurus as a "creature," "beast," "animal," and, occasionally, "cadaver." The show presents the scientists as thoroughly convinced that they are really dissecting a dead dinosaur. Near the end of the show, Matt looks straight into the camera and says, "I'm used to working with dry, dead bones. I have never gotten this filthy, dirty, and disgusting in my entire life. Climbing in a *Tyrannosaurus rex* cadaver—that sort of close proximity—it's indescribable." Yet *T-Rex Autopsy* does not try to deceive viewers into believing the object at the center of the show is the dead body of a dinosaur.

From the opening scene, the National Geographic special is framed by a pretend "as if" in which a *Tyrannosaurus rex* corpse exists in the present. The voiceover reminds viewers throughout the hour-long special that they have to *imagine* that the object was a recently deceased dinosaur. While Matthew was critical of several aspects of the final version, he was satisfied that it "could not have made it more clear" that the dinosaur's body "is not real." The science writer Riley Black concurred: "Some suspension of disbelief is required to enjoy the show, but at least the program is forthright that such an admission ticket is required."[31] Unlike Sampson's ordeal making *Walking with Dinosaurs*, Matthew's experience was a good one: National Geographic's producers kept their promise to him. Matthew even worried that because the show repeated so many times that the *Tyrannosaurus rex* body was not real, it verged on "insulting" the intelligence of the audience. At the same time, National Geographic presents the four human stars as both knowledgeable experts on and passionate fans of *Tyrannosaurus rex*. The scientists explain the current theories about species' morphology and ethnology, like talking heads in science documentaries. Yet they transmit the feel and smell of the dead dinosaur, like chefs on a cooking show. This paradoxical framing—in which the scientists conducting the dissection consider the dinosaur cadaver to be real, while the viewers know it is not—makes *T-Rex Autopsy* fall more into the genre of reality television than science documentary. Where does this place the show and its dead dinosaur on the paleontological continuum? Is its *Tyrannosaurus rex* cadaver comparable to a bone-clone or an anatomical model? Or is it a "giant rubber Muppet," as Matthew jokingly called it?

National Geographic claimed that the object at the center of its television show is the most accurate anatomical model of *Tyrannosaurus rex* ever built. While the scientists on screen never call this object a model, the voiceover narrator explains, "This groundbreaking creature took six months and over ten thousand man-hours to build. It was *modeled* on the most complete T. rex skeleton ever found with scientific input from the world's leading experts" (emphasis added). National Geographic's spokespeople elaborated in media interviews and press releases that it was created using casts and scans of Stan (BHI 3033) and Sue (FMNH PR 2081), along with the latest research about *Tyrannosaurus rex*. These are the same things that scientists and technicians would use to create an anatomical model for a science museum. The feathered exodermis of the National Geographic model suggests that it is more accurate than most anatomical models of *Tyrannosaurus rex* in science museums, which were made decades ago and so have lizard-like scales. In a video on *T-Rex Autopsy*'s website, Dave Hone, a University of London paleontologist who was one of the show's off-camera science consultants, elaborated:

There are obviously plenty of unknowns in paleontology, or areas where there are several plausible conditions. We're sure Tyrannosaurus was feathered, for example . . . From the fossil record there often appears to be a ridge of feathers along the profile outline, looking like a mohawk, but during the fossilization process the feathers get crushed together forming a denser ridge.

To replicate the feather ridge, Hone explained, the production team "stripped the feathery surface from approximately twenty thousand turkey and goose feathers to leave the quills which were then dyed a bluish black and inserted one by one" into a skin made out of polyurethane. Like the fossils prepared in paleo labs, the materiality of this cadaver was composed of an amalgam of animal remains and plastic. In this case, the animal matter was from modern analogues, not prehistoric bones-turned-stone. Unlike Sampson with his Hadrosaur vertebra, Hone could not connect any of the materials in National Geographic's *Tyrannosaurus rex* to an individual prehistoric animal. The dinosaur's exodermis thus indicates that National Geographic's cadaver belongs on the human side of the becoming axis in the paleontological continuum, along with other models, casts, and toys.

Where does this dinosaur belong on the employment axis? The scene focused on the cadaver's exodermis illustrates how National Geographic's 3D *Tyrannosaurus rex* model communicates what leading paleontologists considered accurate knowledge about the species at the time of the show's production. Tori yanks a feather from the Tyrannosaurus's skin; Steve takes it from her, and, bending it between his fingers, comments:

> When you look at this, I mean, it kind of has some give to it. It's kind of like a porcupine quill, or like the hair we have. And that's not what we think of when we think of bird feathers. We think of a quill pen type of thing with a long shaft and all the fluffy bits. And, of course, in birds they use feathers to fly. That's not what we're seeing here. This is a much simpler arrangement and this is what we call "proto-feathers." These are the ancestral start of feathers. A lot of dinosaurs, maybe even all dinosaurs, had feathers. And we know that because of fossils.

The screen shows images of these fossils as Steve continues to explain the evidence. "We have thousands of dinosaurs from China fossilized covered in feathers. And then, because we have so many fossils, we can trace how feathers change. Those feathers start forming wings on long arms in a small animal with big muscles, and that's how flight evolves." The scene switches to a video of a modern owl flying in a snowy forest as the voiceover states, "Bird feathers evolved so they interlock, forming a flexible wing. At seven tons, T. rex was never going to fly. But birds don't just use feathers for flight. They're important in display and attracting a mate. Scientists now think T. rex used its feathers in the same way." As the camera returns to the autopsy theater, Tori concludes, "These proto-feathers on T. rex give us more evidence of that bird-dinosaur relationship." This scene illustrates how

T-Rex Autopsy blends the testimony of scientists, the materiality of the dinosaur model, and video footage of modern birds to convey current theories about the anatomy of *Tyrannosaurus rex* and its evolutionary connection to birds. This scene shows the cadaver and its human interpreters to be adept science teachers who communicate the theory of evolution, and some evidence of it, in an approachable and engaging manner. These were familiar ideas for the people sitting around me at Paleo Joe's pub, but unfamiliar to myriad viewers watching at home.[32]

While the scene focused on *Tyrannosaurus rex* feathers reinforces National Geographic's claim that its show features a scientifically accurate anatomical model, another scene in which Matt and his costars dissect the dinosaur's heart challenges it. As synthetic blood gushes and splatters in every direction, Luke, the veterinarian, explains how *Tyrannosaurus rex*'s heart worked. As Luke runs his blood-soaked hands over the heart, he remarks that the left ventricle wall is "thick and muscular" while the right one is "much thinner." He explains that "the reason" for this is the left ventricle needs to pump blood "at massive, massive pressure into the biggest artery in this guy's body all the way up through the aorta." The blood, once oxygenated, "goes into all the muscles, fueling them, giving them the energy they need to do their job." The dinosaur's "job," viewers learn from the animation that appears on the screen, is to run through the prehistoric forest in pursuit of prey. Unlike Steve's discussion of the epidermis, Luke's explication of the dinosaur's heart does not reference fossil evidence. While rib fossils indicate the heart's size, they do not indicate its form. National Geographic's team modeled its four-chamber heart on those of modern birds. However, bird hearts are typically 1 percent of their body size, but the dinosaur's heart had to be made smaller in order to fit into its rib cage. Riley Black notes that this scene is "more speculative than others." In spite of this, Black concludes that the show "offers a fast and furious short course on vertebrate anatomy."[33] I would instead argue that it "offers a fast and furious short course" on evolution.

National Geographic's televised mock dissection neither represents scientific research techniques nor helps generate new scientific knowledge or theory. Although its model *Tyrannosaurus rex* could never be used for research, the show is more than an entertaining spectacle. The superstar of *T-Rex Autopsy* encourages viewers to reconceive dinosaurs as once-living animals rather than as monsters that lived in a mythical world. This brings me back to the widespread concern with creationism that permeated my fieldwork. The television show does not exhort its viewers to believe in evolution, but instead subtly and inoffensively demonstrates the evolution of species through its amalgamation of the materiality of its dinosaur model, images of modern animals, and scientists' movements and words.[34]

While some people are convinced of evolution by sensing the weight of a stone fossil in their hands, others are persuaded by fingering the spongy texture at the end of a femur cast, and still others are swayed by gazing up at a mounted skeleton that fuses fossils, reconstructions, and casts. Of course, some people are unconvinced by all paleontological objects, discourse, and images of evolution. Their belief in creationism is unshakable. As we saw with charismatic people and objects in chapter 2, objects' power of communication depends on context and audience. However, lessons about evolution that are based in the dominant secular ontology in the United States may be undermining their aim by reproducing the binaries of real and unreal, nature and culture, specimen and artwork, science and fiction. Defending the real-unreal dichotomy between fossils and their others, as Scott Sampson did, can inadvertently result in aligning evolution and creationism as beliefs.

Objects that provocatively combine science and fiction, like National Geographic's dinosaur "Muppet," may be the best advocates for evolution. At the same time that *T-Rex Autopsy* uncritically celebrates gore and violence, it models scientific inquiry through both its human and nonhuman participants. The dinosaur cadaver joins myriad other paleontological objects in embodying the creativity, imagination, and pleasure of doing science, albeit in different ways and to different extents. It also materializes theory. This makes National Geographic's "Muppet" an effective science communicator that has the potential to reach a much larger audience than the mounted fossils at the Denver Museum and the bone-clones in the Morrison Museum can. Precisely because *T-Rex Autopsy* is framed as entertainment, not a science documentary, it may be seen by people who will never enter a science museum. It thus has the potential to incite the curiosity of creationists and their children, whom even the most articulate and charismatic paleontologists, science educators, museum docents, and park guides have little chance of swaying. A binary approach to paleontological objects does not allow for this, but an ontology that conceives of the material world in a multidimensional, processual, and contextual manner—as the paleontological continuum does—encourages curiosity. Furthermore, the paleontological continuum reflects how paleontologists and others actually relate to matter in the course of their work. A continuum could prove to be a more powerful framework for countering creationism, if it is allowed to spread beyond a few idiosyncratic sites, such as the Morrison Museum. I am sure these ideas are threatening to some people, but I believe that for others it is comforting, maybe even liberating, to give words to what they have long been doing: collaborating with a range of humans and nonhumans to make fossils and scientific knowledge together. The final chapter of *Making Our Beasts* suggests additional ways that we can rewrite the stories we tell about the evolution of life on earth and remake our relationship to the prehistory of our planet.

Conclusion

Remaking Our Beasts

The foreboding clouds of anthropogenic climate change have a thin silver lining: the melting snow, receding waters, hurricane winds, and rockslides are exposing traces of the dead. Ancient humans and prehistoric animals are being disinterred from the earth.[1] Each newly exposed remain opens up possibilities for new knowledge and storytelling.[2] I end this book with a call for greater thoughtfulness and deliberateness in what we—scholars, educators, students, consumers, and citizens of an endangered world—do with these traces of the dead, both their physical remains and their life stories. Becoming more aware of how dead-yet-present animals have shaped our lives and our world today helps us to think carefully and critically about how we would like to live with animals—including our fellow humans—in the future.

Making Our Beasts has demonstrated that long dead animals have not stayed put, in the earth or within the physical and conceptual buildings people have built to house them. Their bodies are torn apart and their limbs scattered; they get carried down streams and buried in mud; they become entombed under layers of earth; their bones turn to rock and dust. If some of them make it into the present and are found by people who know how to care for them, they move indoors to labs then to collection shelves and, maybe just maybe, to display stands. A small group of privileged dead dinosaurs—the charismatic, the spectacular, the beastly—spend time in museums, universities, schools, homes, and entertainment sites. These beasts influence people's experiences—of life and time, of gender and race, of violence, and of what is and is not real. In the previous chapters, I have challenged the taken-for-granted notion that interspecies and intraspecies violence is universal and timeless, and I have suggested instead that both violence and caring have histories. They also have futures that we can shape.

Earlier in the book, I argued that people cannot make the dead conform to their dreams. We can question long-standing assumptions about their violence, but we cannot make prehistoric beasts into the pacific, nurturing, and collaborative animals we might wish them to be. We must contend with powerful non-human forces that do more than stymie human creativity. Such forces are also creative, though in different ways. Today's beasts have come into being through a more-than-human process that involves not only the animals that once roamed the earth but also scientists, fossil preparators, curators, volunteers, educators, students, media-makers, and consumers. It also involves rock, wind, rain, plastics, shovels, airscribes, and CGI. Although people alone cannot remake the physical matter or determine the meanings of these dead animals, we can work toward guiding them in certain directions, and away from others. Specifically, I contend that we need to disassemble the gendered and racialized power dynamics in paleontology's institutional structures and everyday practices in order to rewrite narratives about the prehistoric world and the evolution of life in ways that denaturalize violence and hierarchy. In other words, we need to disarm some of our beasts. I conclude the book by suggesting how scholars, educators, and students might work in new ways to help produce the objects, knowledge, and stories of the prehistoric world in the future.

THE HIERARCHY OF LIFE ITSELF

One could characterize the central objective of paleontology as mapping the evolution of life on earth. Through meticulous research, paleontologists work to reconstruct how current life forms came to be, as well as the trajectories of those species that have gone extinct. The resulting maps of life have taken different forms, but hierarchical arrangements have long dominated. Charles Darwin is credited with being the first to describe life on earth as a "great branching tree" at the base of which is the common ancestor of all living things.[3] This evolutionary tree replaced the older notion of "the great chain of being" (*scala naturae* in Latin), a ladder of increasing perfection that European scholars adopted from ancient Greek philosophers. On the lowest rung of this chain sat stone, earth, mineral, and other nonliving matter. Its middle rungs comprised animate life, tiered from plants to vertebrate animals. The vertebrates, too, were ranked, with "man" at the top. Medieval Christians added angels above humans and placed the Christian god at the pinnacle. As previous chapters mention, Linnean taxonomy challenged this "great chain" when it placed humans together with "beasts" in the kingdom *fauna*. Darwinian evolution replaced this chain with a tree of life that looks like an upside-down pyramid with still-growing branches.

In the last few decades, a ranked taxonomic approach to evolution, based on the number of shared characteristics among organisms, has been largely replaced by a phylogenetic one, based on those characteristics that indicate common ancestry.[4] Branches and trees are now called clades and phylograms, and they are constructed using computer calculations of large data sets. As one textbook notes, molecular and DNA research, combined with novel computing technologies, has led to "a revolution in the ways in which paleontologists interpret evolutionary aspects of the fossil record." Yet, arboreal imagery with hierarchical organization still abounds in the contemporary sciences and beyond.[5] As the authors argue, "nothing in biology or paleobiology means anything without a tree."[6] The idea that life is hierarchically and teleologically ordered did not disappear as paleontology and the other evolutionary sciences developed. This tree metaphor is as hierarchical as the "Great Chain" that it replaced. The highest-reaching branches are not only later in time but are also conceived as more complex and advanced than those below them.

My fieldwork showed me how a hierarchical understanding of life remains deeply entrenched in paleontology research and education. I first noticed the ranking of organisms in a paleontology course I attended while preparing for fieldwork. In one class exercise, the professor used simple fasteners, such as nails and screws, as an analogy for organisms, and the students were asked to order them in evolutionary succession. We soon learned that the smallest and simplest nail (a small rod with a slightly pointed tip on one end) belonged at the bottom of our chart and the largest and most elaborate screw (a bigger rod with an intricate head, a threaded neck, and a sharp tip) belonged at the top. This exercise suggested that organisms, or specific features of them, can be temporally ordered based on how many parts they have and how big and elaborate they are. The class exercise thus presented evolution as moving toward greater size and complexity over time, implying that the more intricate an organic structure is, the more advanced it is. The most popular introductory paleontology textbook in the United States, which was assigned for the course, reinforced the class exercise: "Just as trees may branch many times, and branches then branch, and so on, clades of organisms exist in a hierarchy of scales."[7]

This vision of evolution as a movement from simple to complex also appears in museum exhibitions.[8] For example, the Denver Museum's *Prehistoric Journey* is physically laid out to carry visitors on a "trail through time" from the Ancient Sea-floor diorama, where "the first living things were simple and lived underwater" six hundred million years ago, to the fossil preparation laboratory, where "scientists and technicians . . . use modern technology and human dexterity to uncover the mysteries of the fossil record." Walking from the dark halls representing the Paleozoic Era to the brightly lit fossil prep lab at the end of the Cenozoic, how could you not conclude that you had just traced evolution from simplicty to complexity, or that the people behind the lab windows represents its zenith? Across scientific publications, popular articles, museum exhibits, and docent explanations, I found evolution repeatedly portrayed, whether explicitly or implicitly, as a ladder in which life forms are ranked with homo sapiens placed at the summit.

A PINNACLE OF WHITE MASCULINITY

The hierarchical organization of nonhuman life forms in the evolutionary sciences mirrors the racial hierarchy that dominates Euro-American societies. The gender and racial dynamics within paleontology reflect the hierarchical ordering of people within the United States and around the world. My ethnographic research supports Sylvia Wynter's argument that the current scientific definition of humanity is gendered and racialized in ways that make white men seem ideal humans and others seem lesser.[9] The scientists at the top of the prestige and remuneration hierarchy are predominantly white men, while the technicians and educators below them are more diverse in gender and, to a lesser extent, race. For most of the history of paleontology, women were explicitly disallowed from fieldwork while people of color were excluded in other ways. Now, people of many races and genders are encouraged to conduct fieldwork, but they are still judged against a rugged, white, masculine ideal. In earlier chapters, I have examined how the paleontological cowboy has been idealized in films, from *Ten Who Dared* to the *Jurassic Park* series, and even in amusement park rides, such as Disney's DINOSAUR. But I also have shown that the "image of a paleontologist as a scruffy white man in a cowboy hat" is not the invention of the entertainment industry.[10] Paleontology's white cowboy masculinity has a long history in the everyday practice of the science too.[11] The life stories of Arthur Lakes and Barnum Brown exemplify the fossil hunters who took part in "pioneering" scientific exploration and colonialism in North America and other places around the world. Today, commercial, museum- and university-based vertebrate paleontologists, as well as the middle-class urbanites and suburbanites who volunteer for fossil-collecting expeditions, continue this cowboy science in more subtle ways. One paleontologist I met during my fieldwork had a reputation for testing his field-workers by seeing how they reacted when they were attacked by his dog. Another tried out new field assistants by seeing how long they could break up rock with a pickaxe in extreme heat without resting. Even under less excessive circumstances, women and minoritized others continue to receive greater scrutiny because the "underlying assumption" is that they "aren't as skilled, tough or driven" as their white male counterparts.[12] Those who successfully embody white cowboy masculinity are more likely to be rewarded with positions as department heads, chief curators, and distinguished professors, as well as with movie, television, and social media stardom.

Paleontology is far from the only field in which white masculinity has been rewarded. My own discipline of anthropology is hardly immune to this tendency.[13] Yet US vertebrate paleontology is marked by a particular form of scientific virility.

The Kaiparowits Plateau, a remote and inhospitable paleontological excavation site in the Grand Staircase-Escalante National Monument of southern Utah (see map), provides a vivid example of how the heroic cowboy-scientist ideal continues to exclude many from pursuing paleontology. Since 2000, paleontologists, fossil preparators, and volunteers from the Denver Museum have been part of a large collaboration among museums, universities, and government agencies that has uncovered exceptionally well-preserved remains of more than sixteen new species. The Denver Museum's website described the Kaiparowits as "1.9 million acres of extremely rugged, largely roadless terrain" that is "America's last great (relatively) unexplored 'dinosaur boneyard.'" This twenty-first century description of "virgin land" continues the long-standing pioneer discourse of imperialism and colonialism. Scott Sampson described the Kaiparowits this way:

> Life in the field lacks the romance and action of an Indiana Jones film. Most field seasons consist of weeks living in a tent in some desolate corner of the planet. Let's face it: These places are called Badlands for a reason. But what keeps you going is knowing that you may just walk around the next hill and find something that no one has ever seen before, something completely new to the world.[14]

Sampson countered the "lone genius" trope by acknowledging that "the discoveries I've been involved with were not made solo" but have involved "several teams of terrific people." Yet he reproduced the narrative of heroic pioneering, wilderness survival, and discovery of the unknown that exemplifies paleontology's celebration of white cowboy masculinity. Moreover, in idealizing fieldwork, he also devalues lab work. He does not mention all the labor that must come after "find[ing] something that no one has ever seen before" at a dig site and before publishing a scientific article or naming a new species. As chief curator at the Denver Museum, Sampson was more often in a suit than a cowboy hat when I saw him. But he nonetheless perpetuated the rugged heroism of the fossil hunter willing to risk his well-being in pursuit of his prey. As Riley Black has argued, "the rogue-heroic-scientist-makes-amazing-discovery storyline" is not one among many stories told about successful paleontologists; it is "the only one we ever seem to get."[15]

I heard heroic stories of fieldwork at Kaiparowits from many people at the Denver Museum, some told with pride and others with trepidation. Olivia, a white female college student and aspiring paleontologist, told me it "is considered to be the Iron Man of paleontology digs." She experienced Kaiparowits as a site where she was tested to see if she had "what it takes" to become a full-fledged fossil hunter. After she returned from a few weeks volunteering at Kaiparowits, she said that her experience there lived up to its reputation: "It was really tough. Every day, you hike out [from the campsite to the quarry along] a narrow ridge . . . almost straight down, almost straight up . . . It is pretty intense." On her first day of fieldwork, Olivia worked with a paleobotanist and another volunteer, both white men. She recounted:

We headed down these ridges and into a valley. We spent the whole day using pick-axes to pull rocks out of the ridge and then using rock hammers to split them open to look. I was sitting out there for about eight hours. It was either "Oh my god, there's a fossil leaf!" or "No, I can't get anything in here," and toss it down the side. Once we were all done for the day, we had to hike back out. I didn't realize this, but I was semi-dehydrated because I hadn't been drinking enough. Also, I hadn't acclimated to the altitude and I had improper footwear. I didn't bring proper hiking boots . . . I thought, "I'll be fine with my horseback-riding boots because they have treads on the bottom." No way! Climbing out, you put your foot down to kind of get over a vertical part and the whole thing would fall out from underneath so you'd slide back down. I made it about halfway. [The paleobotanist] and the other volunteer were zipping up. They're like mountain goats. I'm sitting there gasping for breath, and then I started to feel sick. I thought to myself, "There is no way I am throwing up in front of my mentor who has stuck his neck out for me, inviting me to come on this dig." What was I going to do? There was no way I was going to get out of this except by climbing up.

Olivia blamed herself for being unprepared, yet her dehydration was partially caused by the difficulty women and others face when relieving themselves in the field. While men simply walk a short distance and turn their backs on their companions, women are expected to go far away to find a spot where they won't be seen. This often leads them to drink insufficient water and to work harder to make up for their longer "bathroom breaks." Furthermore, since Olivia did not know anyone who had volunteered at Kaiparowits before, no one told her what kind of boots to bring. Olivia's attempt to conform to the cowboy dress code by wearing her horse-riding boots hindered her ability to navigate the Kaiparowits's steep terrain. Olivia concluded the story of her initiation into paleontological field research at Kaiparowits thus:

> I finally made it to the top of the ridge. But it really shook my confidence a lot. I thought that if this is what being a paleontologist entails, I don't think I can do this. I feel ashamed about it now, but I said to myself, "if I'm still feeling this way come Wednesday, when we were moving to another camp that was even more remote, I'm going to find a way to bail."

Olivia did not abandon the expedition, but she did abandon her goal of becoming a paleontologist. Following her experience at Kaiparowits, Olivia shifted her ambition from getting a doctorate in paleontology to becoming a fossil lab technician. She thus joined the long line of white women, Native Americans, and other people of color who have occupied the least well-paid and least respected positions in paleontology.[16] Her story illustrates how gender-based exclusion can be more subtle but no less impactful than in the past.

Even though I had no ambitions to prove myself in paleontology, I too experienced the pressure to perform feats of physical strength and endurance to show that I was as capable as my male companions. On my first expedition in the Dakota Badlands, I too was sent by the lead fossil preparator to the top of a butte to see how long I could shovel dirt in the blazing sun. I soon learned that many of the paleontologists, graduate students, and professional excavators—a group of several white men and one woman—followed grueling days of excavations in the field with long nights of masculine comradery at the nearest bar. In a space that evoked an ol' Western saloon, I witnessed their heavy drinking, bug-eating contests, and incredible tales of daring and dangerous adventures during previous expeditions. The one other woman in the group joined her male colleagues in bonding through masculine performance, entering in the banter and competitions with gusto late into the night. While some relish this aspect of fieldwork, other paleontologists, curators, fossil preparators, lab technicians, and volunteers have moved into specialties that do not require them to conduct field excavations so they can avoid situations like these. But there are consequences. In paleontology and other field sciences, fieldwork provides important moments for socialization and professionalization that advance the careers of those who can participate effectively. For volunteers, these moments build comradery and supply "war stories" that encourage their return for multiple field seasons. Paleontological excavations under harsh conditions, like those in the Kaiparowits and the Dakota Badlands, breed a virulent form of white masculinity that reproduces exclusions and inequities, despite policies and programs promoting inclusion.

A less rugged and more cerebral form of scientific white masculinity is inculcated among children in science museums. I observed innumerable moments in which boys, some as young as two years old, were taught that they should be interested in dinosaurs and could become paleontologists, or "a doctor of dinosaurs." Boys' passion for dinosaurs is not merely stimulated, but channeled into pursuing the highest rungs of paleontology as a career, frequently by adults who are not themselves scientists. Carol, a long-time volunteer in the Denver Museum's *Prehistoric Journey* and a retired scientist, described interactions with thousands of visitors that confirmed what I too had observed: Boys "tend to be the ones that are really, really focused" on dinosaurs because their parents actively encourage them. Carol told me that she was repeatedly approached by mothers who wanted to report that their sons were "just so interested in dinosaurs." These moms would tell her, "He wants to become a paleontologist. It's just amazing how much he knows." Carol noted that the mother frequently "implies that she doesn't know much," even suggesting that she is proud that her son knows so much more than she does. These mothers reinforced the idea that paleontology is a masculine career by contrasting their sons' knowledge with their own ignorance. I observed parents and educators explicitly encouraging girls to take an interest in paleontology, yet they often perpetuated scientific masculinity by making clear to girls that they would be "pioneers" venturing into a male field. I also noticed that the children who were encouraged to pursue paleontology as a career were predominantly white. While not as explicitly white and masculine as the cowboy scientist persona, the prodigy child with encyclopedic knowledge of dinosaurs maintains the long-standing

Euro-American association of elite white masculinity with scientific knowledge of and mastery over nature.[17]

Paleontology does not follow other natural sciences in which masculinized scientists explore feminized nature.[18] On the contrary, the gendering of the people in paleontology is largely mirrored in the gendering of paleontology's objects of study. Genuses, species, and individual animals are routinely masculinized in scientific naming practices and colloquial discourse. *Tyrannosaurus rex*, the "tyrant king," epitomizes the gendering of an entire species. But the exception proves the rule: One Late Cretaceous Hadrosaur was named Maiasaurus, meaning the "good mother lizard," because its first specimens were found in a mass site containing thousands of juveniles, hatchlings, and eggs, some with embryonic bones. This "nesting" site is interpreted as evidence that Maiasaurus cared for its young.[19] This herbivorous genus, with its nurturing relational femininity, is the foil to the assumed violent masculinity of the carnivorous dinosaurs. For most animals, male is the expected unmarked norm, while female is the marked exception used only when animals seem to display mothering attributes. This essentializes gender by implying that feminine and masculine characteristics are natural and ancient innate features while ignoring the fact that most vertebrate species include male and female individuals.

Cowboy masculinity is so deeply entrenched that I witnessed scientists and educators who purposefully used gender-neutral language most of the time slip into masculinized pronouns in telling moments. Take, for instance, a tour of Dinosaur Ridge led by a guide named Panola, who identified as a feminist and actively worked to encourage girls to pursue paleontology. Wearing a cowboy hat, khaki vest, cargo pants, and boots that matched her male counterparts, Panola acted out a predation scene for a group of students. She began, "Allosaurus had his mouth way open, and he runs at [the Apatosaurus's] belly." She stopped to ask the students, "What do you think he does?" One offered, "He bites it." "Yeah," Panola affirmed, "he gets just a chunk of meat" and then runs away. Despite her advocacy for women in paleontology, she dressed her white female body in a scientific cowboy outfit and used masculine pronouns to gender predation as male. To be taken seriously as dinosaur experts, women like Panola not only have adopted the masculine dress and speech of acclaimed paleontologists but have also reproduced the gendering of the field's species and specimens. This is because there has not been another place for them in the long-established paleontological drama in which scientific heroes conquer prehistoric beasts.

A PATH TOWARD INTERDEPENDENCY

Transforming paleontology not only involves drawing attention to the field's colonial legacies, exclusionary practices, hierarchies, and essentialism, but also involves imagining alternatives to the established ways of doing science. Better yet, it involves nurturing the nascent alternatives already budding in the spaces around the dominant storyline. I had the privilege of getting to know one such "trailblazer" in paleontology: a fossil excavator, preparator, and lab technician in the Denver Museum's paleontology department named Natalie Toth. After earning a bachelor's degree in geology and a master's in paleontology, Natalie first worked in science education. She was then employed by the Denver Museum's paleontology department, working on fossil preparation, prospecting, and excavation both in the museum's laboratory and at its excavation sites. After only a few years, she became manager of the museum's earth sciences labs. She now supervises hundreds of volunteers and student interns, teaches fossil preparation courses, and makes the most difficult decisions about fossil repair and conservation. While some women have avoided the fieldwork issues described above by forging careers focused on labwork, Natalie has persisted in doing both. In fact, she told me that fieldwork continues to be her favorite part of the job. Despite being in a leadership position at excavation sites, Natalie has faced numerous instances when men have assumed she knew nothing and treated her disrespectfully. Once, a volunteer tried to teach her how to use a shovel. Rather than denounce their actions, Natalie has tried to correct their misconception of her by showing them what she can do. Moreover, she seeks out opportunities—at the Denver Museum, at field stations, and in the media—for a wider audience to witness her paleontological expertise in action. For instance, she appeared on CBS television news alongside the Denver Museum's paleontologists, explaining their excavations at a local construction site. The segment showed her digging in the dirt, wearing a hardhat and safety vest emblazoned with the museum's logo, the same as the men. But unlike male paleontologists, who use the voice and body language of the authoritative scientist on camera, Natalie showed an effervescent excitement about paleontology as she explained what she was doing to the reporter. She demonstrated herself to be her male colleagues' equal in expertise, but not their mimic. Having forged her career in paleontology with few female role models to follow, Natalie has now become one herself, and rather than imitating the men around her, she is working to shift the paradigm.

Scientists and educators have wrung their hands over the fact that the geosciences, and US vertebrate paleontology in particular, have continued to be dominated by white, cisgendered, heterosexual men.[20] In recent years, there have been some important institutional and policy changes that have addressed discrimination and exclusion in the field. As a result, white women have started to move up US paleontology's institutional ladders.[21] Yet there is still an enormous amount of work to do for the community of US paleontologists to reflect the diversity of the national population—much less that of the world at large—and for paleontology to reap the benefits of having a diverse community of scientists. People of color, nonbinary people, and those facing intersecting forms of exclusion continue to be left out of paleontology in both subtle and not so subtle ways.[22] To address this, change needs to encompass much more than official policies and procedures, new programs, and initiatives. The practice of the science itself also needs to be transformed, including how fieldwork and lab work are conducted, how species are named and described, how research is funded and communicated, and how earthly matter is treated. Changes also need to be carried out in the usually unnoticed politics of what people wear, what pronouns they use for scientists and prehistoric animals, and what images they create and display. Too much of the burden of bringing about change has been placed on those who are female, nonbinary, or people of color, and who are asked to represent all those who have been excluded from the geosciences. It is past time for the white men who serve as paleontology's charismatic figureheads to take off their cowboy hats and apply their strength to breaking down the long-sedimented power dynamics in paleontology. Transforming the field's gender dynamics is only a first step in this process.

We can see the transformation of paleontology beginning in other ways as well. Some scientists have begun questioning the established conception of evolution as an arboreal structure, proposing a nonhierarchical web in its place.[23] It takes a long time for changes to be made to textbooks and museum exhibits, so I was surprised to find hints of this alternative approach to the evolution of life at the Denver Museum's Bone Bar. I noticed it in a discussion between a knowledgeable and enthusiastic boy and Norman, one of the docents described in the previous chapter. The boy asked Norman about the replica of an open eggshell with a hatchling inside, which was affixed to the top of the Bone Bar. "That is an Oviraptor," Norman answered. "Now you will be shocked at what he looks like," Norman continued as he took out an illustration of one. He asked, "Does he look like a dinosaur?" "No," the child responded, "he looks like a bird." Norman affirmed, "It does, doesn't it? Tail and all. But I'm telling you, he's a real dinosaur. Because he's got that hip structure that makes a dinosaur, and he lived back in the early Cretaceous period." Norman asked the boy, "So dinosaurs have feathers. Really? Are you sure?" The boy answered with an enthusiastic "Yes!" and then demonstrated that he knew about the evolutionary connection between theropods, such as Oviraptors, and modern birds, such as turkeys, saying, "we have dinosaurs that exist today." Norman continued, "You know your buddy T. rex? When he's a baby, you know what he looks like?" This the child did not know. Norman explained that when a "baby T. rex" was born "he was covered in down feathers." Referencing the Denver Museum's famous curator, Scott Sampson, Norman joked that when the young *Tyrannosaurus rex* is finished playing with Dr. Scott on *Dino Train*, "he walks across the hall to *Sesame Street* and he gets on stage with Big Bird." The child laughed as Norman hit his punchline, "You know why? T. rex is fully feathered like a bird." Norman elaborated, "In the last twenty years, they're finding a lot of remnants that point to the fact that most of the dinosaurs that walked on two legs were heavily feathered." Norman then handed the fossil of a *Tyrannosaurus rex* lower leg bone to the boy, noting that he has the same bone in his body. In this conversation, Norman created a familiar, possibly even familial, connection between dinosaurs, birds, and humans. Although we do not all have feathers, we all have leg bones, he suggested, and begin life as babies who resemble our parents in certain ways but not in others. Feathers illustrated the evolutionary relationship between theropods and modern birds in a way that did not emphasize the superiority of birds' flight feathers over *Tyrannosaurus rex*'s downlike proto-feathers. Despite gendering *Tyrannosaurus rex* male, the docent and visitor emphasized interspecies connections across time and species, rather than placing species or individual animals on a ladder of increased complexity or improvement. Furthermore, in discussing *Tyrannosaurus rex* as a "buddy," Norman acknowledged that affection for one's object of study does not impede scientific knowledge.

In highlighting nonhierarchical ways of discussing relationships among prehistoric animals, modern birds, and humans, I join other scholars from across the social and physical sciences and humanities who are developing alternative epistemologies that expand the acceptable kinds of relationships between humans and nonhumans within science.[24] By *epistemology*, I mean a theory about what knowledge is, and a process for producing, evaluating and recognizing it. A key tenet of the much critiqued but still dominant Euro-American scientific epistemology asserts that researchers must be objective and detached from their objects of study in order to produce accurate knowledge. However, as feminist science scholars have shown, science does not, and actually cannot, produce objective and universally valid knowledge that is free from culture. While many scientists are hesitant to acknowledge "the passions that power" science, the vertebrate paleontologists I have worked with are frank about their love of prehistoric animals, their fascination with the prehistoric world, and their excitement for research.[25] They are eager to share these passions with the public. Many of the research practices I observed in paleontological laboratories and at field sites can offer a model for how a wide range of humans and nonhumans collaborate in the production of scientific knowledge. The problem, as I see it, is that the practice of vertebrate paleontology is not accurately reflected in its theory and discourse. The language of "discovery" misrepresents the actual practice of paleontology by discounting more-than-human collaborations among people and matter and by depicting paleontological research as a process in which energetic scientists investigate inert nature. While vertebrate paleontology acknowledges that research involves passion, it allows only one kind: virile passion. I do not want paleontologists tame their passion; rather, I want them to expand its range far beyond the white masculine form of adventure-science.

In this book, I have tried to elucidate an epistemology of paleontology that recognizes the expertise and labor of scientists, curators, excavators, fossil preparators, lab technicians, students, and volunteers, and the importance of the diverse skill sets they contribute to producing scientific knowledge. Furthermore, I have shown that paleontological knowledge is manifest not only in scientific texts but also in scientific objects. I have highlighted the creative power of diverse ways of seeing, touching, moving, assembling, and representing, which are not often recognized as scientific. This alternative epistemology also recognizes the crucial contributions of materials, such as rock, water, metal, and plastic, as well as the power of social, cultural, biological, chemical, and geological forces, in creating scientific knowledge. The epistemology of science must expand its definition of knowledge to include knowledge as matter alongside the conventional conception of knowledge as abstract and disembodied ideas, represented after the fact in human-made text and image. People do not simply make sense of things. Materials and material forces shape human sense-making.

I cannot end this book without expressing my gratitude to all the people who enabled me to write it. Above all else, I am deeply indebted to the hundreds of paleontologists, paleontology students, fossil excavators, fossil preparators, lab technicians, science educators, museum docents, park guides, students, and visitors at my various field sites who so generously shared their knowledge, ideas, and passions with me. At the Morrison Natural History Museum, I especially want to thank Fritz Gottron, Doug Hartshorn, Doug Klein, and Zack Neher. The museum's director, Matthew Mossbrucker, has been an amazing teacher and sounding board throughout this project. He has shaped this book in many more ways than he realizes. Samantha Sands, Tyler Lyson, Ian Miller, Karla Zelvis, Natalie Toth, and Scott Sampson were particularly important to my research at the Denver Museum of Nature and Science. I remember Mike Getty, who died tragically near the end of my fieldwork there. At Dinosaur Ridge, I want to thank Erin LeCount, Kermit Shields, Alan Founie, Bobbi Kilgore, and Tom Moklestad. I also thank John Hankla for inviting me to participate in his programs. I am grateful to owners of the "dinosaur hotel," Meredith and Greg Tally, and indeed the whole Tally family for their enthusiastic hospitality at the hotel and beyond. There are several others, particularly the staff of the paleontological research station I call "the Badlands Paleontological Station," whom I must thank anonymously. Please know that I am extremely appreciative of them.

This project has benefited tremendously from the hard work of the Colgate students who have been my research assistants: Hailey Biscow, Vanessa Escobar, Angela Jang, Kristen Kennefick, Olivia Miller, Caitlin Mooney, Jolene Patrina, Angelica Smith, and Trey Spadone. Emma Dexter was an amazing editor in the making with a keen anthropological perspective. In addition, I very much appreciate all the logistical, administrative, and life support I have received from my department administrators, Karen Austin and Kayla Snow Smith.

I have been sustained throughout the research and writing process by thoughtful and encouraging colleagues at several institutions. At Colgate, I appreciate the insights I have gained from Michelle Bigenho, Chris Henke, and Nancy Ries. I thank Connie Soja for first teaching me paleontology, and Ryan Hall for sharing his deep knowledge of Indigenous and colonial history in the United States. The year I spent as an external fellow at the University of Rochester's Humanities Center was crucial for this project. I thank Joan Rubin, Kristin Doughty, Bob Foster, and Brian Jacobson for making my year in Rochester such a fabulous experience. I am also grateful for my sabbatical at the Institute for Advanced Study at the University of Minnesota, despite the tragedies of the 2019–20 year. I thank Jennifer Gunn, Susannah L. Smith, and Juliet Burba for everything they did to make IAS such a special place. It was a pleasure to be part of the IAS community with Fernando Burga, Adam Coon, Enid Logan, Jenn Marshall, Hana Maruyama, Jimmy Patiño, Ioanna Pribiag, and Hannah Ramer. I thank Karen Sue Taussig, David Valentine, and Aisha Ghani for welcoming me into the U of MN's anthropology community.

I received valuable feedback from discussants, copanelists, and audience members at numerous conferences and workshops. I received particularly helpful comments on early drafts of chapters from the 2016–17 University of Rochester Humanities Center work-in-progress seminar, the 2019–20 IAS Fellows Seminar Series at the University of Minnesota, and the STS Underground: Investigating the Technoscientific Worlds of Mining and Subterranean Extraction workshop at the Colorado School of Mines. I received wise comments on parts of this book from Pablo Aguilera Del Castillo, Jennifer Hamilton, Lisa Messeri, Lukas Rieppel, and especially Jessica Smith. Three anonymous reviewers, and Rebecca Lave, coeditor of the Critical Environments series, gave me very thoughtful feedback on the whole manuscript.

Writing is always a solitary task, but doubly so during a pandemic. I am enormously grateful to my writing partners who kept me going through the isolation and other challenges of the last few years, especially Ynesse Abdul-Malak, Janel Benson, Georgia Frank, Hannah Lau, Ben Lennertz, Jenn Marshall, and Navine Murshid. Virtual walks together with Julia Schaffer and Donald Moore nurtured me, enabling my return to the computer screen.

I have had the good fortune of having worked with Naja Pulliam Collins at the University of California Press. She has been a fabulous editor every step of the way. Special thanks to her, Rebecca Lave, and Julie Guthman for making my dreams for the book (open access! color photos!) come true. In addition, I greatly appreciate the whole team that has worked hard behind the scenes at the University of California Press for the care they have taken with this book. My writing has benefitted greatly from the editing acumen of Aaron Carico, Flora Spiegel, and Gabriel Bartlett. I thank Robert A. Ripps, my photographer-brother, for helping me choose and prepare the photographs for publication, Lara Scott for her creativity and patience in making the crazy map I envisioned, and Erin Gredell at the Yale Peabody Museum for her help with Arthur Lakes's paintings. Special thanks to Mark Dion for allowing me to use a photo of his sculpture on the cover, and for his inspirational residency at Colgate early in this project. The image is courtesy of the artist and Tanya Bonakdar Gallery, New York / Los Angeles. I appreciate the help I received from Lisa E. Smith at Tanya Bonakdar.

The research for this project received indispensable financial support from Colgate University's Research Council and Social Science Division, and the University of Rochester's Humanities Center. Portions of the introduction and chapter 1 are published in "Becoming Stone: On the Coming-into-Being of Fossils in the American West," in the *Anthropological Quarterly* 93, no. 3. Parts of chapter 2 appear in "More-than-Human Charisma, Iconic

Fossils, and Paleontologists in the Western United States," in the *Journal of the Royal Anthropological Institute* 31, no. 2. I appreciate the editors for granting me permission to reprint these here.

Support for this research has come in so many forms. My father, David Ripps, first took me to see the paleontology exhibits at the American Museum of Natural History, but my interest in them did not develop until decades later. I am saddened that he did not live to see the seed he planted reach fruition in this book. My mother, Sylvia (née Shever) Ripps, has unflaggingly supported me throughout my academic journey in innumerable ways, particularly in recent years with her loving care for my children. This project could not have happened without many friends and family members who helped me through the long process of researching and writing. I am eternally grateful for your love, labor, good humor, and, at times, for your forgiveness of my absences in body or in spirit. I cannot imagine this book, or my life, without the partnership of Jonathan Levine. My debts are beyond reckoning. Ariella and Daniel put up with my trips away from them, or my dragging them with me, despite not being "dinosaur kids." They have kept me from being too serious while also helping me stay hopeful for the future.

INTRODUCTION

1. I have changed the personal names of some of the paleontologists, preparators, and volunteers, and concealed the names and locations of some research institutes and quarries where I conducted fieldwork. I am careful to avoid using pseudonyms unreflexively, as "an inherited disciplinary custom" (Weiss and McGranahan [2021]), or to exploit them to protect myself from criticism by the very people on whom my fieldwork depended (May [2010]). I use pseudonyms, in certain cases, in order to protect paleontologists who are early in their careers, and fossils that are vulnerable to poaching. Anonymity was also part of the terms of access to two of my field sites. There are a couple of people who have double lives in the book. In some places I identify their names to give them credit for their ideas and work, while in other places I use a pseudonym to ensure that I do not disclose something that could harm their career or reputation. I shared drafts of the chapters that quote or describe people using their real names with those people for them to read and comment on. None expressed concern with how I portrayed them.

2. *Objectivity* has meant different things in different times and places. Its use here combines two meanings that are central to its modern Euro-American use: "empirical reliability" and "emotional detachment" (Daston and Galison [1992]; see also Daston and Galison [2007]).

3. Black (2018); Berta and Turner (2020).

4. Van Dooren (2014, 10).

5. The cultural studies approach to paleontology tends to take this view that the prehistoric world has been created in the image of the contemporary one.

6. I began fieldwork in 2013 and continued it—in the summers, during research leaves and a sabbatical, and amid teaching—through 2018, with the majority of it conducted in 2014 and 2016–17. Over the course of this time, I formally interviewed more than 70 scientists, staff, volunteers, and visitors to my fieldwork sites. I carried out innumerable hours

of participant-observation. I did not explicitly ask the people I worked with demographic questions, but observed that the paleontologists working at my numerous field sites were all white men, while the staff and volunteers were also largely white, but not exclusively male. Many educators, fossil preparators and other technicians were female. From the surveys my student research assistants conducted, and the information shared with me by the administration of my field sites, I know that the visitors to these places were more representative of the population of their location in terms of gender and social class, but whites were over-represented, with the exception of students on class trips. The people I spoke with ranged in age from teenagers to nonagenarians. The vast majority of them spoke English as their first language, but I did have a few conversations in Spanish.

7. Prothero (2013, 7).

8. As O'Reilly (2017) notes, "This is not new; science has never existed in a vacuum, despite serious attempts to protect science as a domain of objectivity and a sense of integrity, free from politics" (7).

9. The phrase "culture of no culture" comes from Traweek (1988, 162), while Franklin (1995b) makes the more general point.

10. This is true across the physical and life sciences. Traweek's (1988) ethnography of high-energy physics laboratories in Japan and the United States in the 1980s demonstrates that physicists' national contexts shaped their scientific understanding of time, space, and matter. In the same period, US weapons scientists used birthing metaphors to discuss nuclear weapons (Gusterson [1998]), and immunologists used battlefield metaphors to describe the human immune system (Martin [1994]). Other scholars have followed suit and identified the influences of domains, including state agencies, business enterprises, religious institutions, artistic movements, activists, and advocacy organizations, on a range of sciences.

11. Martin (1998, 30).

12. Martin (2021).

13. Scientists are "made rather than born," wrote anthropologist Hugh Gusterson in his 1998 ethnography of nuclear weapons development at the Lawrence Livermore laboratory (1), riffing on scientist-philosopher Keller's (1995) famous statement that women, men, and science are all "made rather than born" (3).

14. Wylie (2021) explores how paleontology produces a particular kind of technician, the fossil preparator. Parreñas (2018) is one of several works that examine volunteers in science.

15. Vetter (2008; 2016).

16. There is a growing number of ethnographies examining volunteer-tourism, but most are focused on foreign volunteers in the Global South (e.g., Parreñas [2012]; Redfield [2012]; McLennan [2014]). I add to this body of work an examination of volunteer-tourism in relation to legacies of settler colonialism in the Global North.

17. The term "community of practice" is from Lave and Wenger (1991). Traweek's *Beamtimes and Lifetimes* (1988) was among the first studies to explore a scientific field as a culturally specific community.

18. The term "boundary-work" is from Gieryn (1983; 1999).

19. Pandora and Rader (2008, 351).

20. Dewsbury and Naylor (2002, 258).

21. Rieppel (2015); Vetter (2016); Kjærgaard (2012).

22. Noble (2016); Marsh (2019).

23. Coronil (1997).

24. Fiffer (2001); Yates and Peacock (2024).

25. Kjærgaard (2012, 341).

26. For an overview of extractivism, see Jacka (2018). It is important to recognize that the materiality of distinctive kinds of fossils makes a difference in their entanglements in capitalism (Richardson and Weszkalnys [2014]). For example, compare coal (Rolston [2013]) and petroleum (Weszkalnys [2013]; [2015]).

27. For example, David H. Koch, co-owner of the oil refining and distribution company Koch Industries, donated $35 million to the National Museum of Natural History, in Washington, DC, to renovate and expand its paleontology exhibitions. High (2019; 2022) has carefully examined the Colorado oil and gas industry's complex relationship with philanthropy.

28. Yusoff (2024) provides a treatise on how geology is extractivist, exploitative, and racist. Stewart (1993) provides a seminal discussion of the Euro-American fascination with enormity.

29. The increasingly strong links between science and industry over the course of the twentieth century have been a major force in propelling the shift toward technoscience (Shapin [2008]). The shift away from science and society also recognizes that society cannot be separated from empire and colonialism (Anderson [2009]).

30. The notion that a technology is any object or procedure that extends human capacities to intervene in the world comes from Gell (1988) while the idea that technologies can be people's "full partners in . . . the dance of world-making" comes from Haraway (2008, 249).

31. Fischer (2007, 556–57).

32. Detwiler provides a powerful illustration of this in her analysis of the Atacama Large Millimeter/Submillimeter Array (ALMA), one of the most complex astronomical instruments on earth. ALMA had seemed a fully automated scientific instrument operated by supercomputers and scientists far from the Chilean Atacama Desert—that is, until a labor strike halted ALMA operations. Even the most high-tech and automated instruments still depend on sweaty labor, in this case to clean dust off ALMA's antennas (Detwiler [2022]).

33. In several respects, the region is more isolated from services and infrastructure now than it was in the early twentieth century, when cross-continental trains made regular stops there.

34. Wylie (2015, 32).

35. Important works on touch in science include Barad (2012), Myers (2015), and Ballestero (2019c).

36. Barad (2012, 208) points out that measurement too "is a form of touching."

37. Keller (1983) provides an important early study that makes this point.

38. Lorimer (2015, 9).

39. Meyers (2015, 2) engages with Barad's (2006) concept of "entanglement" and Haraway's notion of "body-fullness" (2001) when she describes life science laboratory research as a "a full-bodied practice" (2). In doing this full-bodied research, I join other feminist scholars of science who are "materially immersed" and intimately "engage[d] with the science no less than with the laboratory workers, modelers, theorists, technicians, and technologies. I share their "commitment to be in the science, not to presume to be above or

outside of it" (Barad [2012, 207]). I contribute to this scholarship a greater scrutiny of the blurry line between science and nonscience, particularly education and entertainment.

40. I purposely refuse to enter into the continual debates about which of the sciences are more, or less, rigorous than others. There is no universal set of criteria that can be used to rank all the sciences.

41. Turkle (2007, 6).

42. For discussion of dodos, see Van Dooren (2014, 3).

43. Sanz (2002, xii).

44. Sanz (2002, 47).

45. A study of book purchasing found that Americans across the political spectrum read (or at least bought) the same books about dinosaurs, while there was little overlap in their other book choices (Shi et al. 2017). This is surprising, given the politicization of evolution in the United States.

46. Argentine scientists, including brothers Carlos and Florentino Ameghino, began collecting fossils in Patagonia in the mid-nineteenth century. The Ameghinos exhibited some of their collections at the 1878 Paris Exposition. They eventually entered the American Museum's collections, along with fossils collected by US teams in Patagonia (Lanham [2012, 210–11]). However, when Argentine paleontologists argued that vertebrates first developed in South America, the idea was so threatening to scientists from the Global North that they immediately set out not only to disprove it but also to discredit the work of the scientists who proposed it. The American banking tycoon J. P. Morgan funded Princeton University to conduct a series of expeditions to Patagonia to collect fossils to show otherwise (Rainger [1991, 190–91]).

47. Ingold (2011) has argued that the scholarship on materiality "seems to have hardly anything to say about materials," that is "the stuff that things are made of" (20). I bring the concept of materiality back down to earth.

48. Richardson and Weszkalnys (2014, 16).

49. Haraway (2003a, 15–25) gives a brilliant explanation of the multiple meanings of species beyond biological taxonomy. Van Dooren (2014, 13–14, 21–43) elaborates a conception of species as a more-than-human achievement, rather than an accident of evolution. O'Connor (2007, 8) points out that the term "prehistoric" only began to mean "prehuman" in the early twentieth century.

50. The characterization of dinosaurs as "big, fierce and . . . scary" comes from Gould (1995, 51), as discussed below.

51. The quotation comes from Martin (1991, 501).

52. The quotation comes from Sansi Roca (2005, 150). Ingold (2007; 2011) elaborates this argument.

53. For dominant American ideas about nature, see Stewart (1993); for masculinity, see Willis (1999); for colonialism and savagery, see Haste (1993) and Semonin (2000); for authenticity, see Tanner (2000); for progress, see O'Connor (2007); for capitalism, see O'Neill (1996) and Lauter (2001); and for modernity and postmodernity, see Mitchell (1998; 2001).

54. Willis (1999, 186–87).

55. Brannigan (1980) provides an early overview of the sociology of science that led to this statement. See Brannigan (1981) for elaboration.

56. The notion that discovery is a fairy tale comes from Latour (1987, 134).

57. This understanding draws on the conception of nature as the Other to humanity, or more specifically "that part of the environment which we [humans] have had no hand in creating" (Soper [1995, 16]).

58. It is important to note that I use "fossil" to refer to the fossilized remains of an organism (body fossils). Some of what I describe does not pertain to trace fossils, like footprints.

59. Hallam and Ingold (2014, 8).

60. I join Van Allen (2017; 2018) in arguing that the physical labor of making specimens is a crucial part of knowledge-making, not prior to it. Yet I understand this labor as more than human.

61. Hallam and Ingold (2014).

62. Sanz (2002, 46).

63. Mitchell (1998, 14) argues it began in England in the 1850s, while Gould (1995) asserts it started in the United States in the 1970s. Furthermore, Gould makes a similar argument for the rise of dinomania as he does for evolution, albeit on a vastly different timescale. Gould argues that the evolution of life on earth does not progress evenly but goes through bursts of rapid change, in which the forms of organisms change quickly, followed by periods of little change. He calls this "punctuated equilibrium." Gould (1995) similarly claims that, after the initial public fascination with dinosaurs at the end of the nineteenth century, it was not until the 1970s that dinosaurs "vaulted to a steady level of culturally pervasive popularity" (222) in the United States.

64. Shatto (1976); Debus (2016, 25). There is some controversy over the exact date, but I follow Torrens (1993). O'Connor (2007, 9) notes that the word *scientist* was coined only nine years earlier.

65. Semonin (2000).

66. Semonin (2000); Thomson (2008); Dugatkin (2009).

67. Semonin (2000); Cohen (2002).

68. Rieppel (2012, 461).

69. Rieppel (2012, 468).

70. Rieppel (2019).

71. Gould (1995, 225).

72. Gould (1995, 229).

73. Dinomania expanded the existing "two-way traffic" (Beer [1983] 2000) between scientific and fictional writing, which began in nineteenth-century England. It extended it from writing to film, television, and then the internet. Buckland (2013) makes this point for geology in general in nineteenth century England, while Tattersdill (2017, 4) explores it in terms of dinosaurs in particular.

74. The American dinosaur movie genre is distinct from the Japanese one, despite some overlaps (Sanz [2002]; Debus [2016]).

75. Sanz (2002).

76. Bramwell and Peck (2008).

77. Sanz (2002, 98–103).

78. Sanz (2002, 18–19).

79. Debus (2016, 6).

80. Scholars have also revealed how the geological sciences have influenced popular literary styles from Victorian fantasy to contemporary science fiction (Tattersdill [2017]).

81. Torrens (1993, 257).

82. O'Connor (2007).

83. Sanz (2002, 43–44).

84. Allmon and Ross (2000); Allmon (2006).

85. Jaffe (2000, 292–95).

86. Jaffe (2000, 253–55).

87. Jaffe (2000, 24–31).

88. Jaffe (2000, 253–55).

89. Liggett et al. (2018).

90. Liggett et al. (2018).

91. In 2023, Zion National Park received 4,623,238 visitors and Dinosaur National Monument received 326,529 visitors. See "Park Reports," NPS Stats, https://irma.nps.gov/Stats/Reports/Park/DINO.

92. Gould (1995, 51).

93. Gould (1995, 51).

94. The inseparability of materiality and sociality has been explored by numerous scholars in slightly different ways. Law and Mol (1995) describe it as mutual production, whereas Pels, Hetherington, and Vandenberghe (2002, 2) call it coperformance, while Keane (2003), following Peircean semiotics, writes of copresence. Baldacchino's (2010) analysis of beach sand and Rogers's (2012) analysis of petroleum are good illustrations of how materiality and sociality are intertwined.

95. Yusoff (2018, 72).

96. De León's *The Land of Open Graves* documents how the heat, hills, and animals of the Sonoran Desert disintegrate human bodies with incredible speed. Both the environment in which we conducted fieldwork and our analyses of its agency are similar. Yet the intentionality with which the US government "outsource[s]" the labor of border enforcement to its "perfect silent partner," the desert, with "systematic logic" does not have a parallel in my research (De León [2015, 60–61]). In the Badlands, the desert environment tends to work at cross-purposes to human efforts to preserve prehistoric animal remains.

97. What counts as a "natural resource" is historically mutable and contested (e.g., Davidov [2014]). While Ferry and Limbert (2008) include "diversity" and "history" as natural resources, I use the term to refer to matter extracted from the earth's surface and subsurface.

98. Within the extensive work on natural resource extraction, there are several ethnographies that examine the production of petroleum under capitalism, including my own *Resources for Reform* (Shever [2012]). The smaller body of ethnographies on gemstones include Ferry (2013), Walsh (2010), and Bell (2023).

99. This Marxian approach was pioneered by Nash ([1979] 1993) and Mintz (1986). The "cultural biography of things" method was forged by Kopytoff (1986) and other contributors to Appadurai (1986).

100. For instance, Ferry (2008; 2013) shows how gems are remade to match expectations of them as scientific specimens, commodities, gifts, and religious offerings. Ferry (2010) further explicates four different "regimes of nature" (Escobar [1999]) that underlie changes in the presentation and reception of minerals and gems at North American natural history museums. Yet I observe a common understanding of stones as unchanging and timeless "nature" across these regimes, which intersect with other displays, such as animal

dioramas (Haraway [1989]). In addition, fossils, unlike gems, foreground the transformation of organic matter into inanimate objects that still in some ways embody the animate beings they once were. Fossils, therefore, highlight the blurry boundary between the animate and inanimate, the organic and inorganic, more than other stones do.

101. For dinosaur bones, see Fiffer (2001); for diamonds, see Proctor (2001) and Kurin (2006); for sapphires, see Walsh (2010).

102. The quotation is from Bennett (2010, 21), yet this conception of agency is not hers alone. She draws significantly from the work of Spinoza, Foucault, Deleuze, Guattari, Latour, and others.

103. For door-closers, see Latour (1988); for keys, see Latour (1992; 2000).

104. Some scholars have adopted Heidegger's (1968) distinction between "an object" and "a thing" (e.g., Brown [2001]). For instance, W. J. T. Mitchell asserts that an object is what people perceive—that is, a stable bounded piece of matter "with a name, an identity, a description, a use or function, a history, a science." In contrast, he states, a thing is either "an amorphous, shapeless, brute materiality awaiting organization by a system of objects" or "the excess, the detritus and waste when an object becomes useless, obsolete, extinct, or (conversely) when it takes on the surplus of aesthetic or spiritual value, the *je ne sais quois* of beauty, the fetishism that animates the commodity" (Mitchell [2005, 156]; see also Mitchell [2001]). I do not see this as a meaningful distinction for fossils, since they are often in both these states. Moreover, the focus on people and objects/things neglects the material entities that are neither (de la Cadena [2015a, 440, 462]). Therefore, I focus more on matter in flux, and use the words *object* and *thing* interchangeably to mean a moment in a continual process of material transformation.

105. The quotation is from Miller (2010, 5).

106. Patchett's (2021) study of the ostrich plume, which she bought at a vintage store in New York, provides an excellent example of the blurring boundary between growing and making. Human and avian work come together to "co-constitute" this object of fashion, not science. In a similar vein, Lyons (2014; 2016) shows how some farmers in the Colombian Amazon collaborate with *selva* to grow food in ways that entangle living and dying, and, in the process, challenge the state's soil science paradigm that separates "nature" from "culture."

107. The term "existents" comes from Povinelli (2016). Anthropologists have explored them in a wide variety of contexts—from watery landscapes (McLean [2011]; Ballestero [2019b]), to urban infrastructures (Chu [2014]), to the earth's atmosphere (Howe [2015]), to landmines (Kim [2016]), and beyond.

108. Descola (2014) is an exception. He notes that more-than-human anthropology has included other lifeforms but "leaves a great many nonhumans unaccounted for and expelled beyond the limits of an anthropology-beyond-the-human." He offers "the stones upon which I stumble" as an example of what has been "expelled" from anthropology beyond the human (271–72).

109. This was not always so. In medieval Europe, stones were widely recognized as having "quasi-animate powers of motion and action" and even souls (Robertson [2012, 92–93]). Stones healed people, imparted moral lessons to them, and otherwise possessed their own creative agency that affected humans.

110. Bennett (2010, 57–58).

111. Paton (2013; 2014).

112. Paton (2014, 235).

113. Paton and DeSilvey (2014, 226).

114. Ingold (2011, 26).

115. Haraway initially uses "become *with*" in *When Species Meet* (2008), then further develops the idea of "companion species" in *The Companion Species Manifesto* (2003a). For dogs specifically, see also Haraway (2003b).

116. Sager (2017).

117. I see some precedent for twenty-first-century posthumanist scholarship's imposition of the qualities of life onto nonlife emerging in the late twentieth-century anthropological studies of the social life of things (e.g., Appadurai [1986]).

118. There are significant differences between this vibrant materiality and vitality, which Bennett (2010) and several others who interpret her work (e.g., Tolia-Kelly [2012]) elide in some instances.

119. Povinelli (2016, 45).

120. The phrase "dance of agency" comes from Pickering (2010).

121. Like the aquifers that Ballestero (2019a) has studied, a fossil is "not an orderly articulation between existing entities" but "a kind of movement where differentiating" subjects from objects is near impossible.

122. Kirshenblatt-Gimblett (1998, 2, 17–18).

123. Williams (1985).

124. Barad (2006); Bennett (2010).

125. Eliade ([1949] 2005, 4).

126. For comparison of *their* magic and *our* science, see Malinowski (1948).

127. For a critique of how anthropologists portrayed people who have relationships with stones and mountains as premodern atavists with false beliefs, see Bird-David (1999, 68).

128. Sansi Roca (2005, 144), quoting Gell (1998).

129. Sansi Roca (2005, 150–51).

130. De la Cadena (2015b), Povinelli (2016), and Ballestero (2019b) each make this point in different ways.

131. In addition to Sansi Roca, de la Cadena (2015b), Li (2015), Povinelli (2016), and, to an extent, Raffles (2020) explore alternative ontologies of the earth's materiality.

132. An excellent introduction to this scholarship is found in *Cultural Anthropology*'s virtual forum on "geological anthropology" (Oguz [2020]).

133. Povinelli (2016); Yusoff (2018; 2024).

134. Paleontology is not the only exception, even within geology. Geobiology is another science that refuses to divide *geos* and *bios* but has been little examined by STS scholars (but see Roosth [2018]). Keller (2016) makes a similar argument to mine about the intersection of physics and biology within active matter science. Another important discipline that bridges the study of life and nonlife is geography. As Whatmore (2006) notes, "the vital connections between the geo (earth) and the bio (life)" is "amongst the most enduring of geographical concerns" (601). Lave (2015) highlights recent work in critical physical geography that turns this concern into practice.

135. This resembles Povinelli's (2016) description of the study of viruses (18–19).

136. The concept of "the inorganic slot" is from Ferry (2020).

137. I have found inspiration in exceptions to this trend, particularly Traweek (1988), Ferry (2013) and Messeri (2016). In addition, Noble (2016) examines the construction of paleontology museum exhibits and Wylie (2021) explores the labors of lab technicians in paleontology labs.

138. One important example is Helmreich (2009), particularly his scrutiny of the changing meanings and form of life for biologists. Helmreich carefully dissects the distinction between *bios* and *zoë*, yet he leaves the distinction between these and *geos* unexamined despite exploring microbes that blur the organic/inorganic dichotomy, astrobiology, and the seafloor.

139. Lévi-Strauss (1963).

140. Thanks to my student assistant Emma Dexter for explaining "beast mode" to me.

141. Palmatier (1995, 20).

142. Wynter (2003); McKittrick (2015).

143. Braverman (2017) and Wolfe (2013) call attention to numerous instances in which groups of people are considered nonhuman beasts. Braverman (2017) remarks that "the division of sentient beings into the bifurcated categories of animal and human is not solely a behavioral or biological division but also a place-based, context-specific construction that changes over time" (193–94). As Burton (2023) notes, "across the longue durée of Western modernity, Black and other colonized populations have functioned as the nonhuman Other existing beneath and beyond humanity's normative paradigm" (51). The category of the beast has been repeatedly used in this way to dehumanize Indigenous, Black, and other peoples, going as far as to categorize them as separate species from Europeans.

144. Mayor (2000); Asma, 26–30).

145. Mayor (2005).

146. Lockley is credited with turning ichnology (the study of fossil footprints) from "an esoteric side branch of vertebrate paleontology" into an active area of research (Houck and Lucas [2015, 134]).

1. BECOMING STONE

1. Mitchell (2019).

2. It is important to note that in this chapter I use *fossil* to refer to the fossilized remains of an organism. Some of what I describe does not pertain to trace fossils, such as footprints.

3. MacLeod (2009, 38).

4. In comparing the connotations of "discovery" and "exploration," the historian of science MacLeod (2009) writes, "The history of discovery is one of uniqueness, serendipity, initial encounter, and personal recognition . . . 'Discovery,' moreover, traditionally has a metropolitan referent . . . Discovery is exclusive. By the act of discovery, we lay claim to possession" (35). In the nineteenth century, MacLeod continues, "discovery became the ambition of the scientific traveler," who had "the unremitting desire to be 'first,'" while "the indigenous inhabitant remained, all too often, an artifact; perhaps an opportunity, at most a distraction" (38).

5. MacLeod (2009, 35).

6. This hegemonic view is captured in Shannon's foreword to Marsh's (2019) ethnography of the fossil halls and dinosaur exhibits at the Smithsonian's National Museum of

Natural History. Shannon writes that "while the fossils, the evidence, have stayed the same, how they have been interpreted has changed over time according to new scientific discoveries and new ways of thinking about museums and how they should engage the public" (xiv). I disagree; fossils have not stayed the same.

7. Like the prehistoric human remains found in bogs, a dinosaur skeleton "is not only preserved but also transformed" in "both its appearance and chemical composition" (McLean [2008, 305]). Unlike human bodies in bogs, however, dinosaur bodies are not preserved *in* earthly matter; they become part of the rock itself.

8. The Badlands Station was founded with an endowment from private donors, and it operates through individual donations and fees paid by volunteer participants. It has not received major state or institutional support.

9. Vetter (2008, 288).

10. This formation goes by other names in other places mentioned in the book. It is called the Lance Formation in Wyoming and the Kaiparowits in southern Utah. Foster (2007) provides a clear explanation of the complicated geology and paleontology of this region.

11. Hayden (1874, 20).

12. Jones (2020).

13. Lane (1989). After several similar bills failed in the 1990s, PRPA passed into law in response to the legal lacuna that the battle over ownership of the *Tyrannosaurus rex* "Sue" publicly exposed (Cronin [2014]). PRPA (2009) is the first law to explicitly and comprehensively govern paleontological resources on federal land. It states that fossil collecting permits are granted only to paleontologists with graduate degrees in relevant fields, scholarly research projects, and contracts with approved curation facilities that will store the specimens in perpetuity. PRPA stipulates that amateurs are allowed to collect scientifically insignificant fossils as long as they do not sell them.

14. Vetter (2016, 15, 7).

15. Lorimer (2015, 38).

16. Parreñas (2012, 680). At a rehabilitation center for orangutans in Borneo, mostly white British women undertake the strenuous labor of caring for apes belonging to endangered charismatic species (Parreñas, 2018).

17. Kilroy-Marac (2016, 447). In addition, Lorimer (2015) explores similar dynamics in "scientific ecotourism," arguing that volunteers' desires shape environmental conservation projects.

18. For West Papua, see Stasch (2014).

19. The quotation is from Wrobel (2002, 25). Tsing (2003) describes frontiers as "zones of not yet" that "wrest landscape elements from previous livelihoods and ecologies to turn them into wild resources, available for the industries of the world"—that is, capitalist exploitation (5100). As in the case in Indonesia that Tsing explores, the military has played an important role in creating resource frontiers in the Western United States. This is true not only for economic resources, such as coal and gold, but also for primarily scientific ones, such as fossils and botanical specimens.

20. I am deliberately vague about the team's members and work in order to preserve Darrell's anonymity.

21. Lorimer (2015) describes a similar situation in environmental conservation projects, in which scientists try to "reconcile research, conservation and entertainment" by

"dramatiz[ing] the conduct of science" for volunteers looking for spectacle and adventure in exotic locations (151).

22. Ray's expertise was highly respected by Darrell and the other scientists at the Badlands Station, and he moved between volunteer and staff roles as funding permitted. Ray received recognition for his work in the acknowledgments of academic publications, but has not been listed as an author.

23. There is a large body of anthropological and archeological scholarship on crafts and craft-making around the world. It is ably surveyed by Goody (2001) and Chibnik (2018). With few exceptions, this scholarship examines how novice artisans learn the particular "synthetic reason" (Paxson [2013]) of their craft from expert artisans but overlooks how they learn from matter as well.

24. Grasseni (2010; 2022). I singularize Grasseni's (2010) term "skilled visions" (5) in referring to the specificities of contemporary paleontology in the Badlands. Like Grasseni, I understand paleontological vision as closely linked to other sensory practices. I see sensing, and the connections among senses, as socially produced, rather than natural (cf. Grasseni [2022]). Wylie's (2017) analysis of paleontologists' reluctance to replace fossil excavation with digital imaging techniques provides an illustration of how the linkages between particular ways of seeing and touching can reinforce situated labor relations and power dynamics in paleontology.

25. The argument that ways of seeing are learned comes from Goodwin (1994).

26. Myers (2015, 14). Myers (2015) ethnography of protein crystallographers shows that "bodily knowledge is integral to science" (15). I follow her and Latour (2004a) in arguing that sensing is not done by humans onto passive matter, but "takes shape between bodies and worlds" (Myers [2015, 21]).

27. As Braun and Whatmore (2010) put it, we did not have "fully formed hand[s] with innate capacities apart from the objects that in a sense shape such a hand or afford it its capabilities, even if these objects are such basic things as sticks and stones" (xviii).

28. The phrase "current of materials" is from Ingold (2011).

29. Paton (2013); Hallam and Ingold (2014); Paton and DeSilvey (2014).

30. Van Allen (2018, 404–5).

31. See Van Allen (2017; 2018) for an excellent discussion of the labor involved in bringing natural history specimens into being.

32. Kirshenblatt-Gimblett (1998, 2–3, 17–18).

33. Richardson and Weszkalnys (2014, 13).

34. Rieppel (2012, 461).

35. Rieppel (2019).

36. Johnson et al. (2013).

37. Rader and Cain (2014); Marsh (2019).

38. Rader and Cain (2014, 241–43).

39. Butler (2010, 247).

40. Haraway (1989) provides a critical examination of how dioramas work as "meaning machines" that teach people to see the natural world in a static and unpeopled manner.

41. Wylie (2015, 32).

42. Noble (2016, 207–9).

43. For hotels, see Goffman (1959); for cell phone repair shops, see Bell et al. (2018); for gemstone prep labs, see Walsh (2010).

44. Unlike paleontologists, biologists have explicitly identified their scientific work in terms of making. See Franklin (1995a); Rabinow (1996); Myers (2015).

45. The fossil preparation performed at the lab window was more for engaging museumgoers than for generating scientific knowledge, although the line between the two was blurred as, for instance, visitors' admissions fees helped pay for the laboratory's operation.

46. Some preparators have explained the repair process using Michelangelo's apocryphal statement that a sculptor uncovers the statue already in the stone (Wylie [2015]). This analogy gets closer to capturing an understanding of fossils as more-than-human creations than the puzzle one. Yet it does not fully recognize fossils as made rather than found. It also denies the vibrancy of stone because it fails to recognize that a statue continues to undergo transformation after it leaves the sculptor's studio.

47. Ingold and Hallam (2014, 8).

48. Van Allen (2018) discusses the archiving of life at natural history museums.

49. Ferry (2013, 153–54).

50. This debate has raged within, and beyond, anthropology for a long time. Hartigan (2021) has recently asserted that Geertz (1972) interprets Balinese cocks "merely as a representational screen for human cultural concerns and obsessions" and not as animals with their own social relationships (853).

51. In particular, the contributors to Hallam and Ingold's *Making and Growing* (2014) expand this insight to a wide range of crafts.

52. Paton (2013).

53. Carpenter (2007a).

54. For gold, uranium, and oil, see Limerick (1987).

55. Dussias (1996, 115–16).

56. Schuller (2016).

57. Goetzmann (1966, 304).

58. The first quote is from Wrobel (2002, 10); the second one from Tauxe (1993, 5).

59. Brinkman (2010); Rieppel (2015; 2019).

60. Government support has included military escorts from the War Department, funding from the Geological and Geographic Survey and the National Science Foundation, and access to land from the Bureau of Land Management (BLM).

61. Schuchert and LeVene (1940, 72–93); Colbert (1984, 67–69).

62. Schuchert and LeVene (1940, 284).

63. Semonin (1997, 175); Rieppel (2015, 8).

64. Rainger (1991).

65. Rieppel (2015, 2).

66. Rieppel (2015, 18–19).

67. Plants provide a clear example of prehistoric life that long was neglected in the hunt for charismatic megafauna. Several paleobotanists have recounted to me how the narrow focus on dinosaurs repeatedly has led to the loss of floral fossils, and therefore valuable information about prehistoric environments.

68. Kjærgaard (2012, 344); Rader and Cain (2014, 249–53). For parallel changes in the United Kingdom, see Macdonald (2002, 32–33).

69. The connections between the petroleum industry and vertebrate paleontology run deep. Sinclair Oil, for instance, sponsored fossil hunting expeditions and a dinosaur

exhibition at the Chicago World's Fair (1933–34). The company continues to deploy the long-neck sauropod known as Brontosaurus as its logo.

2. MAKING CHARISMATIC VIOLENCE

1. Johnson et al. (2013, 249).

2. Prothero (2015, 200).

3. Martin (2006, 256).

4. Alberti (2011) provides a collection of fascinating biographies of famous named animals who have spent their "afterlives" in museums.

5. Osborn commissioned Knight to paint watercolor portraits of the dinosaurs the American Museum was excavating, studying, and mounting, including *Tyrannosaurus rex*. Knights' painting of AMNH 5027—which were reproduced in newspapers, magazines, and books—amplified *Tyrannosaurus rex*'s aggressive and combative nature even further than the mounted skeletons did. These paintings represented them "in active, dramatic situations of combat and struggle for survival" (Rainger [1991, 99]). In contrast, others have held Knight responsible for promoting the understanding of dinosaurs as slow-moving leviathans. See Mitchell (1998, 141).

6. Chambers (2005).

7. Prothero (2015).

8. I draw my conception of coming into being, or becoming, from Deleuze and Guatarri's *A Thousand Plateaus* (1987). My reading of it is influenced by the elaboration of becoming in multispecies ethnography. Kirksey (2015), for one, deploys Deleuze and Guatarri's "becomings" to reconceive species as a nonhierarchical grouping of organisms that recognize each other as alike and others as different. Species, he argues, continue to exist not only through their own doings, such as reproduction, but also through the actions of others, such as pollination. He writes that "species emerge with becomings of animate beings who are entangled in ecological, political and economic networks" (773). While Kirksey acknowledges the importance of human practices, he does not fully account for the power imbalances in which scientists hierarchically order life forms in ways that have profound effects on the ranked organisms, including which live and which die. In extending the notion of becoming from nonhuman animals to no-longer-living ones, I pay attention to the unequal power dynamics involved in the coming-into-being of things.

9. Weber ([1921] 2019, 374).

10. Worsley ([1957] 1968, ix).

11. Charisma has been used by scholars to explain how religion worked as a social force in everyday life, and thus was far more than an ethereal belief. Anthropologists have used the concept to analyze shamanism and religious-social movements, from "cargo cults" (Worsley [(1957) 1968]) to Pentecostal Christianity (Shipley [2009]). Weber famously generalized charisma to designate a secular, as well as religious, form of authority based on a leader's ability to inspire devoted followers with his magnetic personality. A charismatic leader, Weber asserted, is an individual whom other members of society recognize as possessing genuinely exceptional qualities and therefore give him the authority to rule them. In Weber's formulation, charismatic authority differs from the other two types he identified, traditional and legal-rational, because it resides in the unusual emotional appeal

of an individual, rather than in the established patterns of society (tradition) or procedural norms (law) that usually govern leadership (Weber [(1921) 2019], 374–89). Weber's influential formulation of charisma thus accounts for how a person's extraordinary personality or abilities translate into social power.

12. Franklin (1995a, 63).

13. Lawrence and Brown (2016, 153).

14. Robinson (2006, 94–95).

15. Noble (2016, 103). Noble's argument draws on Haraway's (1989) analysis of big game hunting in Africa.

16. Wilson (1995); Brusatte (2018, 278).

17. The term "scientific prophet" is from Caduff (2014).

18. The English word "individual" comes from the Latin *individuus*, meaning something that cannot be further divided into parts. Like atoms are to the physical world, individual humans are understood as the basic building blocks of the social world. In this conception, people create society, and only secondarily are shaped by it. Furthermore, the individualized conception of personhood regards people as unique and autonomous agents who each own their own bodies, thoughts, and talents, and intentionally use them to advance their interests and desires (Shever [2012], 193–94). Anthropologists have demonstrated the exceptionality of this conception of personhood by documenting the myriad times and places where being human is enacted otherwise. Furthermore, they have illuminated how an individual conception of personhood ignores the profound ways in which people are enmeshed in complex webs of interconnection that create their very being, as well as their interests, talents, tastes, and desires. By assuming individual personhood, the Weberian conception of charismatics ignores how their magnetic qualities have been created by the society in which they are embedded.

19. Rieppel (2012, 461).

20. Mullen (2010).

21. Leader-Williams and Dublin (2000, 60). See also Caro and O'Doherty (1999, 810).

22. McCormick (1985).

23. Barua (2016, 725–26).

24. Lorimer (2015, 144–45).

25. Lunney (2012). Moreover, the attention to charismatic megafauna has led to the vilification of the humans—so often impoverished rural people of color—who live and work in close proximity to charismatic megafauna.

26. Whiteley (2012, 223).

27. Ritvo (2010, 2).

28. For the care of modern charismatic megafauna, see Parreñas (2012).

29. This scholarly tradition extends from Wallace (1956) to Lindholm (2013) and Dulin (2021).

30. Alongside the large corpus of scholarship exploring the charisma of religious leaders, a few anthropologists have examined scientists as charismatics (Traweek [1988]; Caduff [2014]; O'Reilly [2017]). Caduff argues that microbiologists have become charismatic figures who offer authoritative prophesies about human health. These and allied studies show that scientists' charisma is an "interactional accomplishment" (Thorpe and Shapin [2000]) involving numerous people and historical, political-economic, and social forces. While their analyses demonstrate that charisma is not the achievement of uniquely gifted individuals, their analyses remain humanist.

31. Sansi Roca (2005, 150).

32. The term "nonhuman charisma" is from Lorimer (2007). For the charisma of computers, see Ames (2019); for scientific data, see O'Reilly (2017).

33. Helpful discussions of prehistoric animal taxonomy can be found in Holtz (1997) and Prothero (2013, 77–87).

34. Bowker and Star (1999) explore taxonomy as a contested political practice in a range of other fields that study life, from viruses to nursing work.

35. Mayor (2000; 2005); Asma (2009, 26–30).

36. Asma (2009, 13).

37. Asma (2009, 123–25); Ritvo (2010, 5, 34).

38. Before Europeans' acceptance of Darwinian evolution, species were seen as naturalists' best approximations of the fixed groups of organisms created by God. Variation among organisms within a species was understood as deviation from a divinely created ideal (Prothero [2013, 49]).

39. Torrens (1992, 40–42).

40. Gideon Mantell based his analysis on a single tooth found by Mary Anne Mantell, his wife (Emling [2009, 170]). Gideon Mantell was far from the only gentleman scientist of his time to take scientific credit for fossils that women found. One exceptional female fossil-hunter was Mary Anning, who supported her family by collecting and preparing fossils along the southern coast of England in the first half of the nineteenth century. Anning sold them to scientists, who gave her little to no credit for her discoveries, which included Ichthyosaurus and Plesiosaurus (Emling [2009]; Prothero [2015, 166–73]). Anning's biographer describes her first encounter with Richard Owen this way:

> A tall, thin, imposing figure with a big forehead and even bigger ego, Owen was also a charismatic speaker who could hold any audience rapt for hours. . . . But Mary doesn't seem to have been intimidated by her visitor's reputation . . . Indeed, she might have been irate that, only a year and a half earlier, Owen had slighted her when he described in great detail to the Geological Society of London her remarkable Pleisiosaurus microcephalus—without giving her credit (Emling [2009, 174]).

At least three taxa have been named for her (Berta and Turner [2020, 268]).

41. Torrens (1997, 175).

42. Prothero (2015, 172).

43. Krishtalka (1989, 29).

44. Asma (2009, 129). Ritvo (1997) makes a similar argument.

45. Osborn (1905, 259).

46. Horner and Lessem (1993, 62).

47. Osborn (1917, 762).

48. Noble (2016, 69).

49. Osborn (1917, 765).

50. For a clear explanation of dinosaur dentition and the analysis of its implication for diet, see (Martin [2006, 232–36]). The dentition of carnivores is highly consistent across species, continent, and time, and indicates convergent evolution of unrelated modern groups (Prothero [2013, 542]).

51. Noble (2016, 69); Rainger (1991).

52. Former view from Rieppel (2020).

53. Rainger (1991, 98).

54. *New York Times* (1905). The journalist and poet Langdon Smith was most likely the writer that covered the discovery of this new species. Although the article has no byline, a few lines from Smith's most famous poem, *Evolution: A Fantasy* (1909), provided an epigraph for the article.

55. Derrida (2009) identifies similarities between the concepts of the beast and the sovereign—in particular, that both are outside the law.

56. Quote from Asma (2009, 6)

57. Brown (1915, 322).

58. Brown (1915, 323).

59. *New York Times* (1905).

60. Brinkman (2010, 27–28). Although Osborn had global ambitions for the American Museum of Natural History, its principal rivals were all in the United States: the Carnegie Museum in Pittsburgh and the Field Museum in Chicago. All three were well-funded by millionaire philanthropists (Brinkman [2010, 2–3]).

61. Brinkman (2010, 34).

62. Brown (1915, 322).

63. Rainger (1991, 94–8). No skeleton is ever found entirely complete. When AMNH 5027 was mounted, missing bones (such as the humerus and femur) were filled in using casts of the holotype (AMNH 973) that Brown found in 1902 (Osborn [1917, 763]).

64. Rainger (1991, 69).

65. I conceal the identity of this paleontologist, including the institution where he works and the focus of his research, in order to protect him from any negative repercussions to his career or standing in the field from my writing.

66. "Black Hills Institute Catalog," Black Hills Institute of Geological Research, Inc., http://www.bhigr.com/catalog. Unsurprisingly, Stan's biggest teeth are the body parts that have been replicated most often.

67. Ferry (2013, 153–54).

68. Rieppel (2019, 1).

69. Shever (2020).

70. Rainger (1991, 158).

71. Matthews quoted in Rainger (1991, 158).

72. Rainger (1991, 160).

73. Osborn (1913, 92).

74. Brown (1915, 322).

75. Asma (2009, 6).

76. Rainger (1991, 160).

77. See "Tyrannosaurus rex," American Museum of Natural History, https://www.amnh.org/exhibitions/permanent/saurischian-dinosaurs/tyrannosaurus-rex.

78. Likewise, when children and young adults were asked to draw a *Tyrannosaurus rex*, the majority drew one that closely resembled the portrait of AMNH 5027 that Osborn wrote and William Diller Matthews illustrated in 1905 (Ross et al. [2013, 145]).

79. For instance, a study of theropod shed teeth used dozens of variables to establish which species they belong to with a high level of statistical confidence (Smith [2005]).

80. Rayfield (2004).

81. Osborn (1917, 762).

82. Mitchell (1998, 145). Other media scholars have countered biological essentialist explanations of dinosaurs' popularity with a cultural constructionist argument that is equally indefensible. This position has been most prominently (and entertainingly) put forth by Mitchell (1998). With a nod to Williams's "Ideas of Nature" (1980), he writes that the dinosaur is "a kind of blank slate on which every kind of collective and individual fantasy can be projected" (63). On the one hand, paleontologists "construct their picture of the dinosaur from extraordinarily skimpy evidence" (63). On the other hand, paleo-artists and media-makers create a "schizosaur"— that is, "a shape-shifting, transitional figure that can seem to mean almost anything one minute and almost nothing the next" (145). He asserts that the green color of dinosaur models in the 1900s to 1960s simultaneously represented "the 'colored' racial other, the savage, primitive 'natural man'" and the camouflage of the "modern armored military vehicle" (147). For Mitchell, dinosaurs first reflected American ideas about modernity and, later, postmodernity. Other scholars have argued that dinosaurs embody ideas about authenticity (Tanner [2000]), capitalism (Lauter [2001]), colonialism and savagery (Semonin [2000]), lawlessness (Morris [2007]), nature (Stewart [1993]), and progress (O'Connor [2007]). These readings of dinosaurs as infinitely malleable symbols disregard the materiality of fossils and the rigor of paleontological research. Cultural constructionist explanations of "dinomania" assume that people alone create nature/culture. They do not consider nonhuman forces in the making of either scientific specimens or popular spectacles. They seem to forget that "one cannot make a potent symbol out of just anything" (Meneley [2008, 308]). The same can be said about a potent fact or form (Latour [2004b]; Ingold [2013]). Paleontologists do not solely create the meaning or value of fossils. It must be in their past and in their materiality.

3. FORGING CONNECTION ACROSS MILLIONS OF YEARS

1. Foster (2007, xiv).

2. Arthur Lakes is immortalized at the Colorado School of Mines, the successor to the school where he taught, which has named the library after him and stores his archives.

3. Whites only began to see the Rocky Mountain landscape as "a paragon of scenic excellence" in this period of resource boom and settler colonization. Prior to this moment, European visitors saw the Rockies as "hideous, likening them to boils and other skin disorders" (Wilson [1991, 242]).

4. Lakes (1997, 9).

5. Hunt et al. (2002).

6. Lakes (1997, 10).

7. Hunt et al. (2002).

8. Ogden (2021, 24).

9. The dermatoglyphs of Tierra de Fuego's Indigenous adults and children in the early twentieth century, fingerprints made by the American explorer Charles Wellington Furlong, similarly affect their descendants because their materiality brought "the past . . . forcibly into the present." Ogden (2021), who studied Furlong's dermatoglyphs together with contemporary Indigenous Fuegians, called this "affective resonance" (27). Photographic reproductions of these dermatoglyphs did not have the same effect as the originals on either the Fuegians or the anthropologist.

10. For further discussion of affective attachment to objects of study in scientific research, see Myers (2015).

11. Ochs et al. (1994) and Myers (2009) describe similar use of gesture in the lab meetings of scientists.

12. Daston and Mitman (2005, 2); OED (2016).

13. Ritvo (2010, 8).

14. Crist (1999, 13); see also Beer ([1983] 2000). Darwin's descriptions of plants tended to be mechanistic rather than anthropomorphic. In his study of orchids, for example, as Hustak and Myers (2012) put it, he "delighted in his ability to discern the mechanical functionality and utility of orchid flowers and took this as a demonstration that even beautiful forms had utilitarian, adaptive value" (74–75). However, "his functionalist accounts of adaptation were sometimes muted by stories of affinities, attractions, and intimacies" (79).

15. Crist (1999, 38); Despret and Buchanan (2016, 38–39).

16. S. Mitchell (2005).

17. Wynne (2007) is one prominent scientist who has argued there is no place for anthropomorphism in scientific research.

18. De Waal (1999).

19. In developing an ethologically informed multispecies ethnography, Hartigan (2021) not only surveys the ethological scholarship on animal cultures but uses it to argue that "perhaps Darwin's emphasis on the continuity between humans and nonhumans in mental and emotional terms needs to extend to sociality, as well; it too may involve differences of degree, not kind, contrary to cultural anthropologists' assumptions." In doing so, he narrows culture to an individual's ability "to be differently inscribed by social settings" (855). This definition of culture may fit the horses he studied, but it excludes so much that humans do, especially meaning-making.

20. S. Mitchell (2005, 114–15).

21. Daston and Mitman (2005).

22. Candea (2012, 128).

23. For identification with the organisms, see Mitman (2005) and Paxson (2013). For molecules, see Myers (2015). Myers's (2015) ethnography of protein modeling further shows how even mechanical conceptions are full of anthropomorphism (chapter 7).

24. Despret and Buchanan (2016, 39).

25. Daston and Mitman (2005, 4).

26. Hartigan (2021, 853).

27. Cahill (2013, 79).

28. The quote is from Cahill (2013, 76).

29. Sampson et al. (2013).

30. It resembles the ethogram that anthropologist Hartigan (2021) describes having used in his fieldwork with wild horses in Spain (851). Yet there are important differences. The horse ethogram was created through direct observation of horses, while the paleontologists' emotional grid was based on observations of modern analogs, not of the dinosaurs themselves. Yet both require extrapolating from animal behaviors, such as eye movements, to dispositions or desires, such as being interested, curious, or sexually aroused. No researcher has unmediated access to their subjects' subjectivity. This includes anthropologists.

31. Prothero (2013, 152–53).

32. There is an interesting resonance between astronomers' efforts to understand distant planets through earthly analogues (Messeri [2017]) and paleontologists' efforts to understand temporally distant life forms through contemporary ones.

33. For examples of biologists who develop affective relationships with animals, see Mitman (2005) and Candea (2010). Kalshoven's (2018a) study of taxidermy provides a resonant instance in which people develop intimate relationships with dead nonhuman bodies at the same time as they separate species and distance the human from the animal.

34. Butler (2004, 30).

4. PALEONTOLOGY, COLONIALISM, AND EDUTAINMENT

1. Zoos benefit from the affective connections between visitors and the charismatic megafauna they house, turning these modern animals into "lively capital" they sell (Barua [2020]).

2. Animal Kingdom is hardly the only zoo to feature prehistoric creatures alongside living ones. Numerous traditional zoos have had permanent or temporary dinosaur exhibits. For many years, the Smithsonian National Zoo's Great Cats exhibit displayed a *Tyrannosaurus rex* skull near its live charismatic megafauna, including lions and tigers (Grazian [2015, 49]). The Smithsonian Zoo, like zoos across the country, has hosted visiting exhibits of animatronic dinosaurs as well.

3. Wrobel (2002, 25).

4. The *Oxford English Dictionary* traces the term's invention to a 1983 *Fortune* magazine article, but the idea is much older than that. On the US East Coast, as mass production and transportation made toys increasingly affordable in the 1870s, growing numbers of families began to buy toys for their children that would "channel" play "into paths of virtue and useful skill" (Heininger [1984, 6]). For a comparative study of contemporary edutainment in Japan and other places, see Creighton (1994). For the history of fusing science and edutainment in Britain, see O'Connor (2007).

5. Marshall (2007); Rieppel (2019, 24).

6. Bramwell and Peck (2008, 71); Peck (2008).

7. Marshall (2007, 291).

8. Rader and Cain (2014, 8–50). George Brown Goode—who first curated the Smithsonian's exhibit for the 1876 Philadelphia Centennial Exhibit and then directed the reorganization of the Smithsonian National Museum of Natural History—argued that the natural history museum must be converted from "a cemetery of bric-a-brac into a nursery of living thoughts" (Rader and Cain [2014, 9]). New York City's American Museum of Natural History exemplified the "New Museum" idea. As Rieppel (2019) has documented, the city's new industrial magnates who underwrote the museum wanted to demonstrate their civic concern by funding a museum that would edify a broad public (43–51). They favored natural history because they believed it would teach disciplined habits of mind and appreciation of divine creation, both of which these men highly valued.

9. Kidd (2004, 274–75).

10. At Yosemite, for instance, Indigenous villages were razed, and predators eliminated, so that white tourists could watch popular animals like mountain sheep graze in a "natural" setting (Spence [1999, 116]).

11. Kidd (2004, 276).

12. The line between corporate and public institutions has long been fuzzy. For instance, the American Museum was founded and funded by tycoons of industry and was itself organized as a corporation, albeit a nonprofit one (Rieppel [2019, 54]).

13. Rieppel (2019, 56–57).

14. This thirteen-part series ran from 1948 to 1960. It won eight Academy Awards (Kurtti [1996, 16]).

15. Von Mueller (2011, 153).

16. Decades before the Florida park was created, Walt Disney folded environmental conservation into his relentless nationalism. He admonished Americans:

> Physical America—the land itself—should be as dear to all of us as our political heritage and our treasured way of life. Its preservation and the wise conservation of its renewable resources concerns every man, woman, and child whose possession it is . . . I urge all citizens to join the effort to save America's natural beauty . . . It's our America—do something to preserve its beauty, strength, and natural wealth. (Quoted in Kurtti [1996, 19])

17. Spence (1999) charts the history of the particular conception and space of "wilderness" in the United States.

18. Von Mueller (2011, 154–55).

19. The Walt Disney Company has been a main beneficiary of these drainage projects ever since it began building its massive entertainment complex in Florida in the 1960s (Wilson [1991, 176–77]; Kidd [2004, 271]).

20. Lorimer (2015, 126) analyses similar scripts in other places.

21. Animal Kingdom also has a great deal in common with menageries, or living curio cabinets that showcased exotic animals from around the world according to their owners' interests and travels. However, while menageries emphasized the bizarre, Animal Kingdom follows zoos in highlighting childlike animals that have bodies that resemble their human visitors (Kidd [2004, 279]).

22. Kidd (2004, 268).

23. Kidd [2004, 279].

24. Critics have pointed out myriad other ways in which Disney's environmentalism is deeply flawed. More than 24,000 acres of everglade and other habitat was destroyed to create the entertainment complex in central Florida (Kidd [2004, 271]). Every day since Disney World opened, the corporation has used an enormous quantity of energy to operate its parks and resorts, and it has encouraged its patrons to do so as well through their travel. It is ironic and fitting that the "Tree of Life" at the center of Animal Kingdom was created from an oil rig. Although visitors are unlikely to see any oil, the park is in fact dripping with it. Hydrocarbon fuels might accurately represent quintessential "nature" in the United States, but Disney deploys dinosaurs for this task. See Huber (2013) for an insightful discussion of the central role of petroleum in the United States.

25. Melson (2001, 154).

26. Hall (1981, 89). As Davis (1997) observed at SeaWorld, this kind of environmentalism promotes "feelings and awareness," not "legislation and regulation," and makes a hefty profit on making people feel virtuous (237). Yet I am wary of the argument that every

element of these parks intentionally and effectively extracts as much money as possible from its patrons. For a nuanced discussion of zoos' relationship to environmental conservation, see Braverman (2013).

27. Kidd (2004, 276–78).

28. There are many accounts of these events in the popular press. One of the most extensive is journalist-lawyer Steve Fiffer's *Tyrannosaurus Sue: The Extraordinary Saga of the Largest, Most Fought Over T. Rex Ever Found* (2001). In a saga in which it was nearly impossible not to take sides, Fiffer tends to be sympathetic to the perspective of Pete Larson, the commercial paleontologist who co-owns the Black Hills Institute and directed the expedition in which Sue was found.

29. For example, Scott (2005).

30. Museums have undertaken the decolonization of their collections by repatriating (or rematriating) human fossils and objects of cultural patrimony to federally recognized Native American tribes. While decolonization projects have included seeds as cultural patrimony, they have not, as far as I know, included nonhuman fossils. Lawrence Bradley (2014b; 2014a) and other Native American scholars and activists have called for the repatriation of fossils removed from Native American reservations and ancestral homelands.

31. Cronin (2014).

32. Some commentators feared that the Field Museum had "sold its soul to gain Sue" and a "scientific specimen [would] be turned into a Happy Meal character or a cartoon star," like the ones already sold at DinoLand (Fiffer [2001, 216]). Instead, Disney and McDonald's made a plan to display an accurate replica at DinoLand, while Sue's fossils remained at the Field Museum. James Abruzzo, an expert in nongovernmental organizations at the Rutgers Business School, argues that "commercializ[ing]" Sue would have "cheapen[ed] the investment" that Disney and McDonald's made. By not turning Sue into a cheap plastic toy, the two corporations "purchased a credibility that is very important" for their "brand [enhancement]" (quoted in Fiffer [2001, 227]). It seems that Disney executives wagered that Sue's scientific aura would be more valuable for Animal Kingdom's profitability than selling Sue "stuffies" would be. In addition, this corporate sponsorship has precedents in the corporate sponsorship of zoo exhibits. It follows McDonald's sponsorship of the North American tour of two pandas from China in the 1980s. One zoo director's statement at the time seemed to foreshadow how Sue would be discussed: He called the two pandas "priceless" because "there are more Rembrandts in the world than there are pandas" (Wilson [1991, 249]).

33. Rieppel (2019).

34. Walsh (2012) provides an interesting discussion of how museums, world's fairs, expositions, and department stores displayed people and objects from around the world as commodities to be "compared, considered, and eventually consumed by the public" (25). The glass-clad laboratories where Sue's fossils were prepared are reminiscent of the mock native villages where Indigenous people demonstrated their crafts.

35. Mullen (2010).

36. Various clues indicate the time frame that DinoLand recreates. One sign states that the Dino Institute was established in 1947, while another indicates *Tyrannosaurus rex* fossils were found there in 1955. The prices at the gas pumps indicate a date between 1953 and 1957 (Veness [2015, 215]).

37. Pretes (2003, 138).

38. Tompkins (1992).

39. Von Mueller (2011, 155).

40. This contrasts with the celebration of the hard work of technological innovation at EPCOT Center, one of the other theme parks in Disney's Florida complex.

41. Hannigan (1995, 187).

42. Bradley (2014a).

43. This is a familiar theme in dinosaur films, thus linking DINOSAUR to the *Jurassic Park* ride at neighboring Universal Studios.

44. Tompkins (1992, 8).

45. Holdzkom (2011, 183–84, 188).

46. This issue is further discussed in chapter 1.

47. In a period in which the geographic ideal of "the frontier" may be replaced by "the wall" (Grandin [2019]), Disney's thrill ride is more than an updated version of the myth of the American West. It is also a revitalization of the ideal of endless frontiers in the face of the celebration of geographic enclosure.

48. Davidson et al. (2010, 3).

49. Davidson et al. (2010, 4).

50. Lockley and Marshall (2014, 28).

51. Davidson et al. (2010, 6).

52. Davidson et al. (2010, 6, 18–19).

53. Young (1997, 15).

54. Blackhawk (2006, 192–205).

55. Blackhawk (2006, 224).

56. Davidson et al. (2010, 5).

57. Davidson et al. (2010, 5).

58. Davidson et al. (2010, 4).

59. Davidson et al. (2010, 12).

60. Davidson et al. (2010, 12–13).

61. The story of the Rooneys and the Ute is not quite that simple. White settlers like the Rooneys provided a useful buffer between the Utes, who increasingly retreated to the mountains, and their native rivals, who were even more crowded on the plains. Settlers along the foothills alerted the Ute to approaching raiding parties from the Arapahoe and other rivals and often took the brunt of these raids (Blackhawk [2006, 190]).

62. Quotation from Simmons and Honda (2009, 427).

63. Vetter (2016, 229–30).

64. Lakes (1997, 11–12; punctuation added for clarity).

65. Davidson et al. (2010, 14). This is not entirely accurate because William Harlow Reed and Oramel Lucas also found dinosaur fossils the same summer—in Como Station, Wyoming, and Canon City, Colorado, respectively (Rieppel [2019, 14–16]).

66. Lakes (1997, 17–21).

67. Mayor (2005, 152; 2007, 248).

68. Mayor (2005, 154–55).

69. Quote from Rieppel (2019, 29).

70. Honda and Simmons (2009, iii).

71. For example, Kohl and McIntosh (1997, 3); Rajewski (2008); Lanham (2012, 170).

72. Cited in Honda and Simmons (2009, iii).

73. Simmons and Honda (2009, 426).

74. Kohl and McIntosh (1997, 4–5); Simmons and Honda (2009, 427–28).

75. Rieppel (2019, 33).

76. Simmons and Honda (2009, 427–28).

77. Lakes (1997, 12–15, 164n11).

78. Lakes (1997, 16).

79. Lakes (2007, 7).

80. Lakes (1997, 18, 145).

81. Hoy (1998). It is most likely that Lakes loaded the wooden boxes full of specimens from his quarries onto the Satanic Mine branch of the DSP&P and C&S lines, since, in 1872, Alexander Rooney leased his coal-rich land to the Satanic Mine company (Davidson et al. [2010, 17–18]).

82. Vetter (2016, 6).

83. Modreski et al. (1998, 332); Simmons and Honda (2009, 427).

84. Shever (2025).

85. Kohl and McIntosh (1997, 5–6).

86. Honda and Simmons (2009, 7, 78–82).

87. Rieppel (2019, 33).

88. Rieppel (2019, 16)

89. It is important to recognize that many wonderful people I know there would respond that visitors come to the park to learn about dinosaurs, and the organization is dedicated to educating them about paleontology and geology. They would assert that I should be commending Dinosaur Ridge's tour guides for mentioning the Utes at all. It is true that Indigenous people are entirely excluded from my other field sites devoted to paleontology and prehistory. Moreover, I met others in the field who defended colonization far more vigorously than the staff and volunteers at Dinosaur Ridge. One was a commercial paleontologist who had decorated the wall of his office with tributes to General George Custer, a key leader in the US military campaign against Native Americans that followed the Sand Creek Massacre. Behind a desk crowded with computer screens, stacks of articles, and boxes of paleontology paraphernalia hung an elegantly framed large portrait of this "war hero" surrounded by related memorabilia. His office thus portrayed a US Army officer who was responsible for the murder of countless Native Americans as a national hero and "the valiant defender of civilization against savage hordes" (White [1991, 622]). The idealized history told at Dinosaur Ridge pales in comparison to this salute to the brutal massacre of native peoples. Tours of Dinosaur Ridge celebrate peace, not violence; friendship, not conquest; and reconciliation, not genocide. Yet by whitewashing Rooney's involvement in events such as the Sand Creek Massacre they cover the violence of settler colonialism with the heroic veneer of paleontological discovery.

90. Goetzmann (1966, 304).

91. Thomson (2008, 4–5).

92. Yusoff (2018, 13).

93. Dussias (1996, 115–16).

94. Schuller (2016).

95. Citing examples from Siberia to Argentina, Rieppel (2019) asserts that "the rapid and often violent expansion of European colonial states overseas thus fed directly into the development of vertebrate paleontology" in the eighteenth century (19). Zizzamia (2019)

adds that, in North America, geological history has long been a "tool of erasure that simultaneously eliminated the continent's deep human history and granted the United States an origin story that tied progress of the nation to a natural and truly ancient past" (132). I see these processes as still ongoing in the present.

96. Thomson (2008); Dugatkin (2009).

97. The idea of American animals' inferiority came from volume 5 of George Louis Leclerc Comte de Buffon's monumental and influential *Histoire Naturelle* (1812), titled *Theory of American Degeneracy* (Dugatkin [2009]). Jefferson ([1801] 2022) did not believe that extinction was possible in God's "Great Chain of Being," so he asserted that mammoths would be found in the "unexplored and undisturbed" Western regions of the North American continent (89).

98. Divorcing Animal Kingdom's celebration of Americana from US colonialism is especially important for Disney because the company's 25,000 acre Florida complex sits on top of traditional Seminole lands, where some of the longest and bloodiest of "the Indian Wars" in United States history were fought. The Seminoles neither surrendered to, nor signed a peace treaty with, the US government at the end of Seminole Wars (1842) and have continued to fight for their land ever since (Cattelino [2008, 22–23]). This history makes the American alligator (*Alligator mississippiensis*), symbol of both the Florida Everglades and the Seminoles, an even less likely mascot of the United States at Animal Kingdom than the buffalo. The "prowlers of the Everglades" of Disney's movies were decimated by drainage projects designed to create space for commercial development, including the land that Disney bought in the 1960s to build an East Coast amusement park. It thus comes as little surprise that crocodiles, and not alligators, are the only living animals in DinoLand U.S.A. Disney's deployment of dinosaurs erases reminders of colonial violence both from the Western United States that it reconstructs and from the land on which the amusement park is built. It is not as easy to ignore colonial history at Dinosaur Ridge as at Animal Kingdom. At Dinosaur Ridge, some colonial buildings still stand, and the Rooney family still owns land. The still-living Ute Council Tree marks the current absence, not the continued presence, of Native Americans in the park.

5. THE PALEONTOLOGICAL REAL

1. I am working with the STS paradigm, which argues that ontologies do not preexist and structure human action, but are generated through more-than-human interactions. Lyons's (2014; 2020) ethnography of soil science research and farming in the Colombian Amazon vividly illustrates this.

2. The term "key symbol" is from Ortner (1973). Creationism is not homogeneous; there are several different varieties, with their own authoritative texts, figures, and institutions. Although it has a long history within Protestantism, the modern US creationist movement was born in the early 1960s. The term "creation science" was coined just prior to a series of influential US Supreme Court decisions about teaching evolution and creationism that "galvanized creationists, who interpreted the decision as evidence that a secular conspiracy to spread the cultural, moral, and spiritual plague of evolution was running rampant" across the country (Bielo [2018, 3–8]; Bielo [2019, 7]). There is an extensive scholarship on creationist movements in the United States. Some of the works most helpful to my thinking include Toumey (1994); Park (2001); Butler (2010); Trollinger (2016); Bielo (2018 and 2019).

3. Rader and Cain (2014).

4. In some other places in the book, I use a pseudonym for Matthew Mossbrucker so that I do not inadvertently write anything that could harm his reputation or career. I use his name here, both because it would be impossible to hide his identity and to give him credit for how significantly he has changed my thinking. I hope this chapter does not harm this smart, thoughtful, kind, and generous person, who has contributed greatly to this project.

5. Butler (2010).

6. Lyons makes a similar argument in showing how interactions between scientists and soil generate the dichotomy between nature and culture, while interactions between Amazonian farmers and this same earthly matter generate "partners in/for life" (2014, 214).

7. During the late nineteenth and early twentieth century, the standard protocol was much more fluid, and preparators often chose to highlight the parts of a specimen that had been manipulated, cast, or sculpted in some way (Lukas Rieppel, personal communication). The curators at the University of Wyoming's museum returned to that approach, using a different color to easily distinguish casts from real fossil bones on their newest large sauropod.

8. Povinelli (2016) writes that "the fossil" is a "geological category" for things that "have once been charged with life" but "have lost that life" (17). Like other scholars, she focuses on those prehistoric organisms that are now fuel for capitalism and contemporary human life.

9. Bigenho (2002) describes a similar experience of what she calls "experiential authenticity," a moment of multi-sensory emotional intensity that seems to elude meaning-making and symbolic representation, in the act of playing music in an ensemble (16–18).

10. Carpenter (2007b).

11. Descola (2013).

12. My argument here draws from Ingold and Hallam's (2014) discussion of the distinction made between organisms and artifacts. They trace this binary pair to eighteenth-century European collectors who systematically categorized the things they collected as either art or nature (10–11). This practice of separating organisms from artifacts was continued in the natural history museums that were founded in the nineteenth century.

13. This analysis of art comes from Benjamin ([1936] 1968).

14. Many thanks to Rebecca Lave for pointing this out to me.

15. Butler (2010).

16. In her presidential address at the 1993 American Anthropological Association meeting, Annette Weiner highlighted the "growing contradictions" between the discourse about the natural world and people's lived experiences of it (quoted in Franklin [1995b, 164]). My research has demonstrated that, despite myriad scholarly and popular critiques, this continues today.

17. Rieppel (2012, 465, 486–90).

18. The problematic objects that do not fit this binary include not only fossils, but also pinned beetles, taxidermized birds, gems, and minerals. Some fascinating studies of these objects include Van Allen (2018) on pinned beetles, Kalshoven (2018b) on taxidermized birds, and Ferry (2013) on gems and minerals.

19. While I am aware that Cartesian coordinate systems reduce the world's messy reality to a neat, static, and ostensibly objective structure, I nonetheless see the paleontological continuum as organized along two axes. I have placed them at an angle, and used line color and texture, to diminish the resemblance of this representation of the paleontological continuum with a Cartesian plane.

20. Haraway (1989, 5).

21. Haraway (1989, 5).

22. Here I diverge from the correlation of scientific care with economic interests that Lyons identifies in soil science (2014, 214).

23. I did not conduct fieldwork at this site but draw my knowledge from Butler's (2010) fascinating ethnography of it, as well as the analyses offered by Bielo (2019) and Trollinger (2016).

24. *Jurassic World* became the first film to gross 500 million dollars. It eventually earned 1.6 billion dollars worldwide (https://www.imdb.com/title/tt0369610 and https:// en.wikipedia.org/wiki/Jurassic_World, accessed March 3, 2020).

25. Huls (2013).

26. The term "boundary-work" is from Gieryn (1983; 1999).

27. There are some strong resemblances between the process by which Benjamin Waterhouse Hawkins "revivified" dinosaurs for the Crystal Palace in the mid-nineteenth century (Peck [2008, 68–71) and National Geographic's process in the early twenty-first century. See chapter 4 for discussion of the Crystal Palace.

28. Ian McBride quoted in Paget (1997).

29. Lorimer (2015, 137).

30. In analyzing the Discovery Channel television program *Meerkat Manor*, Candea (2010) similarly argues that animal documentaries are not made by "Machiavellian producers creating spurious engagement out of brute beasts for a gullible audience" (254). Candea illustrates the show's coproduction by meerkats "who delivered story after story and were so good at eliciting human engagement," biologists "who offered facts and helped the cameramen see," and filmmakers "who patiently assembled these various inputs into a show that was more than the sum of its parts" (254). In the case of *T-Rex Autopsy*, one should recognize the role of the synthetic dinosaur body that spouted fluid, oozed guts, and stank up the studio, thus inciting big reactions from the people dismembering it.

31. Black (2015).

32. Although it is impossible to know how the viewers watching at home interpreted the show, the conversation on the Twitter chat "#TRexAutopsy" (now on X) supports this assessment. The similarities between *Tyrannosaurus rex* and modern birds were among the most noted aspects of the show.

33. Black (2015).

34. Drawing on the critical analyses of media by Connolly (2002), among others, Lorimer (2015) asserts that media can "evoke[s] and mobilize[s] particular 'affective logics' toward various political ends" (122). While some animals in media harms conservation, others "inculcate a curious sensibility" that encourages support for environmental conservation (121). I argue that dinosaurs similarly encourages support for evolution within the US context, where it has long been politicized.

CONCLUSION

1. Inspired by Soto Laveaga (2023).

2. McGranahan (2020) argues that ethnography is "theoretical storytelling." I want to add that paleontology can be, too. We cannot separate the excavation and lab preparation of fossils from meaning-making about prehistory or nature.

3. Benton and Harper (2020, 184).

4. Brochu and Sumrall (2001) provide an overview of the arguments to replace the ranked Linnean taxonomic system with a phylogenetic system. This article demonstrates that the new system gets rid of rankings, yet continues to naturalize hierarchy. In arguing for the abolition of the mandatory ranking of each new species, the authors assert that "divergence events in phylogeny create natural hierarchical groupings of ancestors and descendents, but the ranks are purely human constructs that have nothing to do with the groups themselves. Ranks express the subjective opinion of an expert of the perceived importance of a group relative to another" (756).

5. I draw my argument here from Deleuze and Guattari's (1987) critique of arboreal metaphors. As the contributors to *The Social Life of Trees* (Rival [1998]) show, trees are widely used as metaphors but stand for different meanings and values in different contexts. In the North American context, distinct tree metaphors have been merged and deployed in political struggles over Indigenous land rights and environmental conservation (Mauzé [1998]). See Whiteley (2012) for further discussion of arboreal metaphors in science (220).

6. Benton and Harper (2020, 191).

7. Cowen (2000, 47).

8. Recent work about the evolution of feathers provide additional examples of how hierarchical thinking is embedded in paleontological theory. Take an article in *Scientific American* (Prum and Brush [2014]), coauthored by an evolutionary ornithologist and an evolutionary biologist, that translates the first author's research on feather evolution (Prum [1999]; Feo et al. [2016]) into less technical terms for a nonspecialist audience. This article explicitly describes feather evolution as being "composed of a series of hierarchical stages" (Prum and Brush [2014, 83]). One of the images included in both the magazine and one of the first author's scholarly articles illustrates the five stages of protofeather evolution, beginning with a simple hollow tube, and then adding increasingly intricate knobs ("barbs" and "barbules") to it. This line drawing is remarkably similar to the drawing of fasteners I had used in the class exercise during my paleontology course. The *Scientific American* article also maps the five evolutionary stages onto a cladogram of carnivorous bipedal theropod species. The article thus uses image and text together to present feathers as forming a hierarchy of increasing complexity over time and through the evolution of species. I found the hierarchical story of the evolution of protofeathers in this *Scientific American* article reproduced in *Prehistoric Journey*. It seems the article was used in writing the training manual for volunteers at the Bone Bar. An exhibit case extended the story by illustrating how modern birds evolved from small carnivorous dinosaurs, not from the Pterosaurs that hung from the hall's ceiling. This display dramatically portrayed the evolution of dinosaurs into birds as a story of progress through both increasingly elaborate plumage and the movement from ground to higher and higher flight.

9. Wynter (2003). The gender and racial hierarchies in paleontology are inseparable from the geological sciences' participation in imperialism, colonialism, and extractivism. Postcolonial science studies has shown that scientists' claims to universal laws and theories are generated through historically, geographically, and socially specific processes (Anderson [2009]). In myriad scientific fields, people of European descent have depended on the knowledge and skills of colonized subjects and spaces to produce the scientific "discoveries" they claimed as their own. In paleontology, studies have documented how Native Americans contributed to the field's development (e.g., Simpson [1942]; Rieppel [2019, 28–30]) and how Indigenous people resisted becoming objects of paleontological study (e.g., Yusoff

[2024, 132–35]). Yet much work remains to be done to understand how paleontological collections and knowledge, particularly in their contributions to the theory of evolution, have been forged through imperial domination and extraction.

10. I disagree with Black (2018) on this point.

11. George McJunkin, the Black cowboy who found the bones of extinct bison near the New Mexico/Colorado border in the beginning of the twentieth century (Germond [1982]), is an early exception that proves the rule. Since the late nineteenth century, the ideal of the American cowboy, including its paleontological version, has been a white man, even if those practicing it have not always been (Horne [2005]). Today, Black cowboys are elaborating their own forms of masculinity (Babers [2022]), yet they remain marked categories in comparison to the unmarked (white) cowboy forms.

12. Black (2018).

13. See Bradford (2022); Nelson et al. (2017); Zurné (2024). Conceptual tools and methodologies developed through anthropology's ongoing self-critique can help scholars critically assess paleontology, especially given the similarities between these two disciplines. Both have long histories of exploiting Indigenous people's expertise, belongings, and land. Both have continued to promote grueling fieldwork as a rite of passage into the profession, despite the evidence that it harms and excludes many people who could otherwise make valuable contributions.

14. Scott Sampson, interview by Sara E. Pratt, *Earth Magazine*, May 19, 2014, https://www.earthmagazine.org/article/down-earth-scott-sampson.

15. Black (2019).

16. Berta and Turner (2020, 10); Marín-Spiotta et al. (2020, 117).

17. There is a deep genealogy of feminist scholarship on the gendering of scientific discourse and practice in which Keller (1995) is one crucial inspiration. More recently, scholars have explored the multitude of different masculinities. In the contemporary United States, white masculinity has transformed from the unmarked normative category into many contingent and mutable particulars, or "reactive" and "labile" masculinities in Carroll's terms (2011)

18. Classic studies include Merchant (1989) and Haraway (1989).

19. Noble (2016, 197–202).

20. Important ones include Berta and Turner (2020), Holmes et al. (2015), Marín-Spiotta et al. (2020), Monarrez et al. (2022), and Ranganathan et al. (2021).

21. Yet women make up less than 6 percent of the curators and less than 15 percent of the highest ranking (full) professors in paleontology (Berta and Turner [2020, 10, 211]).

22. The scholarship documenting this is synthesized well in Marín-Spiotta et al. (2020) and Ranganathan et al. (2021).

23. In commenting on new research in biology, Kirksey (2015) writes, "The tree of life has not completely melted. New modes of enacting species are instead enabling taxonomists to recognize a chaotic and polycentric structure in its place" (775). He joins Helmreich (2009) in arguing that one can understand species in a non-hierarchical fashion without jettisoning a "tree of life" conception of life on earth. I see hierarchy as central to arboreal metaphors for evolution and the relations among organisms.

24. Haraway's extensive scholarship on normative and alternative scientific epistemologies has profoundly shaped my thinking. Other scholars who have influenced me include Lorimer (2015) and Thompson (2015).

25. The phrase "the passions that power" science is from Lorimer (2015, 38).

BIBLIOGRAPHY

Alberti, Samuel J. M. M., ed.
2011   *The Afterlives of Animals: A Museum Menagerie*. University of Virginia Press.
Algar, James, dir.
1957   "Prowlers of the Everglades." *True-Life Adventures*. Walt Disney Productions.
Allmon, Warren D.
2006   "The Pre-Modern History of the Post-Modern Dinosaur: Phases and Causes in Post-Darwinian Dinosaur Art." *Earth Sciences History* 25 (1): 5–35. https://doi.org /10.17704/eshi.25.1.g2687j050u3w1546.
Allmon, Warren D., and Robert Merrill Ross
2000   "An Art Exhibit on Dinosaurs and the Nature of Science." *Journal of Geoscience Education* 48:296–358. https://doi.org/10.5408/1089-9995-48.3.296.
Ames, Morgan G.
2019   *The Charisma Machine: The Life, Death, and Legacy of One Laptop Per Child*. MIT Press. https://doi.org/10.7551/mitpress/10868.001.0001/.
Anderson, Warwick
2009   "From Subjugated Knowledge to Conjugated Subjects: Science and Globalisation, or Postcolonial Studies of Science?" *Postcolonial Studies* 12 (4): 389–400. https://doi .org/10.1080/13688790903350641.
Appadurai, Arjun, ed.
1986   *The Social Life of Things: Commodities in Cultural Perspective*. Cambridge University Press. https://doi.org/10.1017/CBO9780511819582.
Asma, Stephen T.
2009   *On Monsters: An Unnatural History of Our Worst Fears*. Oxford University Press.
Babers, Myeshia C.
2022   "Cowboy Cool: A Professional Black Cowboy's Perspective." *Transforming Anthropology* 30 (2): 150–64. https://doi.org/10.1111/traa.12241.

Baldacchino, Godfrey
    2010   "Re-Placing Materiality." *Annals of Tourism Research* 37 (3): 763–78. https://doi.org
        /10.1016/j.annals.2010.02.005.

Ballestero, Andrea
    2019a "Aquifers (or, Hydrolithic Elemental Choreographies)." *Fieldsights* (blog). Soci-
        ety for Cultural Anthropology, June 27. https://culanth.org/fieldsights/aquifers-or
        -hydrolithic-elemental-choreographies/.
    2019b *A Future History of Water.* Duke University Press. https://doi.org/10.2307/j
        .ctv11smwo3.
    2019c "Touching with Light, or, How Texture Recasts the Sensing of Underground
        Water." *Science, Technology, & Human Values* 44 (5): 762–85. https://doi.org
        /10.1177/0162243919858717.

Barad, Karen Michelle
    2006   *Meeting the Universe Halfway: Quantum Physics and the Entanglement of Matter
        and Meaning.* Duke University Press. https://doi.org/10.2307/j.ctv12101zq.
    2012   "On Touching—the Inhuman That Therefore I Am." *Differences* 23 (3): 206–23.
        https://doi.org/10.1215/10407391-1892943.

Barua, Maan
    2016   "Lively Commodities and Encounter Value." *Environment and Planning D: Society
        & Space* 34 (4): 725–44. https://doi.org/10.1177/0263775815626420.
    2020   "Affective Economies, Pandas, and the Atmospheric Politics of Lively Capital."
        Transactions of the Institute of British Geographers 45 (3): 678–92. https://doi.org
        /10.1111/tran.12361.

Beaudine, William, dir.
    (1960) 2009   *Ten Who Dared.* Walt Disney Productions.

Beer, Gillian
    (1983) 2000   *Darwin's Plots: Evolutionary Narrative in Darwin, George Eliot, and Nine-
        teenth-Century Fiction.* Cambridge University Press. https://doi.org/10.1017/CBO
        9780511755101.

Bell, Joshua A., Joel Kuipers, Jacqueline Hazen, Amanda Kemble, and Briel Kobak
    2018   "The Materiality of Cell Phone Repair: Re-Making Commodities in Washing-
        ton, DC." *Anthropological Quarterly* 91 (2): 603–33. https://doi.org/10.1353/anq
        .2018.0028.

Bell, Lindsay A.
    2023   *Under Pressure: Diamond Mining and Everyday Life in Northern Canada.* Univer-
        sity of Toronto Press. https://doi.org/10.3138/9781487548575.

Benjamin, Walter
    (1936) 1968   "The Work of Art in the Age of Mechanical Reproduction." In *Illuminations,*
        edited by H. Arendt. Translated by H. Zohn. Schocken Books.

Bennett, Jane
    2010   *Vibrant Matter: A Political Ecology of Things.* Duke University Press. https://doi.org
        /10.2307/j.ctv111jh6w.

Benton, Michael J., and David A. T. Harper
    2020   *Introduction to Paleobiology and the Fossil Record.* Wiley Blackwell.

Berta, Annalisa, and Susan Turner
    2020   *Rebels, Scholars, Explorers: Women in Vertebrate Paleontology.* Johns Hopkins Uni-
        versity Press. https://doi.org/10.1353/book.77835.

Bielo, James S.
   2018   *Ark Encounter: The Making of a Creationist Theme Park*. New York University Press. https://doi.org/10.18574/nyu/9781479872305.001.0001.
   2019   "'Particles-to-People … Molecules-to-Man': Creationist Poetics in Public Debates." *Journal of Linguistic Anthropology* 29 (1): 4–26. https://doi.org/10.1111/jola.12205.
Bigenho, Michelle
   2002   *Sounding Indigenous: Authenticity in Bolivian Music Performance*. Palgrave Macmillan. https://doi.org/10.1007/978-1-137-11813-4.
Bird-David, Nurit
   1999   "'Animism' Revisited: Personhood, Environment, and Relational Epistemology." *Current Anthropology* 40 (S1): 67–91. https://doi.org/10.1086/200061.
Black, Riley
   2015   "T. rex Autopsy Goes into the 'Belly of the Beast' and Beyond." National Geographic, June 7. https://www.nationalgeographic.com/science/phenomena/2015/06/07/t-rex-autopsy/.
   2018   "The Many Ways Women Get Left out of Paleontology." *Smithsonian Magazine*, June 7. https://www.smithsonianmag.com/science-nature/many-ways-women-get-left-out-paleontology-180969239/.
   2019   "It's Time for the Heroic Male Paleontologist to Go Extinct." Slate, April 3. https://slate.com/technology/2019/04/what-the-new-yorker-dinosaur-story-gets-wrong.html/.
Blackhawk, Ned
   2006   *Violence over the Land: Indians and Empires in the Early American West*. Harvard University Press. https://doi.org/10.4159/9780674020993.
Bowker, Geoffrey C., and Susan Leigh Star
   1999   *Sorting Things Out: Classification and Its Consequences*. MIT Press. https://doi.org/10.7551/mitpress/6352.001.0001.
Bradford, Danielle J., and Enrico R. Crema
   2022   "Risk Factors for the Occurrence of Sexual Misconduct During Archaeological and Anthropological Fieldwork." *American Anthropologist* 124 (3): 548–59. https://doi.org/10.1111/aman.13763.
Bradley, Lawrence W.
   2014a   "Dinosaurs and Indians: Fossil Resource Dispossession of Sioux Lands, 1846–1875." *American Indian Culture and Research Journal* 38 (3): 55–84. https://doi.org/10.17953/aicr.38.3.w4l1q51m13442202.
   2014b   *Dinosaurs and Indians: Paleontology Resource Dispossession from Sioux Lands*. Outskirts Press.
Bramwell, Valerie, and Robert McCracken Peck
   2008   *All in the Bones: A Biography of Benjamin Waterhouse Hawkins*. Academy of Natural Sciences of Philadelphia.
Brannigan, Augustine
   1980   "Naturalistic and Sociological Models of the Problem of Scientific Discovery." *British Journal of Sociology* 31 (4): 559–73. https://doi.org/10.2307/589790.
   1981   *The Social Basis of Scientific Discoveries*. Cambridge University Press.
Braun, Bruce, and Sarah Whatmore, eds.
   2010   *Political Matter: Technoscience, Democracy, and Public Life*. University of Minnesota Press.

Braverman, Irus
    2013   *Zooland: The Institution of Captivity*. Stanford University Press. https://doi.org
         /10.1515/9780804784399.
    2017   "Captive: Zoometric Operations in Gaza." *Public Culture* 29 (1): 191–215. https://doi
         .org/10.1215/08992363-3644457.
Brinkman, Paul D.
    2010   *The Second Jurassic Dinosaur Rush: Museums and Paleontology in America at the
         Turn of the Twentieth Century*. University of Chicago Press. https://doi.org/10.7208
         /chicago/9780226074733.001.0001.
Brochu, Christopher A., and Colin D. Sumrall
    2001   "Phylogenetic Nomenclature and Paleontology." *Journal of Paleontology* 75 (4):
         754–57. https://doi.org/10.1666/0022-3360(2001)075<0754:pnap>2.0.co;2.
Brown, Barnum
    1915   "Tyrannosaurus, a Cretaceous Carnivorous Dinosaur: The Largest Flesh-Eater
         That Ever Lived." *Scientific American*, October 9, 322–23. https://doi.org/10.1038
         /scientificamerican10091915-322.
Brown, Bill
    2001   "Thing Theory." *Critical Inquiry* 28 (1): 1–22. https://doi.org/10.1086/449030.
Brusatte, Stephen
    2018   *The Rise and Fall of the Dinosaurs: A New History of a Lost World*. William Morrow.
Buckland, Adelene
    2013   *Novel Science: Fiction and the Invention of Nineteenth-Century Geology*. University
         of Chicago Press. https://doi.org/10.7208/chicago/9780226923635.001.0001.
Buffon, Georges Louis Leclerc Comte de
    1812   *Natural History, General and Particular*. Translated with notes and observations by
         W. Smellie and W. Wood. T. Cadell and W. Davies.
Burton, Orisanmi
    2023   *Tip of the Spear: Black Radicalism, Prison Repression, and the Long Attica Revolt*.
         University of California Press. https://doi.org/10.2307/jj.5864806.
Butler, Ella
    2010   "God Is in the Data: Epistemologies of Knowledge at the Creation Museum."
         Ethnos 75 (3): 229–51. https://doi.org/10.1080/00141844.2010.507907.
Butler, Judith
    2004   *Precarious Life: The Powers of Mourning and Violence*. Verso.
Caduff, Carlo
    2014   "Pandemic Prophecy, or How to Have Faith in Reason." *Current Anthropology*
         55 (3): 296–315. https://doi.org/10.1086/676124.
Cahill, James Leo
    2013   "Anthropomorphism and Its Vicissitudes: Reflections on Homme-Sick Cinema." In
         Screening Nature: Cinema Beyond the Human, edited by A. Pick and G. Narraway.
         Berghahn Books. https://doi.org/10.2307/j.ctt9qczx4.9.
Candea, Matei
    2010   "'I Fell in Love with Carlos the Meerkat': Engagement and Detachment in Human-
         Animal Relations." *American Ethnologist* 37 (2): 241–58. https://doi.org/10.1111
         /j.1548-1425.2010.01253.x.

2012   "Different Species, One Theory: Reflections on Anthropomorphism and Anthropological Comparison." *Cambridge Journal of Anthropology* 30 (2): 118–35. https://doi.org/10.3167/ca.2012.300208.

Caro, T. M., and Gillian O'Doherty
1999   "On the Use of Surrogate Species in Conservation Biology." *Conservation Biology* 13 (4): 805–14. https://doi.org/10.1046/j.1523-1739.1999.98338.x.

Carpenter, Kenneth
2007a   "'Bison' Alticornis and O. C. Marsh's Early Views on Ceratopsians." In *Horns and Beaks: Ceratopsian and Ornithopod Dinosaurs*, edited by K. Carpenter. Indiana University Press. https://doi.org/10.2307/j.ctt1zxz1md.20.
2007b   "How to Make a Fossil." *Journal of Paleontological Sciences* 1.

Carroll, Hamilton
2011   *Affirmative Reaction: New Formations of White Masculinity*. Duke University Press. https://doi.org/10.1515/9780822393870.

Cattelino, Jessica R.
2008   *High Stakes: Florida Seminole Gaming and Sovereignty*. Duke University Press. https://doi.org/10.1515/9780822391302.

Chambers, Paul
2005   "Rock Idol: A Lucky Break Transformed T. Rex from Obscure Fossil into the Most Famous Dinosaur of All Time." *New Scientist* 188 (2519): 31.

Chibnik, Michael
2018   "Artisan." In *International Encyclopedia of Anthropology*, edited by H. Callan. John Wiley and Sons. https://doi.org/10.1002/9781118924396.wbiea1394.

Chu, Julie Y.
2014   "When Infrastructures Attack: The Workings of Disrepair in China." *American Ethnologist* 41 (2): 351–67. https://doi.org/10.1111/amet.12080.

Cohen, Claudine
2002   *The Fate of the Mammoth: Fossils, Myths, and History*. University of Chicago Press.

Colbert, Edwin H.
1984   *The Great Dinosaur Hunters and Their Discoveries*. Dover.

Connolly, William E
2002   *Neuropolitics: Thinking, Culture, Speed*. University of Minnesota Press.

Cooper, Merian C., and Ernest B. Schoedsack, dirs.
1933   *King Kong*. RKO Radio Pictures.

Coronil, Fernando
1997   *The Magical State: Nature, Money, and Modernity in Venezuela*. University of Chicago Press.

Cowen, Richard
2000   *History of Life*. Blackwell.

Creighton, Millie R.
1994   "'Edutaining' Children: Consumer and Gender Socialization in Japanese Marketing." *Ethnology* 33 (1): 35–52. https://doi.org/10.2307/3773973.

Crist, Eileen
1999   *Images of Animals: Anthropomorphism and Animal Mind*. Temple University Press.

Cronin, Keith
   2014   "A Bone to Pick: The Paleontological Resources Preservation Act and Its Effect on Commercial Paleontology." *Albany Government Law Review* 7 (1): 268–89.

Dale, Richard, dir.
   2015   *T-Rex Autopsy*. National Geographic.

Daston, Lorraine, and Gregg Mitman, eds.
   2005   *Thinking with Animals: New Perspectives on Anthropomorphism*. Columbia University Press.

Daston, Lorraine, and Peter Galison
   1992   "The Image of Objectivity." *Representations* (40): 81–128. https://doi.org/10.2307/2928741.
   2007   *Objectivity*. Zone Books. https://doi.org/10.2307/j.ctv1c9hq4d.

Davidov, Veronica
   2014   "Land, Copper, Flora: Dominant Materialities and the Making of Ecuadorian Resource Environments." *Anthropological Quarterly* 87 (1): 31–58. https://doi.org/10.1353/anq.2014.0010.

Davidson, John A., Katherine K. Honda, and Beth Simmons
   2010   *The Rooney Ranch*. Friends of Dinosaur Ridge.

Davis, Susan G.
   1997   *Spectacular Nature: Corporate Culture and the Sea World Experience*. University of California Press.

de la Cadena, Marisol
   2015a   "Anthropology and STS: Generative Interfaces, Multiple Locations." *HAU Journal of Ethnographic Theory* 5 (1): 437–75. https://doi.org/10.14318/hau5.1.020.
   2015b   *Earth Beings: Ecologies of Practice across Andean Worlds*. Duke University Press. https://doi.org/10.2307/j.ctv11smtkx.

De León, Jason
   2015   *The Land of Open Graves: Living and Dying on the Migrant Trail*. University of California Press. https://doi.org/10.1525/9780520958685.

Debus, Allen A.
   2016   *Dinosaurs Ever Evolving: The Changing Face of Prehistoric Animals in Popular Culture*. McFarland.

Deleuze, Gilles, and Félix Guattari
   1987   *A Thousand Plateaus: Capitalism and Schizophrenia*. Translated by B. Massumi. University of Minnesota Press.

Derrida, Jacques
   2009   *The Beast & the Sovereign*. Translated by G. Bennington. University of Chicago Press.

Descola, Philippe
   2013   *Beyond Nature and Culture*. Translated by J. Lloyd. University of Chicago Press. https://doi.org/10.7208/chicago/9780226145006.001.0001.
   2014   "All Too Human (Still)." *HAU Journal of Ethnographic Theory* 4 (2): 267–73. https://doi.org/10.14318/hau4.2.015.

Despret, Vinciane, and Brett Buchanan
   2016   *What Would Animals Say If We Asked the Right Questions?* University of Minnesota Press. https://doi.org/10.5749/minnesota/9780816692378.001.0001.

Detwiler, Katheryn

    2022   "Cosmos, Cloud, and Underground: The Production and Circulation of Astronomical Big Data in Chile's Atacama Desert." Paper presented at the conference, Subterranean Anthropology: Locating Power And Politics Below Ground, Puebla, Mexico, December 8.

De Waal, Frans B. M.

    1999   "Anthropomorphism and Anthropodenial: Consistency in Our Thinking About Humans and Other Animals." *Philosophical Topics* 27 (1): 255. https://doi.org/10.5840/philtopics199927122.

Dewsbury, J-D, and Simon Naylor

    2002   "Practising Geographical Knowledge: Fields, Bodies and Dissemination." *Area* 34 (3): 253–60. https://doi.org/10.1111/1475-4762.00079.

Dugatkin, Lee Alan

    2009   *Mr. Jefferson and the Giant Moose: Natural History in Early America.* University of Chicago Press. https://doi.org/10.7208/chicago/9780226169194.001.0001.

Dulin, John

    2021   "Charismatic Christianity's Hard Cultural Forms and the Local Patterning of the Divine Voice in Ghana." *American Anthropologist* 123 (1): 108–19. https://doi.org/10.1111/aman.13523.

Dussias, Allison M.

    1996   "Science, Sovereignty, and the Sacred Text: Paleontological Resources and Native American Rights." *Maryland Law Review* 55 (1): 84–159.

Eliade, Mircea

    (1949) 2005   *The Myth of the Eternal Return, or Cosmos and History.* Translated by W. R. Trask Princeton University Press. https://doi.org/10.1515/9780691238326.

Emling, Shelley

    2009   *The Fossil Hunter: Dinosaurs, Evolution, and the Woman Whose Discoveries Changed the World.* Palgrave Macmillan.

Escobar, Arturo

    1999   "After Nature: Steps to an Antiessentialist Political Ecology." *Current Anthropology* 40 (1): 1–30. https://doi.org/10.1086/515799.

Feo, Teresa J., Emma Simon, and Richard O. Prum

    2016   "Theory of the Development of Curved Barbs and Their Effects on Feather Morphology." *Journal of Morphology* 277 (8): 995–1013. https://doi.org/10.1002/jmor.20552.

Ferry, Elizabeth Emma

    2008   "Geologies of Power: Value Transformations of Mineral Specimens from Guanajuato, Mexico." *American Ethnologist* 32 (3): 420–36. https://doi.org/10.1525/ae.2005.32.3.420.

    2010   "'Ziegfeld Girls Coming Down a Runway': Exhibiting Minerals at the Smithsonian." *Journal of Material Culture* 15 (1): 30–63. https://doi.org/10.1177/1359183510355224.

    2013   *Minerals, Collecting, and Value across the U.S.-Mexico Border.* Indiana University Press. https://doi.org/10.2979/6800.0.

    2020   "The Inorganic Slot." In *Theorizing the Contemporary* (blog). Society for Cultural Anthropology, September 22. https://culanth.org/fieldsights/the-inorganic-slot/.

Ferry, Elizabeth Emma, and Mandana E. Limbert, eds.
   2008   *Timely Assets: The Politics of Resources and Their Temporalities.* School for Advanced Research Press.

Fiffer, Steve
   2001   *Tyrannosaurus Sue: The Extraordinary Saga of the Largest, Most Fought over T. Rex Ever Found.* W. H. Freeman.

Fischer, Michael M. J.
   2007   "Four Genealogies for a Recombant Anthropology of Science and Technology." *Cultural Anthropology* 22 (4): 539–615. https://doi.org/10.1525/can.2007.22.4.539.

Foster, John Russell
   2007   *Jurassic West: The Dinosaurs of the Morrison Formation and Their World.* Indiana University Press.

Franklin, Sarah
   1995a   "Romancing the Helix: Nature and Scientific Discovery." In *Romance Revisited,* edited by L. Pearce and J. Stacey. New York University Press.
   1995b   "Science as Culture, Cultures of Science." *Annual Review of Anthropology* 24:163–84. https://doi.org/10.1146/annurev.anthro.24.1.163.

Gell, Alfred
   1988   "Technology and Magic." *Anthropology Today* 4 (2): 6–9. https://doi.org/10.2307/3033230.
   1998   *Art and Agency: An Anthropological Theory.* Clarendon Press.

Geertz, Clifford
   1972   "Deep Play: Notes on the Balinese Cockfight." Daedalus 101 (1): 1–37.

Germond, Mary F.
   1982   "George Mcjunkin." In *Dictionary of American Negro Biography,* edited by R. W. Logan and M. R. Winston. W. W. Norton.

Gieryn, Thomas F.
   1983   "Boundary-Work and the Demarcation of Science from Non-Science: Strains and Interests in Professional Ideologies of Scientists." *American Sociological Review* 48 (6): 781–95. https://doi.org/10.2307/2095325.
   1999   *Cultural Boundaries of Science: Credibility on the Line.* University of Chicago Press.

Goetzmann, William H.
   1966   *Exploration and Empire: The Explorer and the Scientist in the Winning of the American West.* Vintage.

Goffman, Erving
   1959   *The Presentation of Self in Everyday Life.* Doubleday.

Goodwin, Charles
   1994   "Professional Vision." *American Anthropologist* 96 (3): 606–33. https://doi.org/10.1525/aa.1994.96.3.02a00100.

Goody, Esther N.
   2001   "Anthropology of Craft Production." In *International Encyclopedia of the Social & Behavioral Sciences,* edited by J. D. Wright. 2nd ed. Elsevier. https://doi.org/10.1016/b0-08-043076-7/00834-2.

Gould, Stephen Jay
   1995   *Dinosaur in a Haystack: Reflections in Natural History.* Harmony Books.

Grandin, Greg
    2019   *The End of the Myth: From the Frontier to the Border Wall in the Mind of America.* Metropolitan Books.

Grasseni, Cristina, ed.
    2010   *Skilled Visions: Between Apprenticeship and Standards.* Berghahn Books. https://doi.org/10.1515/9780857455666.
    2022   "More Than Visual: The Apprenticeship of Skilled Visions." *Ethos*: 1–19. https://doi.org/10.1111/etho.12372.

Grazian, David
    2015   *American Zoo: A Sociological Safari.* Princeton University Press. https://doi.org/10.1515/9781400873616.

Gusterson, Hugh
    1998   *Nuclear Rites: A Weapons Laboratory at the End of the Cold War.* University of California Press.

Hall, Stuart
    1981   "Notes on Deconstructing 'the Popular.'" In *People's History and Social Theory*, edited by R. Samuel. Routledge.

Hallam, Elizabeth, and Tim Ingold, eds.
    2014   *Making and Growing: Anthropological Studies of Organisms and Artefacts.* Ashgate. https://doi.org/10.4324/9781315593258-1.

Hannigan, A. John
    1995   "Theme Parks and Urban Fantasy-Scapes." *Current Sociology* 43 (1): 183–91. https://doi.org/10.1177/001139295043001015.

Haraway, Donna Jeanne
    1989   *Primate Visions: Gender, Race, and Nature in the World of Modern Science.* Routledge.
    2001   "From Cyborgs to Companion Species: Kinship in Technoscience." Keynote lecture presented at the conference Taking Nature Seriously: Citizens, Science, and Environment, University of Oregon, Eugene, February 25.
    2003a   *The Companion Species Manifesto: Dogs, People, and Significant Otherness.* Prickly Paradigm Press.
    2003b   "For the Love of a Good Dog: Webs of Action in the World of Dog Genetics." In *Race, Nature, and the Politics of Difference*, edited by D. S. Moore, J. Kosek, and A. Pandian. Duke University Press. https://doi.org/10.1515/9780822384656-010.
    2008   *When Species Meet.* University of Minnesota Press.

Hartigan, John, Jr.
    2021   "Knowing Animals: Multispecies Ethnography and the Scope of Anthropology." *American Anthropologist* 123 (4): 846–60. https://doi.org/10.1111/aman.13631.

Haste, Helen
    1993   "Dinosaur as Metaphor." *Modern Geology* 18:349–70.

Hayden, Ferdinand Vandeveer
    1874   "The Hayden Expedition." *College Courant* 14 (2): 18–21.

Heidegger, Martin
    1968   *What Is a Thing?* Translated by W. B. Barton Jr. and V. Deutsch. H. Regnery.

Heininger, Mary Lynn Stevens

1984   *A Century of Childhood, 1820–1920.* Margaret Woodbury Strong Museum.

Helmreich, Stefan

2009   *Alien Ocean: Anthropological Voyages in Microbial Seas.* University of California Press. https://doi.org/10.1525/9780520942608.

High, Mette M.

2019   "Projects of Devotion: Energy Exploration and Moral Ambition in the Cosmo-economy of Oil and Gas in the Western United States: Projects of Devotion." *Journal of the Royal Anthropological Institute* 25:29–46. https://doi.org/10.1111/1467-9655.13013.

2022   "Articulations of Ethics: Energy Worlds and Moral Selves." In *The Palgrave Handbook of the Anthropology of Technology,* edited by M. H. Bruun, A. Wahlberg, R. Douglas-Jones, C. Hasse, K. Hoeyer, D. B. Kristensen, and B. R. Winthereik. Palgrave Macmillan. https://doi.org/10.1007/978-981-16-7084-8_31.

Holdzkom, Marianne

2011   "A Past to Make Us Proud: U.S. History According to Disney." In *Learning from Mickey, Donald, and Walt: Essays on Disney's Edutainment Films,* edited by A. B. Van Riper. McFarland.

Holmes, Mary Anne, Suzanne O'Connell, Kuheli Dutt, and Ann E. Austin, eds.

2015   *Women in the Geosciences: Practical, Positive Practices toward Parity.* American Geophysical Union. https://doi.org/10.1002/9781119067573.

Holtz, Thomas R., Jr., and M. K. Brett-Surman

1997   "The Taxonomy and Systematics of the Dinosaurs." In *The Complete Dinosaur,* edited by J. O. Farlow and M. K. Brett-Surman. Indiana University Press.

Honda, Katherine K., and Beth Simmons

2009   *The Legacy of Arthur Lakes.* Friends of Dinosaur Ridge.

Horne, Gerald

2005   *Black and Brown: African Americans and the Mexican Revolution, 1910–1920.* New York University Press. https://doi.org/10.18574/nyu/9780814769720.001.0001.

Horner, John R., and Don Lessem

1993   *The Complete T. Rex.* Simon & Schuster.

Houck, Karen J., and Spencer G. Lucas

2015   "Martin G. Lockley: Premier Student of Fossil Footprints: An Introduction to the Special Issue." *Ichnos* 22 (3–4): 133–35. https://doi.org/10.1080/10420940.2015.1059336.

Howe, Cymene

2015   "Life above Earth: An Introduction." *Cultural Anthropology* 30 (2): 2039. https://doi.org/10.14506/ca30.2.03.

Hoy, James F.

1998   "Rocky Mountain Oysters." In *The Taste of American Place: A Reader on Regional and Ethnic Foods,* edited by B. G. Shortridge and J. R. Shortridge. Rowman & Littlefield.

Hoyt, Harry O., dir.

1925   *The Lost World.* First National Pictures.

Huber, Matthew T.

2013   *Lifeblood: Oil, Freedom and the Forces of Capital.* University of Minnesota Press. https://doi.org/10.5749/minnesota/9780816677849.001.0001.

Huls, Alexander
  2013   "The Jurassic Park Period: How CGI Dinosaurs Transformed Film Forever." *Atlantic*, April 4.
Hunt, Adrian, Martin Lockley, and Sally White
  2002   *Historic Dinosaur Quarries of the Dinosaur Ridge Area*. Friends of Dinosaur Ridge.
Hustak, Carla, and Natasha Myers
  2012   "Involutionary Momentum: Affective Ecologies and the Sciences of Plant/Insect Encounters." *Differences: A Journal of Feminist Cultural Studies* 23 (3): 74. https://doi.org/10.1215/10407391-1892907.
Ingold, Tim
  2011   *Being Alive: Essays on Movement, Knowledge and Description*. Routledge. https://doi.org/10.4324/9781003196679.
  2013   *Making: Anthropology, Archaeology, Art and Architecture*. Routledge. https://doi.org/10.4324/9780203559055.
Ingold, Tim, and Elizabeth Hallam
  2014   "Making and Growing: An Introduction." In *Making and Growing: Anthropological Studies of Organisms and Artefacts*. Ashgate. https://doi.org/10.4324/9781315593258-1.
Irwin, Lee
  1994   *The Dream Seekers: Native American Visionary Traditions of the Great Plains*. University of Oklahoma Press.
Jacka, Jerry K.
  2018   "The Anthropology of Mining: The Social and Environmental Impacts of Resource Extraction in the Mineral Age." *Annual Review of Anthropology* 47 (1): 61–77. https://doi.org/10.1146/annurev-anthro-102317-050156.
Jaffe, Mark
  2000   *The Gilded Dinosaur: The Fossil War between E. D. Cope and O. C. Marsh and the Rise of American Science*. Crown.
Jefferson, Thomas
  (1801) 2022   *Notes on the State of Virginia: An Annotated Edition*, edited by R. P. Forbes. Yale University Press. https://doi.org/10.2307/j.ctv2g5917f.
Johnson, Kirk, Betsy Armstrong, Chip Colwell-Chanthaphonh, Frances Kruger, Kristine Haglund, and Frank-Thorsten Krell
  2013   *Denver's Natural History Museum: A History*. Vol. 4. Denver Museum of Nature and Science.
Jones, Elizabeth D.
  2020   "Assumptions of Authority: The Story of Sue the T-Rex and Controversy over Access to Fossils." *History and Philosophy of the Life Sciences* 42 (1): 2–2. https://doi.org/10.1007/s40656-019-0288-4.
Kalshoven, Petra Tjitske
  2018a   "Gestures of Taxidermy: Morphological Approximation as Interspecies Affinity." *American Ethnologist* 45 (1): 34–47. https://doi.org/10.1111/amet.12597.
  2018b   "Piecing Together the Extinct Great Auk: Techniques and Charms of Contiguity." *Environmental Humanities* 10 (1): 150–70. https://doi.org/10.1215/22011919-4385507.
Keane, Webb
  2003   "Semiotics and the Social Analysis of Material Things." *Language & Communication* 23 (3–4): 409–25. https://doi.org/10.1016/s0271-5309(03)00010-7.

Keller, Evelyn Fox

1983   *A Feeling for the Organism: The Life and Work of Barbara Mcclintock*. W. H. Freeman.

1995   *Reflections on Gender and Science*. Yale University Press.

2016   "Active Matter, Then and Now." *History and Philosophy of the Life Sciences* 38 (3): 1–11. https://doi.org/10.1007/s40656-016-0112-3.

Kidd, Kenneth B.

2004   "Disney of Orlando's Animal Kingdom." In *Wild Things: Children's Culture and Ecocriticism*, edited by S. I. Dobrin and K. B. Kidd. Wayne State University Press.

Kilroy-Marac, Katie

2016   "A Magical Reorientation of the Modern: Professional Organizers and Thingly Care in Contemporary North America." *Cultural Anthropology* 31 (3): 438–57. https://doi.org/10.14506/ca31.3.09.

Kim, Eleana J.

2016   "Toward an Anthropology of Landmines: Rogue Infrastructure and Military Waste in the Korean D.M.Z." *Cultural Anthropology* 31 (2): 162–87. https://doi .org/10.14506/ca31.2.02.

Kirksey, Eben

2015   "Species: A Praxiographic Study." *Journal of the Royal Anthropological Institute* 21 (4): 758–80. https://doi.org/10.1111/1467-9655.12286.

Kirshenblatt-Gimblett, Barbara

1998   *Destination Culture: Tourism, Museums, and Heritage*. University of California Press.

Kjærgaard, Peter C.

2012   "The Fossil Trade: Paying a Price for Human Origins." *Isis* 103 (2): 340–55. https:// doi.org/10.1086/666365.

Kohl, Michael F., and John Stanton McIntosh

1997   "Introduction." In *Discovering Dinosaurs in the Old West: The Field Journals of Arthur Lakes*, edited by M. F. Kohl and J. S. McIntosh. Smithsonian Institution Press.

Kopytoff, Igor

1986   "The Cultural Biography of Things: Commoditization as Process." In *The Social Life of Things: Commodities in Cultural Perspective*, edited by A. Appadurai. Cambridge University Press.

Krishtalka, Leonard

1989   *Dinosaur Plots and Other Intrigues in Natural History*. William Morrow.

Kurin, Richard

2006   *Hope Diamond: The Legendary History of a Cursed Gem*. Smithsonian Books.

Kurtti, Jeff

1996   *Since the World Began: Walt Disney World, the First 25 Years*. Hyperion.

Lakes, Arthur

1997   *Discovering Dinosaurs in the Old West: The Field Journals of Arthur Lakes*, edited by M. F. Kohl and J. S. McIntosh. Smithsonian Institution Press.

2007   *Arthur Lakes' "Rambles around the Ridge,"* compiled by B. H. Simmons and K. Warren. Edited by E. Warren and J. Jacobs. Friends of Dinosaur Ridge.

Lane, N. Gary

1989   "Paleontology: The Academy and the Marketplace." *Journal of Paleontology* 63 (3): 259–60. https://doi.org/10.1017/s0022336000019429.

Lanham, Url

2012   *The Bone Hunters: The Heroic Age of Paleontology in the American West*. Dover.

Latour, Bruno

  1987 *Science in Action: How to Follow Scientists and Engineers through Society*. Harvard University Press.

  1992 "Where Are the Missing Masses? The Sociology of a Few Mundane Artifacts." In *Shaping Technology/Building Society Studies in Sociotechnical Change*, edited by W. E. Bijker and J. Law. MIT Press.

  2000 "The Berlin Key or How to Do Words with Things." In *Matter, Materiality and Modern Culture*, edited by P. Graves-Brown. Routledge.

  2004a "How to Talk About the Body? The Normative Dimension of Science Studies." *Body & Society* 10 (2–3): 205–29. https://doi.org/10.1177/1357034X04042943.

  2004b "Why Has Critique Run out of Steam? From Matters of Fact to Matters of Concern." *Critical Inquiry* 30:225–48. https://doi.org/10.2307/1344358.

Latour, Bruno (as Jim Johnson)

  1988 "Mixing Humans and Nonhumans Together: The Sociology of a Door-Closer." *Social Problems* 35 (3): 298–310. https://doi.org/10.1525/sp.1988.35.3.03a00070.

Latour, Bruno, and Steve Woolgar

  1986 *Laboratory Life: The Construction of Scientific Facts*. Princeton University Press.

Lauter, Paul

  2001 *From Walden Pond to Jurassic Park: Activism, Culture, and American Studies*. Duke University Press. https://doi.org/10.1515/9780822380474.

Lave, Jean, and Etienne Wenger

  1991 *Situated Learning: Legitimate Peripheral Participation*. Cambridge University Press.

Lave, Rebecca

  2015 "Introduction to Special Issue on Critical Physical Geography." *Progress in Physical Geography* 39 (5): 571–75. https://doi.org/10.1177/0309133315608006.

Law, John, and Annemarie Mol

  1995 "Notes on Materiality and Sociality." *Sociological Review* 43 (2): 274–94. https://doi.org/10.1111/j.1467-954x.1995.tb00604.x.

Lawrence, Christopher, and Michael Brown

  2016 "Quintessentially Modern Heroes: Surgeons, Explorers, and Empire, c. 1840–1914." *Journal of Social History* 50 (1): 148–78. https://doi.org/10.1093/jsh/shw014.

Leader-Williams, Nigel, and Holly T. Dublin

  2000 "Charismatic Megafauna as 'Flagship Species.'" In *Priorities for the Conservation of Mammalian Diversity: Has the Panda Had Its Day?*, edited by A. Entwistle and N. Dunstone. Cambridge University Press.

Lévi-Strauss, Claude

  1963 *Totemism*. Translated by R. Needham. Beacon Press.

Li, Fabiana

  2015 *Unearthing Conflict: Corporate Mining, Activism, and Expertise in Peru*. Duke University Press. https://doi.org/10.1515/9780822375869.

Liggett, Gregory, S. Terry Childs, Nicholas Famoso, H. Gregory McDonald, Alan L. Titus, Elizabeth Varner, and Cameron L. Liggett

  2018 "From Public Lands to Museums: The Foundation of U.S. Paleontology, the Early History of Federal Public Lands and Museums, and the Developing Role of the U.S. Department of the Interior." In *Museums at the Forefront of the History and Philosophy of Geology: History Made, History in the Making*, edited by

G. D. Rosenberg and R. M. Clary. Geological Society of America. https://doi.org /10.1130/2018.2535(21).

Limerick, Patricia Nelson

1987   *The Legacy of Conquest: The Unbroken Past of the American West.* W. W. Norton.

Lindholm, Charles, ed.

2013   *The Anthropology of Religious Charisma: Ecstasies and Institutions.* Palgrave Macmillan. https://doi.org/10.1057/9781137377630.

Lockley, Martin, and Clare Marshall

2014   *A Field Guide to the Dinosaur Ridge Area.* Friends of Dinosaur Ridge.

Lorimer, Jamie

2007   "Nonhuman Charisma." *Environment and Planning D: Society and Space* 25 (5): 911–32. https://doi.org/10.1068/d71j.

2015   *Wildlife in the Anthropocene: Conservation after Nature.* University of Minnesota Press. https://doi.org/10.5749/minnesota/9780816681075.001.0001.

Lunney, Daniel

2012   "Charismatic Megafauna." In *The Berkshire Encyclopedia of Sustainability: Ecosystem Management and Sustainability*, edited by D. E. Vasey. Berkshire.

Lyons, Kristina M.

2014   "Soil Science, Development, and the 'Elusive Nature' of Colombia's Amazonian Plains." *Journal of Latin American and Caribbean Anthropology* 19 (2): 212–36. https://doi.org/10.1111/jlca.12097.

2016   "Decomposition as Life Politics: Soils, Selva, and Small Farmers under the Gun of the U.S.-Colombian War on Drugs." *Cultural Anthropology* 31 (1): 56–81. https://doi.org /10.14506/ca31.1.04.

2020   *Vital Decomposition: Soil Practitioners and Life Politics.* Duke University Press. https://doi.org/10.1515/9781478009207.

Macdonald, Sharon

2002   *Behind the Scenes at the Science Museum.* Berg. https://doi.org/10.4324/9781003084785.

MacLeod, Roy

2009   "Discovery and Exploration." In *The Modern Biological and Earth Sciences*, edited by J. V. Pickstone and P. J. Bowler. Vol. 6, *The Cambridge History of Science.* Cambridge University Press.

Malinowski, Bronislaw

1948   *Magic, Science and Religion: And Other Essays.* Beacon Press.

Marsh, Diana E.

2019   *Extinct Monsters to Deep Time: Conflict, Compromise, and the Making of Smithsonian's Fossil Halls.* Berghahn Books. https://doi.org/10.1515/9781789201239.

Marshall, Nancy Rose

2007   "'A Dim World, Where Monsters Dwell': The Spatial Time of the Sydenham Crystal Palace Dinosaur Park." *Victorian Studies* 49 (2): 286–301. https://doi.org/10.2979 /vic.2007.49.2.286.

Martin, Anthony J.

2006   *Introduction to the Study of Dinosaurs.* Blackwell.

Martin, Emily

1991   "The Egg and the Sperm: How Science Has Constructed a Romance Based on Stereotypical Male-Female Roles." *Signs* 16 (3): 485–501. https://doi.org/10.1093/oso /9780198751458.003.0008.

1994   *Flexible Bodies: Tracking Immunity in American Culture from the Days of Polio to the Age of AIDS.* Beacon Press.

1998   "Anthropology and the Cultural Study of Science." *Science, Technology, & Human Values* 23 (1): 24–44. https://doi.org/10.1177/016224399802300102.

2021   *Experiments of the Mind: From the Cognitive Psychology Lab to the World of Facebook and Twitter.* Princeton University Press. https://doi.org/10.23943/princeton/9780691230719.001.0001.

Marín-Spiotta, E., Rebecca T. Barnes, Asmeret Asefaw Berhe, Meredith G. Hastings, Allison Mattheis, Blair Schneider, and Billy M. Williams

2020   "Hostile Climates Are Barriers to Diversifying the Geosciences." *Advances in Geosciences* 53:117–27. https://doi.org/10.5194/adgeo-53-117-2020.

Mauzé, Marie

1998   "Northwest Coast Trees: From Metaphors in Culture to Symbols for Culture." In *The Social Life of Trees: Anthropological Perspectives on Tree Symbolism*, edited by L. M. Rival. Berg.

May, Shannon

2010   "Rethinking Anonymity in Anthropology: A Question of Ethics." *Anthropology News* 51 (4): 10–13. https://doi.org/10.1111/j.1556-3502.2010.51410.x.

Mayor, Adrienne

2000   *The First Fossil Hunters: Paleontology in Greek and Roman Times.* Princeton University Press.

2005   *Fossil Legends of the First Americans.* Princeton University Press.

2007   "Place Names Describing Fossils in Oral Traditions." In *Myth and Geology*, edited by L. Piccardi and W. B. Masse. Geological Society.

McCay, Winsor, dir.

1914   *Gertie the Dinosaur.* Box Office Attractions.

McCormick, John

1985   "Saving 'Charismatic' Animals." *Newsweek*, April 22, 10.

McGranahan, Carole

2020   "Anthropology as Theoretical Storytelling." In *Writing Anthropology: Essays on Craft and Commitment*, edited by C. McGranahan. Duke University Press. https://doi.org/10.1515/9781478009160-013.

McKittrick, Katherine, ed.

2015   *Sylvia Wynter: On Being Human as Praxis.* Duke University Press. https://doi.org/10.1515/9780822375852.

McLean, Stuart J.

2008   "Bodies from the Bog: Metamorphosis, Non-Human Agency and the Making of Collective Memory." *Trames* 12 (3): 299–308. https://doi.org/10.3176/tr.2008.3.05.

2011   "Black Goo: Forceful Encounters with Matter in Europe's Muddy Margins." *Cultural Anthropology* 26 (4): 589–619. https://doi.org/10.1111/j.1548-1360.2011.01113.x.

McLennan, Sharon

2014   "Networks for Development: Volunteer Tourism, Information and Communications Technology, and the Paradoxes of Alternative Development." *PoLAR: Political and Legal Anthropology Review* 37 (1): 48–68. https://doi.org/10.1111/plar.12050.

Melson, Gail F.

2001   *Why the Wild Things Are: Animals in the Lives of Children.* Harvard University Press. https://doi.org/10.4159/9780674040922.

Meneley, Anne

2008   "Oleo-Signs and Quali-Signs: The Qualities of Olive Oil." *Ethnos* 73 (3): 303–26. https://doi.org/10.1080/00141840802324003.

Merchant, Carolyn

1989   *The Death of Nature: Women, Ecology, and the Scientific Revolution.* Harper & Row.

Messeri, Lisa

2016   *Placing Outer Space: An Earthly Ethnography of Other Worlds.* Duke University Press. https://doi.org/10.2307/j.ctv11cw9f7.

2017   "Resonant Worlds." *American Ethnologist* 44 (1): 1–12. https://doi.org/10.1111/amet .12431.

Miller, Daniel

2010   *Stuff.* Polity.

Mintz, Sidney Wilfred

1986   *Sweetness and Power: The Place of Sugar in Modern History.* Penguin Books.

Mitchell, Kirk

2019   "It's a Triceratops! Denver Museum Releases New Info on Highlands Ranch Dinosaur Dig." *Denver Post*, June 21, https://www.denverpost.com/2019/06/21 /highlands-ranch-triceratops-bones-denver-science-museum/.

Mitchell, Sandra

2005   "Anthropomorphism and Cross-Species Modeling." In *Thinking with Animals New Perspectives on Anthropomorphism*, edited by L. Daston and G. Mitman. Columbia University Press.

Mitchell, W. J. T.

1998   *The Last Dinosaur Book: The Life and Times of a Cultural Icon.* University of Chicago Press.

2001   "Romanticism and the Life of Things: Fossils, Totems, and Images." *Critical Inquiry* 28 (1): 167–84. https://doi.org/10.1086/449037.

2005   *What Do Pictures Want? The Lives and Loves of Images.* University of Chicago Press. https://doi.org/10.7208/chicago/9780226245904.001.0001.

Mitman, Gregg

2005   "The Media of Science, Politics, and Conservation." In *Thinking with Animals New Perspectives on Anthropomorphism*, edited by L. Daston and G. Mitman. Columbia University Press.

Modreski, Peter J., Norbert E. Cygan, Elizabeth P. Rall, and Robert G Raynold

1998   "A Window on the Mesazoic Era Dinosaur Ridge Morrison, Colorado." *Rocks and Minerals* 73 (5): 330. https://doi.org/10.1080/00357529809602999.

Monarrez, Pedro M., Joshua B. Zimmt, Annaka M. Clement, William Gearty, John J. Jacisin III, Kelsey M. Jenkins, Kristopher M. Kusnerik, et al.

2022   "Our Past Creates Our Present: a Brief Overview of Racism and Colonialism in Western Paleontology." *Paleobiology* 48 (2): 173–85. https://doi.org/10.1017 /pab.2021.28.

Morris, Rosalind

2007   "The Age of Dinosaurs." *New Ohio Review* 1 (1): 175–79.

Mullen, Bill

2010   "Sue Pays Off for Field Museum: 10 Years on, T. Rex Proving to Be a $8.3 Million Bargain." *Chicago Tribune*, May 16.

Myers, Natasha
  2008   "Molecular Embodiments and the Body-Work of Modeling in Protein Crystallography." *Social Studies of Science* 38 (2): 163–99. https://doi.org/10.1177/0306312707082969.
  2009   "Performing the Protein Fold." In *Simulation and Its Discontents*, edited by S. Turkle. MIT Press. https://doi.org/10.7551/mitpress/8200.003.0015.
  2015   *Rendering Life Molecular: Models, Modelers, and Excitable Matter*. Duke University Press. https://doi.org/10.1515/9780822375630.

Nash, June C.
  (1979) 1993   *We Eat the Mines and the Mines Eat Us: Dependency and Exploitation in Bolivian Tin Mines*. Columbia University Press.

Nelson, Robin G., Julienne N. Rutherford, Katie Hinde, and Kathryn B. H. Clancy
  2017   "Signaling Safety: Characterizing Fieldwork Experiences and Their Implications for Career Trajectories." *American Anthropologist* 119 (4): 710–22. https://doi.org/10.1111/aman.12929.

New York Times
  1905   "Mining for Mammoths in the Bad Lands: How the Monster Tyrannosaurus Rex Was Dug out of His 8,000,000-Year-Old Tomb." *New York Times*, December 3.

Nightengale, Neil, and Barry Cook, dirs.
  2013   *Walking with Dinosaurs*. 20th Century Fox.

Noble, Brian
  2016   *Articulating Dinosaurs: A Political Anthropology*. University of Toronto Press. https://doi.org/10.3138/9781442621312.

O'Brien, Willis, dir.
  1919   *Ghost of Slumber Mountain*. World Film.

O'Connor, Ralph
  2007   *The Earth on Show: Fossils and the Poetics of Popular Science, 1802–1856*. University of Chicago Press. https://doi.org/10.7208/chicago/9780226616704.001.0001.

Oguz, Zeynep, ed.
  2020   "Geological Anthropology." In *Theorizing the Contemporary* (blog). Society for Cultural Anthropology, September 22. https://culanth.org/fieldsights/series/geological-anthropology/.

O'Neill, John
  1996   "Dinosaurs-R-Us: The (Un) Natural History of Jurassic Park." In *Monster Theory: Reading Culture*, edited by J. J. Cohen. University of Minnesota Press.

O'Reilly, Jessica
  2017   *The Technocratic Antarctic: An Ethnography of Scientific Expertise and Environmental Governance*. Cornell University Press. https://doi.org/10.7591/9781501708367.

Ochs, Elinor, Sally Jacoby, and Patrick Gonzales
  1994   "Interpretive Journeys: How Physicists Talk and Travel through Graphic Space." *Configurations* 2 (1): 151–71. https://doi.org/10.1353/con.1994.0003.

OED
  2016   "Anthropomorphism." In *Oxford English Dictionary*. Oxford University Press.

Ogden, Laura A.
  2021   *Loss and Wonder at the World's End*. Duke University Press. https://doi.org/10.2307/j.ctv1xg5hso.

Ortner, Sherry B.
  1973   "On Key Symbols." *American Anthropologist* 75 (5): 1338–46. https://doi.org/10.1525/aa.1973.75.5.02a00100.
Osborn, Henry Fairfield
  1905   "Tyrannosaurus and Other Cretaceous Carnivorous Dinosaurs." *Bulletin of the American Museum of Natural History* 21 (14): 259–65.
  1913   "Tyrannosaurus: Restoration and Model of the Skeleton." *Bulletin of the American Museum of Natural History* 32 (4): 91–95.
  1917   "Skeletal Adaptations of Ornitholestes, Struthiomimus, Tyrannosaurus." *Bulletin of the American Museum of Natural History* 35 (43): 733–71.
Paget, Derek
  1997   "Drifting Towards Hollywood: Drama-Documentary Goes West." *Continuum* 11 (1): 23–42. https://doi.org/10.1080/10304319709359416.
Paleontological Resources Preservation Act
  2009   Pub. L. No. 119–1416 U.S.C.S. § 470aaa-1-11.
Palmatier, Robert A.
  1995   *Speaking of Animals: A Dictionary of Animal Metaphors.* Greenwood Press.
Pandora, Katherine, and Karen A. Rader
  2008   "Science in the Everyday World." *Isis* 99 (2): 350–64. https://doi.org/10.1086/588693.
Park, Hee-Joo
  2001   "The Creation-Evolution Debate: Carving Creationism in the Public Mind." *Public Understanding of Science* 10 (2): 173–86.
Parreñas, Juno Salazar
  2012   "Producing Affect: Transnational Volunteerism in a Malaysian Orangutan Rehabilitation Center." *American Ethnologist* 39 (4): 673–87. https://doi.org/10.1111/j.1548-1425.2012.01387.x.
  2018   *Decolonizing Extinction: The Work of Care in Orangutan Rehabilitation.* Duke University Press. https://doi.org/10.1215/9780822371946.
Patchett, Merle
  2021   "Feather-Work: A Fashioned Ostrich Plume Embodies Hybrid and Violent Labors of Growing and Making." *GeoHumanities* 7 (1): 257–82. https://doi.org/10.1080/2373566X.2021.1904789.
Paton, David A.
  2013   "The Quarry as Sculpture: The Place of Making." *Environment and Planning A: Economy and Space* 45 (5): 1070–86. https://doi.org/10.1068/a45568.
Paton, David A., and Caitlin DeSilvey
  2014   "Growing Granite: The Recombinant Geologies of Sludge." In *Making and Growing: Anthropological Studies of Organisms and Artefacts*, edited by E. Hallam and T. Ingold. Ashgate. https://doi.org/10.4324/9781315593258-12.
Paxson, Heather
  2013   *The Life of Cheese: Crafting Food and Value in America.* University of California Press. https://doi.org/10.1525/9780520954021.
Peck, Robert McCracken
  2008   "The Art of Bones." *Natural History* 117 (10): 24–29.
Pels, Dick, Kevin Hetherington, and Frédéric Vandenberghe
  2002   "The Status of the Object." *Theory, Culture & Society* 19 (5–6): 1–21. https://doi.org/10.1177/026327602761899110.

Pickering, Andrew
  2010   "Material Culture and the Dance of Agency." In *The Oxford Handbook of Material Culture Studies*, edited by D. Hicks and M. C. Beaudry. Oxford University Press. https://doi.org/10.1093/oxfordhb/9780199218714.013.0007.
Povinelli, Elizabeth A.
  2016   *Geontologies: A Requiem to Late Liberalism.* Duke University Press. https://doi.org/10.2307/j.ctv11g9857.
Pretes, Michael
  2003   "Tourism and Nationalism." *Annals of Tourism Research* 30 (1): 125–42. https://doi.org/10.1016/s0160-7383(02)00035-x.
Proctor, Robert N.
  2001   "Anti-Agate: The Great Diamond Hoax and the Semiprecious Stone Scam." *Configurations* 9 (3): 381–412. https://doi.org/10.1353/con.2001.0019.
Prothero, Donald R.
  2013   *Bringing Fossils to Life: An Introduction to Paleobiology.* Columbia University Press.
  2015   *The Story of Life in 25 Fossils: Tales of Intrepid Fossil Hunters and the Wonders of Evolution.* Columbia University Press. https://doi.org/10.7312/prot17190.
Prum, Richard O.
  1999   "Development and Evolutionary Origin of Feathers." *Journal of Experimental Zoology* 285 (4): 291–306. https://doi.org/10.1002/(sici)1097-010x(19991215)285:4<291::aid-jez1>3.0.co;2-9.
Prum, Richard O., and Alan H. Brush
  2014   "Which Came First, the Feather or the Bird?" *Scientific American* 23:78–85. https://doi.org/10.1038/scientificamericandinosaurs0514-76.
Rabinow, Paul
  1996   *Making PCR: A Story of Biotechnology.* University of Chicago Press.
Rader, Karen Ann, and Victoria Cain
  2014   *Life on Display: Revolutionizing U.S. Museums of Science and Natural History in the Twentieth Century.* University of Chicago Press. https://doi.org/10.7208/chicago/9780226079837.001.0001.
Raffles, Hugh
  2020   *The Book of Unconformities: Speculations on Lost Time.* Pantheon.
Rajewski, Genevieve
  2008   "Where Dinosaurs Roamed." *Smithsonian Magazine*, May 1.
Rainger, Ronald
  1991   *An Agenda for Antiquity: Henry Fairfield Osborn and Vertebrate Paleontology at the American Museum of Natural History, 1890–1935.* University of Alabama Press.
Ranganathan, Meghana, Ellen Lalk, Lyssa M Freese, Mara A Freilich, Julia Wilcots, Margaret L Duffy, and Rohini Shivamoggi
  2021   "Trends in the Representation of Women among Us Geoscience Faculty from 1999 to 2020: The Long Road toward Gender Parity." *AGU Advances* 2 (3): e2021AV000436. https://doi.org/10.1002/essoar.10506485.2.
Rayfield, Emily J.
  2004   "Cranial Mechanics and Feeding in Tyrannosaurus Rex." *Proceedings of the Royal Society B: Biological Sciences* 271 (1547): 1451–59. https://doi.org/10.1098/rspb.2004.2755.

Redfield, Peter
  2012 "The Unbearable Lightness of Ex-Pats: Double Binds of Humanitarian Mobility." *Cultural Anthropology* 27 (2): 358–82. https://doi.org/10.1111/j.1548-1360.2012.01147.x.
Richardson, Tanya, and Gisa Weszkalnys
  2014 "Introduction: Resource Materialities." *Anthropological Quarterly* 87 (1): 5–30. https://doi.org/10.1353/anq.2014.0007.
Rieppel, Lukas
  2012 "Bringing Dinosaurs Back to Life: Exhibiting Prehistory at the American Museum of Natural History." *Isis* 103 (3): 460–90. https://doi.org/10.1086/667969.
  2015 "Prospecting for Dinosaurs on the Mining Frontier: The Value of Information in America's Gilded Age." *Social Studies of Science* 45 (2): 161–86. https://doi.org/10.1177/0306312715570650.
  2019 *Assembling the Dinosaur: Fossil Hunters, Tycoons, and the Making of a Spectacle.* Harvard University Press. https://doi.org/10.4159/9780674240339.
  2020 "How Dinosaurs Became Tyrants of the Prehistoric." *Environmental History* 25 (4): 774–87.
Ritvo, Harriet
  1997 *The Platypus and the Mermaid, and Other Figments of the Classifying Imagination.* Harvard University Press.
  2010 *Noble Cows and Hybrid Zebras: Essays on Animals and History.* University of Virginia Press.
Rival, Laura M., ed.
  1998 *The Social Life of Trees: Anthropological Perspectives on Tree Symbolism.* Berg.
Robertson, Kellie
  2012 "Exemplary Rocks." In *Animal, Vegetable, Mineral: Ethics and Objects*, edited by J. J. Cohen. Oliphaunt Books.
Robinson, Michael F.
  2006 *The Coldest Crucible: Artic Exploration and American Culture.* University of Chicago Press. https://doi.org/10.7208/chicago/9780226721873.001.0001.
Rogers, Douglas
  2012 "The Materiality of the Corporation: Oil, Gas, and Corporate Social Technologies in the Remaking of a Russian Region." *American Ethnologist* 39 (2): 284–96. https://doi.org/10.1111/j.1548-1425.2012.01364.x.
Rolston, Jessica Smith
  2013 "The Politics of Pits and the Materiality of Mine Labor: Making Natural Resources in the American West." *American Anthropologist* 115 (4): 582–94. https://doi.org/10.1111/aman.12050.
Roosth, Sophia
  2018 "Turning to Stone: Fossil Hunting and Coeval Estrangement in Montana." *Res: Anthropology and Aesthetics* 69–70:62–75. https://doi.org/10.1086/699900.
Ross, Robert M., Don Duggan-Haas, and Warren D. Allmon
  2013 "The Posture of Tyrannosaurus Rex: Why Do Student Views Lag Behind the Science?" *Journal of Geoscience Education* 61 (1): 145–60. https://doi.org/10.5408/11-259.1.

Sager, Mike
  2017   "The Ultimate Summer Camp Activity: Digging for Dinosaurs." *Smithsonian Magazine*, September 13. https://www.smithsonianmag.com/science-nature/ultimate-summer-camp-activity-digging-dinosaurs-180964862/.

Sampson, Scott D., Eric K. Lund, Mark A. Loewen, Andrew A. Farke, and Katherine E. Clayton
  2013   "A Remarkable Short-Snouted Horned Dinosaur from the Late Cretaceous (Late Campanian) of Southern Laramidia." *Proceedings of Biological Sciences* 280 (1766): 1–7. https://doi.org/10.1098/rspb.2013.1186.

Sansi Roca, Roger
  2005   "The Hidden Life of Stones: Historicity, Materiality and the Value of Candomblé Objects in Bahia." *Journal of Material Culture* 10 (2): 139–56. https://doi.org/10.1177/1359183505053072.

Sanz, José Luis
  2002   *Starring T. Rex!: Dinosaur Mythology and Popular Culture*. Indiana University Press.

Schuchert, Charles, and Clara Mae LeVene
  1940   *O. C. Marsh, Pioneer in Paleontology*. Yale University Press.

Schuller, Kyla
  2016   "The Fossil and the Photograph: Red Cloud, Prehistoric Media, and Dispossession in Perpetuity." *Configurations: A Journal of Literature, Science, and Technology* 24 (2): 229–61. https://doi.org/10.1353/con.2016.0008.

Scott, Eric
  2005   "Is Selling Vertebrate Fossils Bad for Science?" *Palaios* 20 (6): 515–17. https://doi.org/10.2110/palo.2005.s06.

Semonin, Paul
  2000   *American Monster: How the Nation's First Prehistoric Creature Became a Symbol of National Identity*. New York University Press.

Shannon, Jennifer A.
  2019   Foreward to *Extinct Monsters to Deep Time: Conflict, Compromise, and the Making of Smithsonian's Fossil Halls*, by Diana E. Marsh. Berghahn Books.

Shapin, Steven
  2008   *The Scientific Life: A Moral History of a Late Modern Vocation*. University of Chicago Press.

Shatto, Susan
  1976   "Byron, Dickens, Tennyson, and the Monstrous Efts." *Yearbook of English Studies* 6:144–55. https://doi.org/10.2307/3506397.

Shever, Elana
  2012   *Resources for Reform: Oil and Neoliberalism in Argentina*. Stanford University Press. https://doi.org/10.1515/9780804783200.
  2020   "Becoming Stone: On the Coming-into-Being of Fossils in the American West." *Anthropology Quarterly* 93 (3): 461–96. https://doi.org/10.1353/anq.2020.0055.
  2025   "More-Than-Human Charisma, Iconic Fossils, and Paleontologists in the Western United States." *Journal of the Royal Anthropological Institute* 31 (4). https://doi.org/10.1111/1467-9655.14263.

Shi, Feng, Yongren Shi, Fedor A. Dokshin, James A. Evans, and Michael W. Macy
    2017   "Millions of Online Book Co-Purchases Reveal Partisan Differences in the Consumption of Science." *Nature Human Behaviour* 1:1–9. https://doi.org/10.1038/s41562-017-0079.

Shipley, Jesse Weaver
    2009   "Comedians, Pastors, and the Miraculous Agency of Charisma in Ghana." *Cultural Anthropology* 24 (3): 523–52. https://doi.org/10.1111/j.1548-1360.2009.01039.x.

Simmons, Beth, and Katherine Honda
    2009   "Arthur Lakes: Founder of the Famed Colorado School of Mines Geology Museum." *Rocks and Minerals* 84 (5): 426–30. https://doi.org/10.3200/rmin.84.5.426-432.

Simpson, George Gaylord
    1942   "The Beginnings of Vertebrate Paleontology in North America." *Proceedings of the American Philosophical Society* 86 (1): 130–88.

Smith, Joshua B.
    2005   "Heterodonty in Tyrannosaurus Rex: Implications for the Taxonomic and Systematic Utility of Theropod Dentitions." *Journal of Vertebrate Paleontology* 25 (4): 865–87. https://doi.org/10.1671/0272-4634(2005)025[0865:hitrif]2.0.co;2.

Smith, Langdon
    1909   *Evolution: A Fantasy.* J. W. Luce.

Soper, Kate
    1995   *What Is Nature?: Culture, Politics and the Non-Human.* Blackwell.

Soto Laveaga, Gabriela
    2023   "Presidential Panel: Knowing on the Edge." Keynote lecture presented at the Society for the Social Studies of Science Conference, Cholula, Mexico.

Spence, Mark David
    1999   *Dispossessing the Wilderness: Indian Removal and the Making of the National Parks.* Oxford University Press.

Spielberg, Steven, dir.
    1993   *Jurassic Park.* Universal Studios.

Stasch, Rupert
    2014   "Primitivist Tourism and Romantic Individualism: On the Values in Exotic Stereotypy About Cultural Others." *Anthropological Theory* 14 (2): 191–214. https://doi.org/10.1177/1463499614534114.

Stewart, Susan
    1993   *On Longing: Narratives of the Miniature, the Gigantic, the Souvenir, the Collection.* Duke University Press.

Tanner, Ron
    2000   "Terrible Lizard! The Dinosaur as Plaything." *Journal of American & Comparative Cultures* 23 (2): 53–65. https://doi.org/10.1111/j.1542-734X.2000.2302_53.x.

Tattersdill, Will
    2017   "Work on the Victorian Dinosaur: Histories and Prehistories of 19th-Century Palaeontology." *Literature Compass* 14 (6). https://doi.org/10.1111/lic3.12394.

Tauxe, Caroline S.
    1993   *Farms, Mines, and Main Streets: Uneven Development in a Dakota County.* Temple University Press.

Thompson, Charis M.
  2015   "Situated Knowledge, Feminist and Science and Technology Studies Perspectives."
          In *International Encyclopedia of the Social & Behavioral Sciences*, edited by
          J. D. Wright. 2nd ed. Elsevier. https://doi.org/10.1016/b0-08-043076-7/03159-4.
Thomson, Keith Stewart
  2008   *The Legacy of the Mastodon: The Golden Age of Fossils in America*. Yale University
          Press.
Tolia-Kelly, Divya P.
  2012   "The Geographies of Cultural Geography Iii: Material Geographies, Vibrant
          Matters and Risking Surface Geographies." *Progress in Human Geography* 37 (1):
          153–60. https://doi.org/10.1177/0309132512439154.
Thorpe, Charles, and Steven Shapin
  2000   "Who Was J. Robert Oppenheimer? Charisma and Complex Organization." *Social
          Studies of Science* 30 (4): 545–90. https://doi.org/10.1177/030631200030004003.
Tompkins, Jane P.
  1992   *West of Everything: The Inner Life of Westerns*. Oxford University Press.
Torrens, Hugh
  1992   "When Did the Dinosaur Get Its Name?" *New Scientist* 134 (1815): 40–44.
  1993   "The Dinosaurs and Dinomania over 150 Years." *Modern Geology* 18 (2).
  1997   "Politics and Paleontology: Richard Owen and the Invention of Dinosaurs." In
          The Complete Dinosaur, edited by J. O. Farlow and M. K. Brett-Surman. Indiana
          University Press.
Toumey, Christopher P.
  1994   *God's Own Scientists: Creationists in a Secular World*. Rutgers University Press.
Traweek, Sharon
  1988   *Beamtimes and Lifetimes: The World of High Energy Physicists*. Harvard University
          Press.
Trevorrow, Colin, dir.
  2015   *Jurassic World*. Universal Studios.
Trollinger, Susan L.
  2016   *Righting America at the Creation Museum*. Johns Hopkins University Press.
Tsing, Anna Lowenhaupt
  2003   "Natural Resources and Capitalist Frontiers." *Economic and Political Weekly* 38
          (48): 5100–06.
Turkle, Sherry
  2007   *Evocative Objects: Things We Think With*. MIT Press.
Van Allen, Adrian
  2017   "Bird Skin to Biorepository: Making Materials Matter in the Afterlives of Natural
          History Collections." *Knowledge Organization* 44 (7): 529–44. https://doi.org
          /10.5771/0943-7444-2017-7-529.
  2018   "Pinning Beetles, Biobanking Futures: Practices of Archiving Life in a Time of
          Extinction." *New Genetics and Society* 37 (4): 387–410. https://doi.org/10.1080/14
          636778.2018.1546573.
Van Dooren, Thom
  2014   *Flight Ways: Life and Loss at the Edge of Extinction*. Columbia University Press.
          https://doi.org/10.7312/vand16618.

Veness, Susan

  2015   *The Hidden Magic of Walt Disney World: Over 600 Secrets of the Magic Kingdom, Epcot, Disney's Hollywood Studios, and Animal Kingdom.* Adams Media.

Vetter, Jeremy

  2008   "Cowboys, Scientists, and Fossils." *Isis* 99 (2): 273–303. https://doi.org/10.1086/588688.

  2016   *Field Life: Science in the American West During the Railroad Era.* University of Pittsburgh Press. https://doi.org/10.2307/j.ctt1gxxqcp.

Von Mueller, Eddy

  2011   "'Nature Is the Dramatist': Documentary, Entertainment and the World According to the True-Life Adventures." In *Learning from Mickey, Donald and Walt: Essays on Disney's Edutainment Films,* edited by A. B. Van Riper. McFarland.

Wallace, Anthony F. C.

  1956   "Revitalization Movements." *American Anthropologist* 58 (2): 264–81. https://doi.org/10.1525/aa.1956.58.2.02a00040.

Walsh, Andrew

  2010   "The Commodification of Fetishes: Telling the Difference between Natural and Synthetic Sapphires." *American Ethnologist* 37 (1): 98–114. https://doi.org/10.1111/j.1548-1425.2010.01244.x.

Walsh, Casey

  2012   "Anthropology and the Commodity Form: The Philadelphia Commercial Museum." *Critique of Anthropology* 32 (3): 223–40. https://doi.org/10.1177/0308275x12449100.

Weber, Max

  (1921) 2019   *Economy and Society: A New Translation.* Translated by K. Tribe. Harvard University Press.

Weiss, Erica, and Carole McGranahan

  2021   "Rethinking Pseudonyms in Ethnography: An Introduction." In *Rethinking Pseudonyms in Ethnography,* edited by E. Weiss and C. McGranahan. American Ethnologist, December 13. https://americanethnologist.org/online-content/collections/rethinking-pseudonyms-in-ethnography/rethinking-pseudonyms-in-ethnography-an-introduction/.

Weszkalnys, Gisa

  2013   "Oil's Magic: Contestation and Materiality." In *Cultures of Energy: Power, Practices, Technologies,* edited by S. Strauss, S. Rupp, and T. F. Love. Left Coast Press.

  2015   "Geology, Potentiality, Speculation: On the Indeterminacy of First Oil." *Cultural Anthropology* 30 (4): 611–39. https://doi.org/10.14506/ca30.4.08.

Whatmore, Sarah

  2006   "Materialist Returns: Practising Cultural Geography in and for a More-Than-Human World." *Cultural Geographies* 13 (4): 600–609. https://doi.org/10.1191/1474474006cgj377oa.

White, Richard

  1991   *"It's Your Misfortune and None of My Own": A History of the American West.* University of Oklahoma Press.

Whiteley, Peter M.

  2012   "Epilogue: Prolegomenon for a New Totemism." In *The Anthropology of Extinction: Essays on Culture and Species Death,* edited by G. M. Sodikoff. Indiana University Press.

Williams, Raymond

    1980   "Ideas of Nature." In *Problems in Materialism and Culture*. NLB.

    1985   *Keywords: A Vocabulary of Culture and Society*. Oxford University Press.

Willis, Susan

    1999   "Imagining Dinosaurs." In *Girls, Boys, Books, Toys: Gender in Children's Literature and Culture*, edited by B. L. Clark and M. R. Higonnet. Johns Hopkins University Press.

Wilson, Alexander

    1991   *The Culture of Nature: North American Landscape from Disney to the Exxon Valdez*. Between the Lines.

Wilson, H. W.

    1995   "Bakker, Robert T." In *Current Biography Yearbook*. H. W. Wilson.

Wolfe, Cary

    2013   *Before the Law: Humans and Other Animals in a Biopolitical Frame*. University of Chicago Press. https://doi.org/10.7208/chicago/9780226922423.001.0001.

Worsley, Peter

    (1957) 1968   *The Trumpet Shall Sound: A Study of 'Cargo' Cults in Melanesia*. Schocken Books.

Wrobel, David M.

    2002   *Promised Lands: Promotion, Memory, and the Creation of the American West*. University Press of Kansas.

Wylie, Caitlin Donahue

    2015   "'The Artist's Piece Is Already in the Stone': Constructing Creativity in Paleontology Laboratories." *Social Studies of Science* 45 (1): 31–55. https://doi.org/10.1177/0306312714549794.

    2017   "Trust in Technicians in Paleontology Laboratories." *Science, Technology, & Human Values* 43 (2): 324–48. https://doi.org/10.1177/0162243917722844.

    2021   *Preparing Dinosaurs: The Work Behind the Scenes*. MIT Press. https://doi.org/10.7551/mitpress/12643.001.0001.

Wynne, Clive D. L.

    2007   "What Are Animals? Why Anthropomorphism Is Still Not a Scientific Approach to Behavior." *Comparative Cognition & Behavior Reviews* 2:125–35. https://doi.org/10.3819/ccbr.2008.20008.

Wynter, Sylvia

    2003   "Unsettling the Coloniality of Being/Power/Truth/Freedom: Towards the Human, after Man, Its Overrepresentation—an Argument." *Community Reading Group* 3 (3): 257–337. https://doi.org/10.1353/ncr.2004.0015.

Yates, Donna, and Emily Peacock

    2024   "T. Rex Is Fierce, T. Rex Is Charismatic, T. Rex Is Litigious: Disruptive Objects in Affective Desirescapes." *International Journal of Cultural Property* 30 (4): 1–23. https://doi.org/10.31235/osf.io/3wzet.

Young, Richard K.

    1997   *The Ute Indians of Colorado in the Twentieth Century*. University of Oklahoma Press.

Yusoff, Kathryn

    2018   *A Billion Black Anthropocenes or None*. University of Minnesota Press. https://doi.org/10.5749/9781452962054.

2024   *Geologic Life: Inhuman Intimacies and the Geophysics of Race.* Duke University Press. https://doi.org/10.2307/jj.12949156.

Zizzamia, Daniel

2019   "Restoring the Paleo-West: Fossils, Coal, and Climate in Late Nineteenth-Century America." *Environmental History* 24 (1): 130–56. https://doi.org/10.1093/envhis /emy092.

Zondag, Ralph, and Eric Leighton, dir.

2000   *Dinosaur.* Buena Vista Pictures.

Zurné, Lise, and Loes Oudenhuijsen

2024   "Sexualised Researchers in Ethnographic Encounters." *Tijdschrift voor Gender-studies* 27 (2/3): 171–89. https://doi.org/10.5117/tvgn2024.2-3.005.zurn.

Abruzzo, James, 209n32

Academy of Natural Sciences (Philadelphia), 19, 137

Actor Network Theory (ANT): "actants," 11; nonhuman agency, 24; object/thing distinction, 195n104

affect and affective relationships: definition of, 12–13; Disney's Animal Kingdom and, 113–14, 118–19, 208n21; media and, 214n34; nonhuman animals and, 110, 207n33; science and, 13, 157
—with fossils: animality and, 91, 96, 98–100, 102; children and anthropomorphism, 102–5; grief and, 90–91, 110, 111–12; modeling and, 110–11; paleontologists and, 13, 182, 191–92nn39–40; preparators and, 112; scientist and, 12, 182; touching fossils and, 149–51, 155, 156, 157, 213n9. *See also* anthropomorphism—in paleontology; touch and the senses in paleontology

affective resonance, 205n9

agency: Actor Network Theory (ANT) and, 24; "dance of agency," 26–27, 196n121; dichotomy between growing and making, 24, 195n106; distributive agency, 23–24, 25, 195n102; "existents" and, 24–25, 195n107–108; fossil making and, 50; fossils as vibrant, 26, 27, 196n118; Tyler Lyson and, 25–26; materiality and, 24–25; more-than-human, 24–25, 195n108; posthumanism and, 26, 196nn117–118; power dynamics and, 24; rock and, 26–29. *See also* fossil making

Alberti, Samuel J. M. M., 201n4

alligator (*Alligator mississippiensis*), 118, 119, 212n98

Allosaurus, 72, 80, 92, 102, 102–5, 136, 138, 146, 179

Ameghino, Carlos and Florentino, 192n46

American Museum of Natural History (New York): dinosaur exhibits, 19, 62–63, 67, 116; founding of, 208n12; funding for, 204n60; Charles R. Knight and, 62–63, 201n5; New Museum movement and, 116, 207n8. *See also* Osborn, Richard Fairfield; *Tyrannosaurus rex*—AMNH 5027

American Philosophical Society (Philadelphia), 19

AMNH 5027. *See Tyrannosaurus rex*—AMNH 5027

Anderson, Warwick, 191n29, 215n9

animality: edutainment and, 170–71; mounted skeletons and, 149; museum visitors and, 92–93, 94–96, *95*, 98–100, 102, 181–82

Anning, Mary, 203n40

Anthropocene, 24, 29, 172

anthropocentrism, 101, 105, 111

anthropology: of charisma and charismatics, 30, 63–64, 66–69, 201–2nn11,18; of dinomania, 22; of "existents," 24–25, 195nn107–108; of extractivism, 191n26; of gemstones, 23, 56, 79, 194–95nn98–100, 213n18; of materiality, 14–15, 24–25, 29–30, 192nn47,52, 194n94, 195n106, 196nn117,127,130–132; of media, 20, 214n30; of the more-than-human, 24–25, 195n108; multispecies, 201n8, 206n19, 216n23; of museum and other objects, 49–50, 193n60, 196n117, 213nn12,18;

anthropology (*continued*)
of natural resources, 194–95nn97–99,100;
of rock, 26–29; of science, 7, 11, 14–15, 29–30,
190nn8–10,13–14, 191–92nn32,39, 197nn137–138,
212n1; of touch, 191n35; of volunteer-tourism,
190n16
anthropomorphism: animalcentric
anthropomorphism, 101; in arts and
literature, 101, 106; in Christianity, 100;
definition of, 91, 100–101; ethology and,
101, 102, 206n17; human exceptionalism
and, 32–33, 101, 105, 111; language and,
102; in scientific research, 101–2, 109–11,
206nn14,19,23
—in paleontology: children and, 102–6; in film,
106–10, 206n30; human exceptionalism and,
111; in park tours, 1–4, 91–92; as pedagogical
tool, 105–9; as research technique, 109–11;
touch and, 102. *See also* affect and affective
relationships—with fossils; dinosaurs:
emotions of
Apatosaurus, 80, 92–93, 96, 97–98, 102, 132, 136,
138, 146, 179
Arapahoe people, 132, 210n61
Argentina: Americanness in, 13–14; Carnotaurus,
127–28; dinosaurs in, 88; paleontologists in,
192n46
Asma, Stephen T., 71, 204n56

Babers, Myeshia C., 216n11
badlands, 39
Bakker, Robert T.: as charismatic paleontologist,
64–66, 82; contributions to paleontology,
65, 82; creationists and, 145; loyal following,
65–66; media and, 9, 66, 164; at Morrison
Museum ("Dr. Bob"), 65–66, 98, 145, 148,
155; mounting of *Tyrannosaurus rex*, 61, 62,
64–67, 82, 148
Baldacchino, Godfrey, 194n94
Ballestero, Andrea, 196n121
Barad, Karen, 191–92n36,39
Barney (purple dinosaur), 122
Barnum, P. T., 117
beasts and beastliness: the Crystal Palace
Exhibition and, 116; dinosaurs associated
with, 31–32, 70–71, 73; enslavement and,
30–31, 197n143; European folktales of, 31–32,
70, 73, 81–82; Linnaean taxonomy and, 31,
70; race and, 30–31, 135, 197n143. *See also*
animality
Bennett, Jane, 23–24, 25, 26, 27, 195n102, 196n118
Bielo, James S., 212n2, 214n23
Bigenho, Michelle, 213n9

bios (life) and *geos* (nonlife), 29–30, 196n134,
197n138
birds: as analogues in paleobiology, 110, 169–70,
214n32; evolution of feathers, 169, 181–82,
215n8; theropod dinosaurs and, 145, 170,
181–82, 215n8
Blackhawk, Ned, 131, 210n61
Black Hills Institute of Geological Research,
120, 128, 209nn28. *See also Tyrannosaurus
rex*—"Stan" (BHI 3033)
Black, Riley, 168, 170, 176
Boas, Franz, 19, 67, 73
body fossils. *See* fossils
"bone wars," 18–19, 21, 57, 71–72, 74, 136
Bowker, Geoffrey C., 203n34
brachiosaur, 164
Bradley, Lawrence, 209n30
Brannigan, Augustine, 192n55
Braun, Bruce, 199n27
Braverman, Irus, 197n143
Brinkman, Paul D., 204n60
Brochu, Christopher A., 215n4
Brontosaurus, 116, 136, 145, 153, 200–201n69
Brown, Barnum: AMNH 5027 and, 81–82;
description of *Tyrannosaurus rex*, 73–74,
133–34; excavations, 75, 204n63; Arthur Lakes
and, 97, 138; masculinity and, 81, 175; Richard
Fairfield Osborn and, 74–75; preparation of
fossils, 75–76; prospecting fossils, 75
Brusatte, Stephen Louis, 167
Brush, Alan H., 215n8
Buckland, Adelene, 193n73
Buckland, William, 70
buffalo (*Bison bison*), 140–41
Buffon, George Louis Leclerc, Comte de, 212n97
Bureau of Indian Affairs, 21
Bureau of Land Management (BLM), 39–40, 54,
200n60
Burton, Orisanmi, 197n143
Butler, Ella, 51, 214n23
Butler, Judith, 111

Caduff, Carlo, 202nn17,30
Cain, Victoria, 207n8
Candea, Matei, 207n33, 214n30
Candomblé, 28–29, 69
Cannon, George, 57
capitalism and paleontology, 10, 19, 40, 50. *See
also* Native Americans, dispossession of;
philanthropists; settler colonialism; United
States—nationalism in
care-work: casts and, 161–62, 214n22; fossils and, 112
Carnegie, Andrew, 58–59, 68

Carnegie Museum (Pittsburgh), 71, 204n60
carnivores: dentition of, 72–73; diet of, 72, 203n50; scientific research on, 82–83, 204n79. *See also Tyrannosaurus rex*
Carnotaurus, 115, 126–28
Carroll, Hamilton, 216n17
casts and casting: definition of, 148; museum docents and, 86; as research specimens, 78, 79–80, 152–53, 161; restoration and, 78–79, 80, 148, 153; scientific accuracy and, 79–80, 88; of skulls, 76, 146, 148; of *Tyrannosaurus rex*, 77–80, *78, 82. See also* mounted skeletons; paleontological continuum; paleontological real; reconstructions
CGI (computer-generated imagery): in edutainment, 108, 110, 162–63, 164; modeling and, 108, 110, 148, 162
charisma and charismatics: anthropology of, 30, 63–64, 66–69, 201–2nn11,18; beasts and, 73; as Christian/religious concept, 63; more-than-human, 68–69, 87–89, 202n30, 203n32; scientists as, 64, 69, 202n30. *See also* charismatic megafauna; charismatic paleontologists; charismatic prehistoric megafauna
charismatic megafauna: environmentalism and, 67–68, 202n25; zoos and, 113–14, 118–19, 208n21. *See also* charismatic prehistoric megafauna
charismatic paleontologists: gender and race and, 181; individual fossils and, 67, 87; masculinity of, 64, 66–67, 81, 138–39; media and, 64, 66, 106; physical appearance, 66; professional success, 64; public attention, 19, 64, 65–66, 106; relationality of, 68–69, 202n30
charismatic prehistoric megafauna, 68–69, 87–89, 202n30; "faces" of, 68; in museums, 68; nicknames of, 62, 67; philanthropists and collectors and, 58–59, 60, 68; research on, 59, 200n67; as symbols of Americanness, 18; volunteer recruitment and, 43–45, 68, 198–99n21. *See also* dinomania; Stegosaurus; Triceratops; *Tyrannosaurus rex*
Cheyenne River Sioux Reservation, and *Tyrannosaurus rex* "Sue," 120, 209nn28,30
Chibnik, Michael, 199n23
Chicago World's Fair (1933–34), 19, 200–201n69
children and childhood, 13, 102–6, 120, 164, 178–79, 207n4, 216n17. *See also* dinomania; education
China, 169
Christianity: anthropomorphism and, 100; charismatics and, 63; great chain of being and, 173, 174, 212n97

climate change. *See* Anthropocene
colonialism. *See* settler colonialism
Colorado: first Triceratops unearthed in (1887), 57; Gold Rush of, 130, 131; Stegosaurus as state fossil, 98, 136
Colorado Rockies mascot, 57
commercial fossil collectors: Disney World and, 128–29; excavations by, 43; opposition to, 9; *Tyrannosaurus rex* "Sue" and, 120, 209nn28,30. *See also* Black Hills Institute; fossil hunting
Connolly, William E., 214n34
consolidants, 42, 43–44, 46, 47–48
Cope, Edward D.: the "bone wars" and, 18–19, 21, 57, 71–72, 74, 136; as charismatic paleontologist, 64; the Great Dinosaur Rush and, 65; Arthur Lakes and, *97, 137*
craftwork, 17, 45, 56, 199n23, 200n51
creationism and creationists, 143, 212n2; conception of dinosaurs, 15, 17, 56, 200n50; edutainment and, 170–71; museums and, 51, 144–45; paleontological continuum and, 157–58, 171; rock-dating methods and, 144; unauthorized tours by, 143
Creation Museum, 162, 214n23
Crichton, Michael, 19
Crystal Palace Exhibition (England), 20, 116, 214n27
cultural studies, 15, 17, 56, 88, 189n5, 200n50, 205n82
Custer, George, 211n89
Cuvier, Georges, 70

Darwin, Charles, 101, 173, 206nn14,19. *See also* evolution
Daston, Lorraine, 101–2
Davis, Susan G., 208–9n26
De León, Jason, 194n96
Deleuze, Gilles, 24, 201n8, 215n5
Denver, Colorado, 5, 34–36, *36,* 57
Denver Museum of Nature and Science: creationists at, 143; "Dinosaur Detective" program, 102–5; fossil preparation laboratory, 51–52, *53,* 54–56, 79–80, 200nn40,45–46,51; founding and financing of, 50, 200n45; Free Days, *107,* 143, 146; historic dinosaur hall, 50–51; *Tyrannosaurus rex* in, 61, *62,* 64–67, 82, 148. *See also* Sampson, Scott
—*Prehistoric Journey* exhibit, 50–51, 144–45, 174–75, 199n40; Bone Bar, 83–87, *84–85,* 92–93, 102, 146–47, 154, 215n8; enviroramas, 51–54, *52–53,* 199n40

Department of the Interior (United States), 21

Dern, Laura, 164

Derrida, Jacques, 204n55

Descola, Philippe, 156, 195n108

DeSilvey, Caitlin, 25

Detwiler, Katheryn, 191n32

De Waal, Frans B. M., 101

DinoLand. *See* Disney World—DinoLand U.S.A. (Animal Kingdom)

dinomania: as bridging political divides, 13, 192n45; definition of, 17–18, 22; demographics of, 18, 178–79, 216n17; entertainment and, 19–21, 193nn73–74,80; museum exhibits and, 19, 50, 116, 207n8; nationalism (US) and, 13–14, 17–19, 21; paleontologists and, 19

Dinosaur Hotel, 142, 143–44, 165

dinosauria, 31, 70–71

Dinosaur National Monument, 32, 194n91

Dinosaur Ridge (Colorado): creationists at, 143; as edutainment, 115–16; footprints, 1–4, 32, 91–92, 98, 152, 197n146; geological formations of, 35–36; land ownership of, 129, 131; Scout Days, 143; settler colonialism and, 115–16, 129–39, 140, 141, 211n89, 212n98; touching and, 129, 156; Ute Council Tree, 131; volunteers, 129

dinosaurs: as beasts, 31–32; coining of term (Owen), 18; emotions of, 107–11, 206n30, 207n32–33; evolution of, 6; symbolism, 36–37

Dinosaur Train (PBS Kids), 5, 106–7, 181

dioramas, 51, 199n40

Diplodocus, 50, 97–98, 138, 146

discovery: agency and, 11, 15–17, 24, 35, 38; definition of, 16, 74, 192n55; DinoLand U.S.A. and, 16, 125; fossil collecting and, 35; individualism and, 35, 38, 74; myth of, 15–16; philanthropy and, 58–59, 60; property ownership and, 35, 74, 136, 197n4; wilderness and, 16, 35, 118, 125, 134, *134–35*, 193n57, 197n4. *See also* fossil making; paleontology—as adventure; settler colonialism

Disney (Walt Disney Company), 117–18, 208n16; purchase of *Tyrannosaurus rex* "Sue" and, 120–21, 128–29, 209nn28,30,32,34; *Ten Who Dared* (1960 film), 66; *True-Life Adventures* series, and "Prowlers of the Everglades" episode, 117–18, 208n14, 212n98

Disney World (Florida), 125, 210n40; Animal Kingdom (zoo), 113–15, 118–19, 121, 127, 208–9nn21,24,26; environmental destruction and, 118, 119, 208nn19,24, 212n98; *Jurassic Park* ride (Universal Studios), 210n43

—DinoLand U.S.A. (Animal Kingdom): American West and, 119–20, 121–22, *122*, 123–25, 209n36; "Boneyard Playground," 16, 36–37, 123–24, *124*; commercial fossil collectors and, 128–29; crocodiles in, 115, 119, 212n98; Dino-Rama, 113, 115, 122–23, *123*; DINOSAUR (thrill ride), 5, 16, 115–16, 117, 119, 124–29, 139–41, 175, 210nn43,47, 212n98; fossil preparation laboratory, 9, 121, 209n34; Native Americans and, 119, 125, 129, 212n98; paleontology and, 140–41, 212n98; settler colonialism and, 115–16, 118, 119, 124–25, 128–29, 139–41, 212n98; *Tyrannosaurus rex* cast "Dino-Sue," 82, 113, *114*, 121, 150, 209n32. *See also* edutainment

dispossession. *See* Native Americans, dispossession of

distributive agency, 23–24, 25, 195n102. *See also* agency

Edmontosaurus, 81, 83–84, 146

education: evolution and, 149–51, 155–56, 157, 213n9; media as "hook" for, 164; race and racism and, 178–79. *See also* children and childhood; edutainment; museums

edutainment, 3, 116, 207n4; creationists and, 170–71; the Crystal Palace Exhibition and, 20, 116, 214n27; environmentalism and, 117–19, 208–9nn16,26; immersive, 115, 117, 127, 128, 140; roadside attractions as, 121; settler colonialism and, 115–16, 118; social class and, 116–19, 207n8, 208–9n26. *See also* Dinosaur Ridge (Colorado); Disney World— DinoLand U.S.A. (Animal Kingdom); media industry; National Park Service; zoos

Eliade, Mircea, 27

Emling, Shelley, 203n40

entertainment industries. *See* edutainment; film; media industry; television

environmentalism: charismatic megafauna and, 67–68; edutainment and, 117–19, 208–9nn16,26; volunteer-tourism and, 198–99nn16–17,21. *See also* charismatic megafauna

epistemology of science, 2, 12, 40, 182, 189n2; alternative, 182–83, 216n24; distinction between *bios* (life) and *geos* (nonlife), 29–30; the earth as passive ground for human history, 22–23; human exceptionalism in, 32–33, 101, 105, 111; Linnean taxonomy and, 70, 203n34; as part of Euro-American tradition, 27; personhood and, 66–67, 68, 202n18; race and, 30–31, 197n143; rock and,

25, 36; universalism in, 70, 182, 203n34,
215n9; vision and, 46. *See also* individualism
and individualization
ethology, 101, 102, 110, 206nn17,19,30
eugenics, 72–73
Europe: conception of species, 203n38; folktales
of beasts and monsters, 14, 31–32, 70, 73,
81–82; medieval recognition of lithic agency,
195n109
evolution: Denver Museum and, 144–45;
education and, 149–51, 155–56, 157, 213n9;
edutainment and, 170–71, 214n34; hierarchy
in, 173–75, 215nn4,8, 216n23; history of,
72–73, 203n38; mounted skeletons and,
149; museums and, 50–51, 144–45, 149,
177; paleontology and, 6, 173; phylogenetic
system, 174, 215nn4,8; popular media
and, 164; punctuated equilibrium, 193n63;
representations of, 173, 174, 181–82, 208n24,
215n5, 216n23. *See also* Linnaean taxonomy
extractivism and paleontology, 1, 10–11, 57–58,
139–40, 191n26, 191n27. *See also* Native
Americans, dispossession of; settler
colonialism

feminist scholarship, 182, 216n17
Ferry, Elizabeth, 56, 194–95nn97,100, 196n136,
213n18
fiction: *Dinosaur Train* and, 106–7; fact versus,
105–6, 164; science and, 142, 145, 150, 156,
163, 171; *T-Rex Autopsy* and, 167–68, 170–71,
214n34
Field Museum of Natural History (Chicago),
204n60. *See also Tyrannosaurus rex*—"Sue"
(FMNH PR 2081)
Fiffer, Steve, 209nn28,32
film: "emotion grid" for animation, 107–8,
109–10, 206n30; *Gertie the Dinosaur* (1914),
20; *Ghost of Slumber Mountain* (1919), 63;
Japanese dinosaur movie genre, 193n74; *King
Kong* (1933), 20, 63, 96; *The Lost World* (1925),
20; *Ten Who Dared* (1960), 66; *Walking with
Dinosaurs* (2013), 107–10, 163. *See also* media
industry; television
—American dinosaur movie genre, 19–20,
193nn73–74, 210n43; as "hook" for science,
164; *Tyrannosaurus rex* AMNH 5027 as
model for, 63. *See also Jurassic Park* film
series
—animal documentaries, 109, 118, 165–67,
214n30; *True-Life Adventures* series, and
"Prowlers of the Everglades" episode, 117–18,
208n14, 212n98

Florida Everglades, 118, 119, 208nn19,24, 212n98.
See also Disney World
footprints. *See* fossils—footprints
fossil hunting, 3, 4, 5, 21, 33, 38, 42–43, 46,
57–58, 64–65, 81. *See also* Brown, Barnum;
commercial fossil collectors; Lakes, Arthur;
Lyson, Tyler; masculinity
fossilization: bacteria and, 153; biological and
geological forces in, 26, 37–38, 46, 55–56, 60,
75–76, 198n7; casting as continuation of, 153;
flattening bones, 38, 76, 77, 88, 148; repairing
distortions of, 78–79, 80, 148, 153
fossil making: overview, 17, 26–27, 49–50, 60,
158, 173; climatic forces in, 39, 46, 54–55, 60;
consolidants in, 44, 47–48, 55, 158; craftwork
and, 17, 45, 56, 199n23, 200n51; discovery
myth and, 15–17, 24, 35, 38; distributive
agency and, 50; fossil excavation and, 45–49,
47, 49, 199nn23–24,26–27; fossil jacketing,
48–49, 49; fossil preparation, 53–56, 79, 158,
200nn46,51; fossil prospecting and, 42, 43–
44, 46; human-nonhuman interactions and,
7–8, 17, 26–27, 193n60; labor hidden, 56; lithic
agency and, 27, 38, 47, 55–56; materialism
and, 27; as more-than-human process, 26,
37–38, 46, 55–56, 60, 75–76, 198n7; museum
visitors and, 53, 200n44; the paleontological
real and, 153–54, 158; philanthropists and
collectors and, 58–59, 60, 200n67; power
dynamics and, 8, 27, 60; preparators and,
17, 53, 54, 56, 79, 200n46; of skulls, 23, 37,
38, 75–80, 78, 82, 88, 146, 148; tools and,
46, 47, 54–55; volunteers' labor and, 60. *See
also* agency; fossilization; paleontological
continuum; paleontology laboratories
fossils: articulation of, 38; as assemblages, 80;
body fossils, 193n58; as both specimen and
artifact, 17, 24, 50, 80, 156–57, 158; fossil
record, 6, 16, 38, 148, 204n63; "frags" (broken
pieces), 154; "isolated," 154; popular view of,
22, 36–37, 197–98n6; preconquest significance
to Native Americans, 135; radiocarbon and
other dating methods, 144, 151, 153; rareness
and, 5, 6; trace fossils, 110, 193n58; as vibrant
objects, 26, 27, 56, 195n104, 196n118. *See also*
fossil making
—footprints: affect and, 98–100, 102, 205n9;
casting of, 161–62; interpretation of, 1–4,
91–92; research on, 32, 197n146; touch and,
96, 99; as trace fossils, 193n58
Foster, John Russell, 198n10
Foucault, Michel, 29
Franklin, Sarah, 64, 66

funding of paleontology: media industry and, 164; by US government, 21, 57–58, 200n60. *See also* philanthropists
Furlong, Charles Wellington, 205n9

Gamble, Luke, 167
Geertz, Clifford, 200n50
gemstones, 23, 56, 79, 194–95nn98–100, 213n18
gender and gendering, 175, 179, 215–16n9; clothing and, 1–2, 3, 177, 179; of objects of study, 179; in paleontology, 175, 179, 180, 215–16n9; science and, 3, 178–79, 216n17; socialization of children, 178–79, 216n17. *See also* masculinity
geology, 18, 29; colonialism and, 139–40, 211–12n95. *See also* extractivism and paleontology; Native Americans, dispossession of; paleontology; settler colonialism
—mining and petroleum: dinosaur as logo for petroleum company, 200–201n69; ethnographies of gemstones, 23, 56, 79, 194–95nn98–100, 213n18; ethnographies of petroleum, 194n98; as extractivist enterprises, 10, 57; philanthropy and, 10, 50, 59, 191n27, 200–201n69; resource rushes, booms, and busts, 10; retirees, 1, 10–11
geontopower (Povinelli), 29
Gertie the Dinosaur (1914 film), 20
Ghost of Slumber Mountain (1919 film), 63
Gieryn, Thomas F., 214n26
Goetzmann, William H., 58, 139
Good, George Brown, 207n8
Goodwin, Charles, 199n25
Goody, Esther N., 199n23
Gould, Stephen Jay, 17–18, 19–20, 22, 63, 64, 193n63
Grasseni, Cristina, 199n24
great chain of being (*scala naturae*), 173, 174, 212n97
Great Dinosaur Rush (1877–92), 61–62, 65, 97, 115
Guattari, Felix, 24, 201n8, 215n5

Hadrosaurs, 19, 113, 149–50, 154, 160, 169, 179
Hallam, Elizabeth, 200n51, 213n12
Hamon, Geneviève, 109
Haraway, Donna, 25, 160, 191–92n39,49, 194–95n100, 196n115, 199n40, 202n15, 216n24
Hartigan, John, Jr., 200n50, 206nn19,30
Hawkins, Benjamin Waterhouse, 214n27
Hayden, Ferdinand V., 18, 39, 58, 97
Heidegger, Martin, 195n104

Hell Creek Formation: conditions in, 11–12, 39, 41–42; geology of, 42, 46, 47, 55, 154; Tyler Lyson and, 25–26; other names of, 198n10
Helmreich, Stefan, 197n138, 216n23
Hermann, Adam, 75–76
Herridge, Victoria, 167
Hetherington, Kevin, 194n94
High, Mette M., 191n27
Holocene era, 30
holotype, 98, 156–57
Holz, Thomas R., Jr., 203n34
Hone, Dave, 168–69
Horne, Gerald, 216n11
Houston Museum of Natural Science, 98
Huls, Alexander, 164
humans: evolution and, 72–73, 93, 94–96, 174–75, 215–16n9; exceptionalism, 101, 105, 111; Linnaean taxonomy and, 70, 173; nature/culture divide and, 156–57, 158, 213nn6,12,16,18
Hustak, Carla, 206n14

ichnology, 197n146. *See also* fossils—footprints
Ichthyosaurus, 203n40
Iguanodon, 70, 115, 126–27, 132
individualism and individualization, 66–67, 68, 202n18; charisma and, 63–64, 66–67, 68, 201–2nn11,18; discovery myth and, 35, 38; knowledge production and, 70; masculinity and, 64, 66–67
Ingold, Tim, 192n47, 199n28, 200n51, 205n82, 213n12
Irwin, Lee, 28

Japanese dinosaur movie genre, 193n74
Jefferson, Thomas, 16, 18, 140–41, 212n97
Johnson, Kirk, 26
Jurassic Park film series, 88, 163–65, 214n24; amusement park ride based on, 210n43; Robert T. Bakker and, 66; CGI in, 162–63, 164; dinomania and, 19–20; paleontology laboratories in, 12, 51; *Tyrannosaurus rex* AMNH 5027 as model for, 63

Kaiparowits Plateau, 12, 176–77, 178, 198n10
Kalshoven, Petra Tjitske, 207n33, 213n18
Keane, Webb, 194n94
Keller, Evelyn Fox, 191n37, 196n134, 216n17
Kennedy, John F., 64
Kilroy-Marac, Katie, 198n17
King Kong (1933 film), 20, 63, 96
Kirkaldy, George Willis, 71

Kirksey, Eben, 201n8, 216n23
Kirshenblatt-Gimblett, Barbara, 49
Knight, Charles R., 62–63, 124, 201n5
Koch, David H., 68, 191n27
Kopytoff, Igor, 194n99
Kurtti, Jeff, 208nn14,16

Lakes, Arthur, 96–97, 133–34; the "bone wars" and, 136; colonialism and, 133, 134, 136, 137–39; Colorado School of Mines and, 133, 136, 137, 138, 205n2; Edward D. Cope and, 97, 137; early career, 97, 136–37; excavations by, 96–98, 133–34, 136, 137–38, 156–57; the Great Dinosaur Rush and, 97, 136; Othniel C. Marsh and Peabody Museum and, 97–98, 136, 137–38, 152, 211n81; masculinity and, 138–39, 175; Morrison Museum and, 98, 152; paintings of, 134–35, 144
Langham, Wallace, 126
Larson, Pete, 209n28
Latour, Bruno, 7, 24, 192n56, 199n26, 205n82
Lave, Jean, 190n17
Lave, Rebecca, 196n134
Law, John, 194n94
Leakey, Mary, 99
Lesquereux, Leo, 97
Lévi-Strauss, Claude, 30
Limbert, Mandana E., 194n97
Linnaean taxonomy, 31, 69–71, 73, 203n34; anthropomorphism and, 102; evolution and, 70; holotypes, 98, 156–57; humans in, 87, 173; naming conventions, 70; new species, 71–72; phylogenetic system and, 174, 215n4; universalism and, 70, 203n34. See also evolution
Linnaeus, Carl, 31, 70. See also Linnaean taxonomy
lithic matter. See rock (lithic matter)
Lockley, Martin, 32, 197n146
Lorenz, Konrad, 102
Lorimer, Jamie, 40, 198–99nn17,21, 203n32, 208n20, 214n34, 216nn24–25
Lost World, The (1925 film), 20
Lucas, Oramel, 210n65
Lunney, Daniel, 202n25
Lyons, Kristina M., 195n106, 212n1, 213n6, 214n22
Lyson, Tyler, 25–26

McDonald's (Corporation), 120–21, 209nn28,30,32,34
McGranahan, Carole, 214n2
McJunkin, George, 216n11

McLean, Stuart J., 198n7
MacLeod, Roy, 35, 197n4
Maiasaurus, 179
mammals, 6, 19, 70; as analogues in paleobiology, 110
mammoth. See mastodon (Mammut americanum)
Manifest Destiny, 129, 136, 139
Mantell, Gideon, 70, 203n40
Mantel, Mary Anne, 203n40
Marsh, Othniel C.: the "bone wars" and, 18–19, 21, 57, 71–72, 74, 136; as charismatic paleontologist, 64; coining theropoda order, 71; the Great Dinosaur Rush and, 65; Arthur Lakes and, 97–98, 136, 137–38, 152, 156–57; Native Americans and, 136; the Peabody Museum at Yale University and, 58, 97–98, 137–38, 152
Martin, Anthony J., 203n50
Martin, Emily, 7, 14–15
masculinity, 177, 178; boyhood and, 86–87, 178–79, 216n17; charisma and, 64, 66–67, 81, 138–39, 175; clothing and, 66, 177, 179; cowboy hat as symbol of, 66; edutainment and, 66, 126, 128, 139; fieldwork and, 175–78, 216n13; paleontology and, 175, 176, 178; race and, 216n11; of Tyrannosaurus rex, 1, 73, 81, 83–85, 86–89, 133–34. See also gender and gendering; race and racism
mastodon (Mammut americanum) (mammoth), 18, 19, 130, 140, 212n97
materiality, 4–5, 14, 27, 192n47
—sociality and, 4–5, 14, 22, 27, 173; inseparability of, 194n94; storytelling and, 4, 172, 214n2; Tyrannosaurus rex and, 14–15, 31, 63, 76, 80, 82–83, 87–89, 192n49. See also paleontological continuum
Matthews, William Diller, 80, 204n78
Mayor, Adrienne, 135
media industry: anthropomorphism in, 107–9; discovery myth in, 14; funding paleontological research, 164; "hook" for science education, 164; paleontologists and, 9, 13, 64, 66, 106, 107–10, 164, 206n30; science in, 9; viewer interpretation of, 20. See also edutainment; film; television
Megalosaurus, 70
Merril, Judith, 160
Mesozoic era, 6
Messeri, Lisa, 207n32
Miller, Daniel, 24
mining. See geology—mining and petroleum
Mintz, Sidney Wilfred, 194n99

250 INDEX

Mitchell, W. J. T., 15, 193n63, 195n104, 205n82
Mitman, Gregg, 101–2, 206n23, 207n33
models and modeling, 148; anatomical models, 148, 168; biomechanical models, 110, 162, 169–70; CGI models, 108, 110, 148, 162–63, 164, 165; in media and entertainment, 108, 110, 162–63, 165, 168–69
modern charismatic megafauna. *See* charismatic megafauna
Mol, Annemarie, 194n94
Montreal World's Fair (1967), 117
the more-than-human, 15; agency, 24–25, 195n108; anthropology of, 24–25, 195n108; charisma, 68–69, 87–89, 202n30, 203n32; fossil making and, 26, 37–38, 46, 55–56, 60, 75–76, 198n7; ontology and, 212n1. *See also* agency; charisma and charismatics; fossil making
Morgan, J. P., 58–59, 192n46
Morrison Natural History Museum, 98, 152; Robert Bakker ("Dr. Bob") and, 65–66, 98, 145, 148, 155; bone-clones (research-grade casts) in, 152–53; casts as pedagogical tools in, 155–56; creationists and, 145, 152; fossil preparation laboratory, 154–55, 166, *166*; Stegosaurus in, 98–100, 152; touch and, 93–94, 154–56; tours of, 93, 94–96, *95*, 98–100, 102; Triceratops in, 155–56, 160; *Tyrannosaurus rex* in, 77–78, *78*, 94, 155. *See also* Mossbrucker, Matthew (director of Morrison Museum); paleontological real— alternative ontology of
Mossbrucker, Matthew (director of Morrison Museum), 77, 93–94, 96, 98–100; creationists and, 145; on the paleontological real, 151–56; *T-Rex Autopsy* (television special) and, 20, 117, 164–71, *166–67*, 214nn27,30. *See also* paleontological real—alternative ontology of
mounted skeletons: casts and reconstructions in, 146, 148, 149, 213n7; as complex assemblages, 80, 156, 158; as embodiment of theory, 148; paleontological continuum and, 160–61; of Stegosaurus, 146; of *Tyrannosaurus rex*, 61, 62, 80–82, 204n78
museums: evolution in, 117, 174–75, 215n8; funding of, 50, 58–59, 60, 200–201n69, 204n60; natural history museums reimagined as science museums, 117; New Museum movement and, 116–17, 120, 207n8; paleontology exhibits, 50; popularity of, 19, 50, 116, 207n8. *See also* American Museum of Natural History (New York); Carnegie Museum (Pittsburgh); Denver Museum of Nature and Science; Field Museum of Natural History (Chicago); Morrison Natural History Museum; Peabody Museum of Natural Science (Yale University); philanthropists; Smithsonian National Museum of Natural History (Washington, DC)
Myers, Natasha, 13, 199n26, 206nn10–11,14,23

NAGPRA (Native American Graves Protection and Repatriation Act), 120, 209n30
Nash, June C., 194n99
National Geographic, *T-Rex Autopsy*, 20, 117, 164–71, *166–67*, 214nn27,30
nationalism. *See* United States—nationalism in
National Park Service, 21, 117, 194n91; removal of Native Americans and animals, 207n10
National Science Foundation, 200n60
Native American Graves Protection and Repatriation Act (NAGPRA), 120, 209n30
Native Americans: early colonists' and field researchers' dependence on, 8, 132, 135–36, 215–16n9; repatriation and, 120, 135, 209n30; *Tyrannosaurus rex* "Sue" and, 120, 209nn28,30. *See also* Native Americans, dispossession of; *specific peoples*
Native Americans, dispossession of, 30–31, 135–36, 197n143; burial grounds robbed, 134; genocidal wars against, 131–32, 135–36, 211n89, 212n98; Manifest Destiny and, 129, 136, 139; Mexican-American War and, 131; National Park Service and, 207n10; representation in the Crystal Palace Exhibition, 116; Seminoles contesting, 212n98; settler colonialism and, 57–58, 131–32, 140; slaughter of the buffalo and, 140–41
—paleontology and, 10, 139–41, 211–12nn89,95,98; colonial infrastructure facilitating, 133, 134, 136, 137–38; Dinosaur Ridge and, 115–16, 129–39, 140, 141, 211n89, 212n98; Disney World DinoLand U.S.A. and, 115–16, 118, 119, 125, 129, 212n98; fossil hunting and, 81; homestead laws used, 133; park tours and, 4, 139; providing geographical data, 57–58, 140; treaties and tribal sovereignty and, 57–58, 140; wilderness narrative and, 16, 35, 118, 125, 134, *134–35*, 193n57, 197n4
natural history museums. *See* museums
nature: nature/culture dichotomy, 156–57, 158, 193n57, 213nn6,12,16,18; state of, 30
Neill, Sam, 164
New Museum movement, 116–17, 120, 207n8

news media, 34–35, 57, 73–74, 180, 204n55
New York Times, 73, 74, 76, 204n54
Noble, Brian, 51, 72, 202n15
nonhuman agency. *See* agency; more-than-human, the
North Dakota Heritage Center, 25

objectivity, 2, 12, 40, 183, 189n2
Ochs, Elinor, 206n11
O'Connor, Ralph, 192n49
Ogden, Laura A., 205n9
ontology: creationist, 145; definition of, 142; more-than-human interactions and, 212n1; subaltern, 28–29. *See also* paleontological real
O'Reilly, Jessica, 190n8
Ortner, Sherry B., 212n2
Osborn, Richard Fairfield (American Museum of Natural History): ambitions of, 74–75, 204n60; Barnum Brown and, 74–75; Charles R. Knight and, 201n5; naming of *Tyrannosaurus rex*, 71–73, 88; *Tyrannosaurus rex* AMNH 5027 and, 80–82, 204n78
Ostrom, John, 65
Oviraptor, 86, 87, 181–82
Owen, Richard, 18, 31, 70–71, 203n40

Pachyrhinosaurus, 107–9, 110
Painlevé, Jean, 109
paleobiology, 110, 169–70, 207n32, 214n32
paleobotany, 200n67
paleontological continuum, 158–63, *159*, 213n19; "SF" and, 160; *T-Rex Autopsy* and, 168–71. *See also* paleontological real—alternative ontology of
paleontological real: definition of, 142–43, 212n1. *See also* paleontological continuum
—alternative ontology of: casting and, 153, 161–62; fossil making and, 153–54, 158; Morrison Museum and, 143; research-grade casts (bone-clones) and, 152–53, 155–56; touch and, 155–56
—Euro-American ontology of: creationism and, 147, 151, 157, 171; geography and, 150, 161; growing/making dichotomy, 145, 146–48, 149, 150, 152, 156–57, 158; human labor and, 147, 153, 157; living/nonliving dichotomy, 150, 213n8; nature/culture dichotomy, 156–57, 158, 213nn6,12,16,18; science/fiction dichotomy, 142, 145, 150, 156, 163, 171; specimen/artifact/artwork division, 156–57, 158; terminology, 147, 152; touch and, 149–51, 155, 156, 157, 213n9
Paleontological Research Institute (PRI), 9

Paleontological Resources Preservation Act (PRPA), 39–40, 198n13
paleontologists: childhood and background of, 40, 63, 106; commercial fossil collectors and, 58, 64; eugenics and, 72–73; gender and, 203n40; Native Americans and, 8, 135–36, 215–16n9; teaching by, 93; technicians and, 40, 175, 178. *See also* charismatic paleontologists
paleontology: *bios* (life) and *geos* (nonlife) and, 29–30; boundary-work of, 8–9, 164; "community of practice," 8, 190n17; as "full-bodied practice," 13, 191–92n39; history of, 18, 58; prestige of, 9, 13, 192n40; promotion of careers in, 64, 106, 178–79, 216n17; research, 6–7, 9, 15, 82–83, 153, 204n79; standards, 79–80, 88; storytelling and, 4, 31–32, 172, 214n2; subfields of, 29. *See also* anthropomorphism—in paleontology; commercial fossil collectors; edutainment; evolution; fossils; funding of paleontology; media industry; museums; paleontological real; paleontologists; paleontology laboratories; philanthropists
—as adventure: overview, 3; colonialism and, 4, 16, 139–40; edutainment and, 3–4, 139–40; fossil hunting and, 14, 16; masculinity and, 64, 66–67, 81, 138–39, 175; media and entertainment industries and, 14; volunteer-tourism and, 41; Western frontier and, 4, 16, 41, 58, 139, 198n19. *See also* discovery; extractivism and paleontology; fossil hunting; masculinity; Native Americans, dispossession of—paleontology and; settler colonialism; United States—nationalism in
paleontology laboratories, 7, 177; at Denver Museum, 51–52, *53*, 54–56, 79–80, 200nn40,45–46,51; high-tech vs. low-tech, 11–12, 191nn32–33; inside DinoLand U.S.A., 9, 121, 209n34; at Morrison Natural History Museum, 154–55, 166, *166*; technicians and, 8, 40, 175, 177, 178; volunteers and, 8, 53–55, 60. *See also* fossil making
Paleozoic era, 6
Parreñas, Juno Salazar, 190n14, 198n16
Patchett, Merle, 195n106
Paton, David, 25
Pawnee people, 21
Paxson, Heather, 199n23, 206n23
Peabody, George, 21, 58–59
Peabody Museum of Natural Science (Yale University), 58, 74, 97–98, 137–38, 152

Peal, Charles Willson, 19
Pels, Dick, 194n94
Petrified Forest (Arizona), 21
petroleum. *See* geology—mining and petroleum
Philadelphia Centennial Exhibit (1876), 207n8
philanthropists: aims of, 207n8; corporate
     sponsorships and, 120–21, 209n32; importance
     for research, 58–59, 60, 200n67; paleontologists
     courting, 59; paleontology used by, 10,
     191n27; prestige of science and, 50, 58–59, 60,
     200–201n69, 204n60. *See also* museums
phylogenetic system, 174, 215nn4,8
Pickering, Andrew, 196n120
Plesiosaurus, 70–71, 203n40
political ecology, 23, 27, 194n97; gemstone
     ethnographies and, 23, 56, 79, 194–95nn98–
     100, 213n18
postcolonial science studies, 215–16n9
posthumanism, 111; charisma and, 69, 203n32;
     STS (science and technology studies) and,
     23, 27, 212n1. *See also* agency
Povinelli, Elizabeth, 24–25, 26, 29, 195n107, 213n8
Powell, John Wesley, 128, 139
private property, 21, 35, 39, 58–59
Prothero, Donald R., 203nn33,38,40,50
PRPA (Paleontological Resources Preservation
     Act), 39–40, 198n13
Prum, Richard O., 215n8
Pteranodon, 94–96, *95*, 106–7, 110
Pterosaurs, 215n8

race and racism: beastliness and, 30–31, 197n143;
     the Crystal Palace Exhibition and, 116; racial
     exclusion in paleontology, 15, 175–76, 178,
     180–81, 216nn11,13; science and, 175; science
     education and, 178–79. *See also* whiteness
Rader, Karen Ann, 207n8
railroads, 137–38, 211n81
Rainger, Ronald, 201n5, 204n63
Rashad, Phylicia, 126
reconstructions, 148. *See also* casts and casting;
     models and modeling; mounted skeletons
Reed, William Harlow, 210n65
Rieppel, Lukas, 138, 158, 207n8, 208n12, 210n65,
     211–12n95, 213n7
rock (lithic matter): agency of, 26–29; Euro-
     American conception of, 22–23, 25, 36;
     gemstones, 23, 56, 79, 194–95nn98–100,
     213n18; humans becoming with, 25; medieval
     European conception of, 195n109; pilgrimages
     to, 27–29; properties of, 23, 194n96; time and,
     23, 25, 26, 56. *See also* agency

Rockefeller, John D., 68
Rocky Mountains: colonial perception of, 205n3;
     excavations, 57; geology of, 55
Rogers, Douglas, 194n94
Rooney family, 130–32, 133, 134, 136, 137–38,
     210n61, 211n81

Sacrison, Stan, 77
Sampson, Scott: on anthropomorphism, 106–9;
     as charismatic scientist, 64, 89, 93, 106;
     contributions to paleontology, 106; Denver
     Museum and, 93; *Dinosaur Train* and, 5,
     106–7, 181; on dinosaur violence, 89; on
     fossils, 148–51, 155, 156, 157, 171; masculinity
     and, 176; the media industry and, 9, 106–10,
     164; *Walking with Dinosaurs* and, 107–10, 163,
     206n30. *See also* paleontological real
Sand Creek Massacre (1864), 132, 211n89
Sansi Roca, Roger, 28–29
Sanz, José Luis, 17–18, 20
science: creativity and imagination and,
     20–21, 162, 171, 193nn73,80; "SF" and,
     160, 171; as social and political practice, 7,
     190nn8,10; teaching and research, 9. *See also*
     epistemology of science; ontology
science-in-action, 7–8
science museums. *See* museums
Scientific American, 73–74, 76, 215n8
scientific ontology. *See* paleontological real
scientists, 7, 8; authority and, 3, 179; as
     charismatic figures, 64, 69, 202n30. *See also*
     charismatic paleontologists
Seminole people, 212n98
settler colonialism, 30–32, 197n143; DinoLand
     U.S.A. and, 119, 124–25, 128–29, 139–41,
     212n98; Dinosaur Ridge and, 115–16, 129–39,
     140, 141, 211n89, 212n98; edutainment and,
     115–16, 118; Manifest Destiny and, 129, 136,
     139; paleontology and, 8, 10, 57–58, 66, 133,
     137–41, 211–12nn89,95,98. *See also* Native
     Americans, dispossession of
Shannon, Jennifer A., 197–98n6
Shapin, Steven, 191n29
Sinclair Oil, 200–201n69
Sioux peoples (Lakota and Dakota), 120, 125, 136,
     209nn28,30
Smith, Langdon, 204n54
Smithsonian National Museum of Natural
     History (Washington, DC), 191n27,
     197–98n6, 207n8
Smithsonian National Zoo (Washington, DC),
     207n2

sociality. *See* materiality—sociality and

social media, 165, 214n32

species, 69, 201n8, 203n38. *See also* evolution; Linnaean taxonomy

specimens vs. artifacts and artworks, 17, 24, 50, 80, 156–57, 158

Spence, Mark David, 208n17

Spielberg, Steven. *See Jurassic Park* film series

"Stan." *See Tyrannosaurus rex*—"Stan" (BHI 3033)

Star, Susan Leigh, 203n34

Stegosaurus, 19, 98–100, 102, 110; Arthur Lakes and, 96–98, 133–34, 136; as Colorado's state fossil, 98, 136; at Denver Museum, 102–5; at Dinosaur Ridge, 132, 133; Othniel C. Marsh and, 98, 138; at Morrison Museum, 93, 94, 98–99, 102; mounted skeletons, 146

Stewart, Susan, 191n28

storytelling, 4, 31–32, 172, 214n2

Stygimoloch, 51

"Sue." *See Tyrannosaurus rex*—"Sue" (FMNH PR 2081)

Sumrall, Colin D., 215n4

Supreme Court (United States), 212n2

Tattersdill, Will, 193nn73,80

Tauxe, Caroline S., 200n58

taxonomy. *See* evolution; Linnaean taxonomy

technicians, 8, 40, 175, 177, 178

television: *Dinosaur Train*, 5, 106–7, 181; *T-Rex Autopsy*, 20, 117, 164–71, *166–67*, 214nn27,30; *True-Life Adventures* series, and "Prowlers of the Everglades" episode, 117–18, 208n14, 212n98. *See also* film; media industry

Tennyson, Alfred Lord, 80

Ten Who Dared (1960 film), 66, 128, 139, 175

Theropods, 61, 71, 204n79; modern birds as descendants of, 145, 170, 181–82, 215n8

Thompson, Charis M., 216n24

Thomson, Keith, 139

Torrens, Hugh, 21

Toth, Natalie, 180

touch and the senses in paleontology, 4; affect and, 90–91; bone-clones and, 155–56; excavations and, 45–46; grief and, 90–91, 110, 111–12; as knowledge, 46, 199nn26–27; measurement and, 191n36; at Morrison Museum, 93–94, 154–56; rock and, 45–46, 149–51, 155, 156, 157, 213n9; as socially produced, 46, 199n24; vision and, 12, 45–46, 47, 199nn24–27

toys, 13, 207n4

Traweek, Sharon, 190nn10,17

tree of life (arboreal symbolism), 173, 174, 208n24, 215n5, 216n23

T-Rex Autopsy (National Geographic television special), 20, 117, 164–71, *166–67*, 214nn27,30

Triceratops, 50, 57; at Chicago World's Fair (1933–34), 19; in Denver Museum, 51, *52*; in DinoLand U.S.A., 113, 115, 122–23, *123*; excavations, 34–35, 36, *36*, 53–54, 57; fossilization of, 37–38, 50; naming, 57; "the Trike," 42, 43–45, 46, 50, 59, 90

Triceratops horridus, 36, *36*

Trollinger, Susan L., 214n23

Tsing, Anna Lowenhaupt, 198n19

Tyrannosaurus rex: as beast, 73, 81, 96; at Chicago World's Fair (1933–34), 19; in DinoLand U.S.A., 115, 123, *124*; feathered exodermis of, 168–70, 181–82, 215n8; fossilization of, 75–76; funding and, 59, 68, 7959; history of excavations, 61–62, 136; Charles R. Knight's illustrations, 62–63, 201n5; masculinity and, 1, 73, 81, 83–85, 86–89, 133–34; media coverage, 73–74, 204n55; mounting of, 61, 62, 80–82, 204n78; museum exhibits of, 61, 62–63, 67, 73, 79, 82, 83–87, *84–85*; naming of, 14, 71–74, 88, 204n55; nicknames, 62; preparation and casting of, 75–80; scientific knowledge of, 72–73, 82, 96, 168–70, 181–82, 215n8; teeth, 12, 72–73, 74, 75–80, *78*, 82–88, 204n79; violence and, 14, 31, 63, 71–74, 86–87, 88–90, 96, 204n55; volunteers and, 68, 79; zoos and, 207n2

—AMNH 5027, 62–63, 67; American Museum of Natural History (NYC) and, 62–63, 67; casts and reproductions of, 62, 75–76, 77, 79–80, 204n63; Denver Museum and, 61, 62, 64–67, 82, 148; excavations of, 75; fossilization of, 75–76; Charles R. Knight's illustrations, 62–63, 201n5; media and, 63; mounting of, 80–82, 204n78

—"Stan" (BHI 3033): Black Hills Institute of Geological Research and, 77, 78–80; casts of, 77–80, 204n66; excavations of, 62, 77; Morrison Museum and, 77–78, *78*, 94, 155; nickname, 77; preparation of, 77–80, *78*; *T-Rex Autopsy* and, 168

—"Sue" (FMNH PR 2081), 62, 67, 150; Disney's "Dino-Sue," *114*, 121, 150; Field Museum of Natural History and, 10, 67, 120–21, 128–29, 198n13, 209nn28,30,32,34; mounting of, 82; preparation of, 82, 121, 209n34; *T-Rex Autopsy* and, 168

United States: Bureau of Indian Affairs, 21; Bureau of Land Management (BLM), 39–40, 54, 200n60; Geological Survey (USGS), 21, 58, 97, 137, 200n60; Interior Department, 21; National Science Foundation, 200n60; Native American Graves Protection and Repatriation Act (NAGPRA), 120, 209n30; Paleontological Resources Preservation Act (PRPA), 39–40, 198n13; Supreme Court, 212n2; Works Project Administration (WPA), 121. *See also* capitalism and paleontology; discovery; extractivism and paleontology; geology—mining and petroleum; National Park Service; Native Americans; Native Americans, dispossession of; settler colonialism; US military
—nationalism in: Argentine paleontologists and, 13–14, 192n46; the "bone wars" (1870s to 1890s) and, 18–19, 21, 57; Disney and, 117–18, 208n16; edutainment and, 140–41, 212n98; funding for expeditions and exhibitions, 21, 57–58, 200n60; Thomas Jefferson and, 16, 18, 140–41, 212n97; naming new species and, 19; paleontologists and, 57–58, 140; prehistoric animals used in, 16, 18, 140–41, 211–12nn95,97
University of Wyoming, 213n7
US Geological Survey (USGS), 21, 58, 97, 137, 200n60
US military: Native Americans and, 131–32, 135–36, 211n89, 212n98; paleontology expeditions and, 8, 21, 57–58, 140, 200n60; resource frontiers and, 198n19
Ute people, 131–32, 134, 135, 211n89; the Colorado Gold Rush and, 131, 210n61; council trees of, 130, 131, 134, 135, 212n98; Rooney family and, 130–32, 210n61

Van Allen, Adrian, 49, 193n60, 199n31, 200n48, 213n18
Vandenberghe, Frédéric, 194n94
Van Dooren, Thom, 4, 192n49
Velociraptor, 164
vertebrates (subphylum), 93
violence: colonization and, 30–31; DinoLand U.S.A. and, 119, 124–25, 139–41, 212n98; DinoLand U.S.A. thrillride and, 125–29, 210nn43,47; Dinosaur Ridge and, 115–16, 129–39, 140, 141, 211n89, 212n98; edutainment and, 115–16, 118; film and, 20;

naturalization of, 81, 115–16, 118, 133–34, 172–73; *Tyrannosaurus rex* and, 1, 73, 81, 86–89, 133–34. *See also* Native Americans, dispossession of; settler colonialism
volunteers: college students as, 8, 38; in fossil preparation laboratories, 8, 53–55, 60; as museum docents, 83–87; recognition of, 40; retirees from commercial geology as, 1, 10–11; socioeconomic status of, 40–41, 60
volunteer-tourism, 8, 38, 41–42, 43–45, 59–60, 178, 190n16, 198–99nn8,17,19,21

Walking with Dinosaurs (2013 film), 107–10, 163
Walsh, Casey, 209n34
Weber, Max, 63–64, 201–2nn11,18
Weiner, Annette, 213n16
Wenger, Etienne, 190n17
Western Interior Seaway (Cretaceous period), 5, 35–36; map of, *xi*
Western United States. *See* capitalism and paleontology; discovery; extractivism and paleontology; geology—mining and petroleum; National Park Service; Native Americans, dispossession of; settler colonialism
Whatmore, Sarah, 196n134, 199n27
whiteness: of charismatic paleontologists, 64, 66–67, 202n18; dinomania and, 18, 178–79, 216n17; evolution and, 72–73; of paleontological staff and volunteers, 40; scientific definition of humanity and, 175, 215–16n9. *See also* gender and gendering; masculinity; race and racism; settler colonialism
Williams, Raymond, 27, 205n82
Willis, Susan, 15
Wolfe, Cary, 197n143
women in paleontology: credit denied to, 203n40; fieldwork and, 175, 177–78; inclusion of, 178, 180–81, 216n21; laboratory work by, 177; masculinity and, 177, 178, 179. *See also* gender and gendering
Woolgar, Steve, 7
Works Project Administration (United States WPA), 121
Worsley, Peter, 64, 201–2n11
Wrobel, David M., 198n19, 200n58
Wylie, Caitlin Donahue, 190n14, 199n24, 200n46
Wynne, Clive D. L., 206n17
Wynter, Sylvia, 30, 175, 215–16n9

Yale University, and paleontology,
    21, 57; Peabody Museum of
    Natural Science, 58, 74, 97–98,
    137–38, 152
Yosemite National Park, 207n10
Yusoff, Kathryn, 29, 191n28

Zion National Park, 194n91
Zizzamia, Daniel, 211–12n95
Zoological Society of London, 71
zoos, 116–17, 119, 208–9nn26,32; charismatic
    megafauna in, 113–14, 118–19, 121, 208n21;
    prehistoric animals in, 115, 207n2

Founded in 1893,
UNIVERSITY OF CALIFORNIA PRESS
publishes bold, progressive books and journals
on topics in the arts, humanities, social sciences,
and natural sciences—with a focus on social
justice issues—that inspire thought and action
among readers worldwide.

The UC PRESS FOUNDATION
raises funds to uphold the press's vital role
as an independent, nonprofit publisher, and
receives philanthropic support from a wide
range of individuals and institutions—and from
committed readers like you. To learn more, visit
ucpress.edu/supportus.